Optimization of Non-thermal Technology in Food Processing

Optimization of Non-thermal Technology in Food Processing

Guest Editors

Milan Houška

Roman Buckow

MDPI

Basel • Beijing • Wuhan • Barcelona • Belgrade • Novi Sad • Cluj • Manchester

Guest Editors
Milan Houška
Czech Agrifood Research Center
Prague
Czech Republic

Roman Buckow
School of Chemical and Biomolecular Engineering
The University of Sydney
Sydney
Australia

Editorial Office
MDPI AG
Grosspeteranlage 5
4052 Basel, Switzerland

This is a reprint of the Special Issue, published open access by the journal *Foods* (ISSN 2304-8158), freely accessible at: https://www.mdpi.com/journal/foods/special_issues/8L7T3Q73BT.

For citation purposes, cite each article independently as indicated on the article page online and as indicated below:

Lastname, A.A.; Lastname, B.B. Article Title. *Journal Name* **Year**, *Volume Number*, Page Range.

ISBN 978-3-7258-7242-8 (Hbk)
ISBN 978-3-7258-7243-5 (PDF)
https://doi.org/10.3390/books978-3-7258-7243-5

Contents

About the Editors . . . . **vii**

Milan Houška and Roman Buckow
Optimization of Non-Thermal Technology in Food Processing
Reprinted from: *Foods* **2026**, *15*, 283, https://doi.org/10.3390/foods15020283 . . . . **1**

Jan Tříska, Naděžda Vrchotová, Jan Strohalm, Milan Houška, Eliška Kovářıková, Pavla Novotná, et al.
Effect of Ascorbic Acid Addition on the Phenolic Compounds Content in Homogenates from Aerial Parts of Spearmint, Fennel, and Thyme
Reprinted from: *Foods* **2025**, *14*, 2165, https://doi.org/10.3390/foods14132165 . . . . **5**

Alejandro Berzosa, Laura Garza-Moreno, Joaquín Quílez, Javier Raso, Ignacio Álvarez-Lanzarote and Juan Manuel Martínez
Permeabilization of *Cryptosporidium* spp. Oocysts in Water, Apple and Carrot Juice by Pulsed Electric Field Technology
Reprinted from: *Foods* **2025**, *14*, 2112, https://doi.org/10.3390/foods14122112 . . . . **16**

Graciela A. Miranda-Mejía, Anaberta Cardador-Martínez, Viridiana Tejada-Ortigoza, Mariana Morales-de la Peña and Olga Martín-Belloso
Modulating Yogurt Fermentation Through Pulsed Electric Fields and Influence of Milk Fat Content
Reprinted from: *Foods* **2025**, *14*, 1927, https://doi.org/10.3390/foods14111927 . . . . **39**

Gyeong-Seo Park, Hyeon Seo, Han-Baek Lee, Ji-Won Lee, Hafiz Muhammad Shahbaz, Se-Ho Jeong and Dong-Un Lee
Enhanced Peelability and Quality of Whiteleg Shrimp (*Litopenaeus vannamei*) Using Pulsed Electric Field (PEF) Treatment
Reprinted from: *Foods* **2025**, *14*, 148, https://doi.org/10.3390/foods14020148 . . . . **53**

Yi Zhou, Huixin Zuo, Zhaoqi Dai, Zonglin Guo, Benjamin W. B. Holman, Yanqin Ding, et al.
Changes to Pork Bacterial Counts and Composition After Dielectric Barrier Discharge Plasma Treatment and Storage in Modified-Atmosphere Packaging
Reprinted from: *Foods* **2024**, *13*, 4162, https://doi.org/10.3390/foods13244162 . . . . **61**

Cristina Arroqui, Sandra Horvitz, María José Noriega, Idoya Fernández-Pan, Francisco C. Ibañez and Paloma Vírseda
Effect of High-Pressure Processing Pretreatment on the Textural Properties of Cooked *Nuovo Maratelli* Rice
Reprinted from: *Foods* **2024**, *13*, 4052, https://doi.org/10.3390/foods13244052 . . . . **77**

Uyen Ha Dao, Jitlada Na Lamphun, Sitthidat Tongdonyod, Sirinya Taya, Suphat Phongthai and Wannaporn Klangpetch
Optimization of High-Pressure Processing for Microbial Inactivation in Pigmented Rice Grass Juice and Quality Impact Assessment during Refrigerated Storage
Reprinted from: *Foods* **2024**, *13*, 2995, https://doi.org/10.3390/foods13182995 . . . . **92**

Breanna Polen, Brahmaiah Pendyala, Ankit Patras and Doris H. D'Souza
Inactivation of Hepatitis A Virus and Feline Calicivirus on Model Food Contact Surfaces by Ultraviolet Light (UV-C) Systems
Reprinted from: *Foods* **2024**, *13*, 2892, https://doi.org/10.3390/foods13182892 . . . . **111**

Philippe Raymond, François St-Germain, Sylvianne Paul, Denise Chabot and Louise Deschênes
Impact of Nanoparticle-Based TiO_2 Surfaces on Norovirus Capsids and Genome Integrity
Reprinted from: *Foods* **2024**, *13*, 1527, https://doi.org/10.3390/foods13101527 . . . . **129**

Miroslava Jandová, Michaela Fišerová, Pavla Paterová, Lucie Cacková, Pavel Měřička, Jan Malý, et al.
High-Pressure Inactivation of *Bacillus cereus* in Human Breast Milk
Reprinted from: *Foods* **2023**, *12*, 4245, https://doi.org/10.3390/foods12234245 . . . . **143**

Marco Dalla Rosa, Santina Romani, Pietro Rocculi, Urszula Tylewicz and Silvia Tappi
Progress in Low-Impact Processing Technologies to Deliver More Sustainable and Healthy Food Tomorrow
Reprinted from: *Foods* **2025**, *14*, 2332, https://doi.org/10.3390/foods14132332 . . . . **160**

About the Editors

Milan Houška

Milan Houška is a senior scientist working at the Department of Food Science at the Agrifood Research Centre. He is interested in Food Science and Technology, especially food engineering. He has led projects focused on high-pressure technology applications in the food industry in the Czech Republic. Novel technologies such as Pulsed Electric Field are of interest to him.

Roman Buckow

Roman Buckow is a leading food scientist and the Director of the Centre for Food Science at La Trobe University. He is internationally recognised for his contributions to sustainable food and beverage manufacturing, bioprocessing innovation, alternative proteins, and advanced food processing technologies. Before joining La Trobe in 2024, he served as Professor of Practice in Food Engineering at the University of Sydney and previously held senior leadership roles at CSIRO, including Principal Scientist and Research Manager for Foods for Health and Food Quality. He also served as Chief Technology Officer at All G Foods, where he led the development of next generation plant-based and precision fermented food products.

Professor Buckow holds an MSc and PhD in Food and Bioprocess Engineering from the Berlin University of Technology. Throughout his career, he has led multidisciplinary teams across academia, government, and industry, delivering strategic research outcomes and commercial innovations in food structure design, novel processing, protein extraction, and fermentation technologies.

A committed educator and mentor, he has supervised over 40 postgraduate researchers and regularly contributes to national and international advisory boards, editorial committees, and professional organisations. His work continues to shape the future of sustainable food systems and drive innovation across the agri food and biomanufacturing sectors.

foods

Editorial

Optimization of Non-Thermal Technology in Food Processing

Milan Houška [1,*] and Roman Buckow [2]

1 Department of Food Science, Czech Agro-Food Centre (CARC), Radiová Street 1285/7, 102 00 Prague, Czech Republic
2 La Trobe Institute for Sustainable Agriculture and Foods, La Trobe University, Bundoora, VIC 3086, Australia
* Correspondence: milan.houska@carc.cz; Tel.: +420-737-287-007

1. Brief Overview of Recent Developments in the Field

Any new food processing technology must first be subjected to extensive research to verify the safety and quality of the food that it creates. Following this, the effects of the various physical parameters of the technology (for example, temperature, pressure, pH, water activity) are able to be examined. When enough publications are made focusing on a particular technology, authors are able to write reviews. High-pressure processing technology [1–3] and pulsed electric fields [4] have gone through this process. Additionally, Reference [4] has provided a comprehensive overview of a number of modern non-thermal technologies [5].

Our Special Issue is dedicated to non-thermal food processing, but it covers only a very small part of this area of food production. These methods have a promising future as consumers are increasingly coming to prefer gentler processing methods to safeguard their health, despite that fact that these products have shorter shelf lives.

Dalla Rosa et al. [C11] provided a review for our Special Issue, considering the impact of environmentally friendly technologies on the bioavailability of various food ingredients and their safety. This review paper also focused on the overall energy consumption for processing, emphasizing that non-thermal pretreatments, such as cold plasma, pulsed electric fields, high-pressure processing and ultrasound processing, combined with mild thermal drying technology, would help reduce energy costs. Any new technology needs some energy input for their application. This has been comprehensively reviewed in Reference [6]. This paper considers the advantages of non-thermal food processing in terms of quality and carbon footprint reduction; see Table 1.

Dynamic and hydrostatic high pressure, vacuum impregnation, ultrasound, pulsed electric fields and cold plasma applications can result in a less negative effect with respect to traditional thermal treatments. These technologies can help extract important substances from secondary raw materials or waste. This ability helps the circular bioeconomy.

In Figure 1, a comparison of the positive and negative properties of the technologies considered is reported.

These treatments can produce structural changes that improve the bioaccessibility and/or the bioavailability of bioactive compounds, such as probiotic microorganisms, to improve food healthiness and the gut microbiome.

Non-thermal low-impact processing technologies could help the food industry to pursue more sustainable practices and offer better food functionality, thereby being seen more favorably in the context of the ultra-processing debate.

However, an interdisciplinary approach among food engineers, microbiologists, food chemists, bio-NMR specialists and nutritionists is necessary to discover the best processing conditions, and to increase and assess the sustainability of processing in the food industry.

 https://doi.org/10.3390/foods15020283

Table 1. Effects on food quality and carbon footprint elements of non-thermal processing (adapted from [6]).

Effects on Food Quality	Carbon Foodprint Reduction
Minimal quality loss	Less wastewater
Increasing bioavailability	Increase energy and water savings
Reduction in processing contaminants	Lower environmental impact
Maintenance of nutritional values	Decreased operational costs
Maintenance of sensorial properties	Decreased electricity
Inactivation of microorganisms	Less time-consuming
Improvement of heat and mass transfer	Inexpensive
Improvement of firmness and texture	Non-hazardous
Decreased color change	Minimal source demands
Increased shelf-life	Simple processing design

Our Special Issue includes works that deal with the effects of high pressure [C1], [C4], [C5] and [C10]; high voltage pulses [C7], [C8] and [C9]; UV-C radiation [C3]; and plasma treatment [C6] and nanoparticle-based TiO_2 [C2] on various types of food and raw materials.

We consider that the most methodologically significant publication on the verification of any new technology against classical pasteurization is Paper [7]. From a methodological point of view, it does not matter that this work is focused on the treatment of human milk. A very useful overview of new non-thermal food processing technologies is provided in Paper [6].

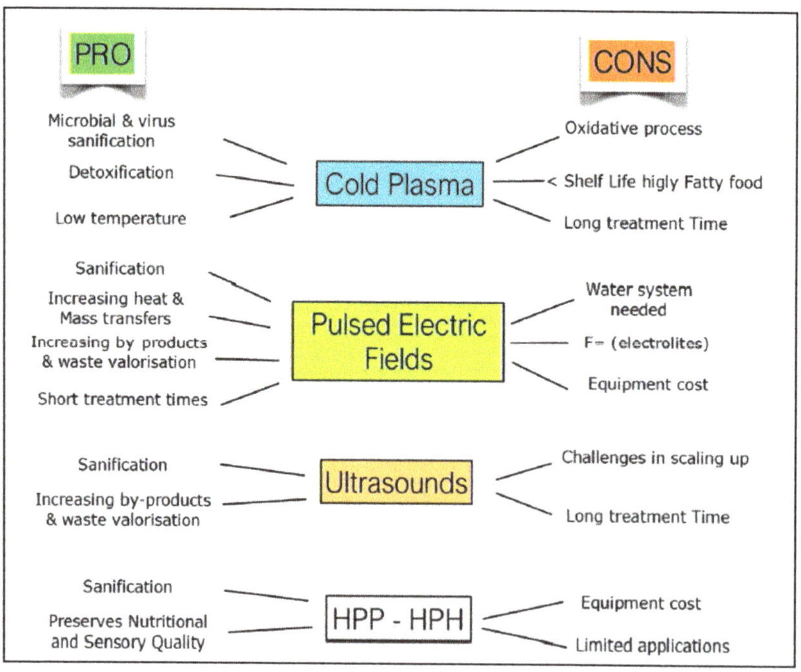

Figure 1. Comparison of pros and cons of investigated technologies (adapted from C11).

https://doi.org/10.3390/foods15020283

2. The Gap in Knowledge and How This Special Issue Has Addressed Those Gaps

Our Special Issue is only a small contribution to the already established knowledge of non-thermal food processing technologies. Therefore, it is very difficult to assess whether there are any gaps in the knowledge of the field on the whole. Each research piece usually reveals new problems to be solved before the studied technology can be introduced into industrial practice (see, for example, the issue of scaling up).

It can be stated that the best application in food practice has so far been achieved by high-pressure treatment technology. The most frequently used products on the market that use this technology are fruit and vegetable juices, in which important significant substances (for example, vitamins and dyes) are able to be preserved. A certain uniqueness is represented by Article [C1], which presents the treatment of human breast milk by this technology.

3. Primary Focus on Future Research That Should Be Considered

The future of non-thermal food processing technologies lies in researching methods that will reliably ensure the safety of products and the preservation of their most nutritionally important and healthy ingredients.

Each method must enable the implementation of a reliable system of hazard analysis and critical control points that will allow any health risks to be significantly reduced or even eliminated.

The costs of their applications in practice must be comparable to the real sales price in order to ensure commercial success.

4. Concluding Remarks

Gentle technologies without targeted heating can enable the food industry to maintain consumer interest in foods that maintain a high content of their original substances, such as vitamins and enzymes.

The application of these technologies requires safeguarding health and ensuring quality within the framework of the HACCP system. It is important to maintain an interdisciplinary approach between food engineers, microbiologists, chemists, machine design specialists, nutritionists and employees of regulatory authorities—all of whom work together to define standards and regulations.

Author Contributions: M.H. writing—original draft preparation, R.B. supervision. All authors have read and agreed to the published version of the manuscript.

Funding: This research received no external funding.

Conflicts of Interest: The authors declare no conflicts of interest.

List of Contributions:

1. Jandová, M.; Fišerová, M.; Paterová, P.; Cacková, L.; Měřička, P.; Malý, J.; Kacerovský, M.; Kovaříková, E.; Strohalm, J.; Demnerová, K.; et al. High-Pressure Inactivation of Bacillus cereus in Human Breast Milk. *Foods* **2023**, *12*, 4245. https://doi.org/10.3390/foods12234245.
2. Raymond, P.; St-Germain, F.; Paul, S.; Chabot, D.; Deschênes, L. Impact of Nanoparticle-Based TiO_2 Surfaces on Norovirus Capsids and Genome Integrity. *Foods* **2024**, *13*, 1527. https://doi.org/10.3390/foods13101527.
3. Polen, B.; Pendyala, B.; Patras, A.; D'Souza, D.H. Inactivation of Hepatitis A Virus and Feline Calicivirus on Model Food Contact Surfaces by Ultraviolet Light (UV-C) Systems. *Foods* **2024**, *13*, 2892. https://doi.org/10.3390/foods13182892.

https://doi.org/10.3390/foods15020283

4. Dao, U.H.; Lamphun, J.N.; Tongdonyod, S.; Taya, S.; Phongthai, S.; Klangpetch, W. Optimization of High-Pressure Processing for Microbial Inactivation in Pigmented Rice Grass Juice and Quality Impact Assessment during Refrigerated Storage. *Foods* **2024**, *13*, 2995. https://doi.org/10.3390/foods13182995.
5. Arroqui, C.; Horvitz, S.; Noriega, M.J.; Fernández-Pan, I.; Ibañez, F.C.; Vírseda, P. Effect of High-Pressure Processing Pretreatment on the Textural Properties of Cooked Nuovo Maratelli Rice. *Foods* **2024**, *13*, 4052. https://doi.org/10.3390/foods13244052.
6. Zhou, Y.; Zuo, H.; Dai, Z.; Guo, Z.; Holman, B.W.B.; Ding, Y.; Shi, J.; Ding, X.; Huang, M.; Mao, Y. Changes to Pork Bacterial Counts and Composition After Dielectric Barrier Discharge Plasma Treatment and Storage in Modified-Atmosphere Packaging. *Foods* **2024**, *13*, 4162. https://doi.org/10.3390/foods13244162.
7. Park, G.-S.; Seo, H.; Lee, H.-B.; Lee, J.-W.; Shahbaz, H.M.; Jeong, S.-H.; Lee, D.-U. Enhanced Peelability and Quality of Whiteleg Shrimp (*Litopenaeus vannamei*) Using Pulsed Electric Field (PEF) Treatment. *Foods* **2025**, *14*, 148. https://doi.org/10.3390/foods14020148.
8. Miranda-Mejía, G.A.; Cardador-Martínez, A.; Tejada-Ortigoza, V.; Morales-de la Peña, M.; Martín-Belloso, O. Modulating Yogurt Fermentation Through Pulsed Electric Fields and Influence of Milk Fat Content. *Foods* **2025**, *14*, 1927. https://doi.org/10.3390/foods14111927.
9. Berzosa, A.; Garza-Moreno, L.; Quílez, J.; Raso, J.; Álvarez-Lanzarote, I.; Martínez, J.M. Permeabilization of *Cryptosporidium* spp. Oocysts in Water, Apple and Carrot Juice by Pulsed Electric Field Technology. *Foods* **2025**, *14*, 2112. https://doi.org/10.3390/foods14122112.
10. Tříska, J.; Vrchotová, N.; Strohalm, J.; Houška, M.; Kovářiková, E.; Novotná, P.; Bednář, J.; Pavela, R. Effect of Ascorbic Acid Addition on the Phenolic Compounds Content in Homogenates from Aerial Parts of Spearmint, Fennel, and Thyme. *Foods* **2025**, *14*, 2165. https://doi.org/10.3390/foods14132165.
11. Dalla Rosa, M.; Romani, S.; Rocculi, P.; Tylewicz, U.; Tappi, S. Progress in Low-Impact Processing Technologies to Deliver More Sustainable and Healthy Food Tomorrow. *Foods* **2025**, *14*, 2332. https://doi.org/10.3390/foods14132332.

References

1. Hendrickx, M.E.; Knorr, D. (Eds.) *Ultra High Pressure Treatment of Foods*; Springer Science & Business Media: Berlin/Heidelberg, Germany, 2001.
2. Ledward, D.A.; Johnston, D.E.; Earnshaw, R.G.; Hasting, A.P.M. *High Pressure Processing of Foods*; Nottingham University Press: Loughborough, UK, 1995.
3. Houška, M.; da Silva, F.V.M. (Eds.) *High Pressure Processing of Fruit and Vegetable Products*; CRC Press: Boca Raton, FL, USA, 2018.
4. Toepfl, S.; Heinz, V.; Knorr, D. Applications of Pulsed Electric Fields Technology for the Food Industry. In *Pulsed Electric Fields Technology for the Food Industry*; Food Engineering Series; Raso, J., Heinz, V., Eds.; Springer: Boston, MA, USA, 2006. [CrossRef]
5. Zhang, H.Q.; Barbosa-Canovas, G.V.; Balasubramaniam, V.M.; Dunne, C.P.; Farkas, D.F.; Yuan, J.T.C. *EDITORS (2011) Nonthermal Processing Technologies for Food*; Blackwell Publishing Ltd.: Hoboken, NJ, USA, 2011; ISBN 978-0-813-81668-5.
6. Yudhistira, B.; Punthi, F.; Gavahian, M.; Chang, C.-K.; Hazeena, S.H.; Hou, C.-Y.; Hsieh, C.-W. Nonthermal technologies to maintain food quality and carbon footprint minimization in food processing: A review. *Trends Food Sci. Technol.* **2023**, *141*, 104205. [CrossRef]
7. Clifford, V.; Walter, L.L.; Klein, L.D.; Wu, T.; Wang, Y.; Bansal, N. A Framework for Evaluation of New Processing Technologies in Human Milk Banking. *Compr. Rev. Food Sci. Food Saf.* **2025**, *24*, e70288. [CrossRef] [PubMed]

foods

MDPI

Article

Effect of Ascorbic Acid Addition on the Phenolic Compounds Content in Homogenates from Aerial Parts of Spearmint, Fennel, and Thyme

Jan Tříska [1,2,*], Naděžda Vrchotová [1,2], Jan Strohalm [3], Milan Houška [3], Eliška Kovaříková [3], Pavla Novotná [3], Jan Bednář [1] and Roman Pavela [3,4]

1 Global Change Research Institute, CAS, Bělidla 986/4a, 603 00 Brno, Czech Republic; vrchotova.n@czechglobe.cz (N.V.); bednar.j@czechglobe.cz (J.B.)
2 Faculty of Agriculture and Technology, University of South Bohemia, Studentská 1668, 370 05 České Budějovice, Czech Republic
3 Czech Agrifood Research Centre, Drnovská 507, Ruzyně, 161 06 Prague, Czech Republic; jan.strohalm@carc.cz (J.S.); milan.houska@carc.cz (M.H.); eliska.kovarikova@carc.cz (E.K.); pavla.novotna@carc.cz (P.N.); roman.pavela@carc.cz (R.P.)
4 Department of Plant Biotechnology, College of Life Sciences and Biotechnology, Korea University, Seoul 02841, Republic of Korea
* Correspondence: triska.j@czechglobe.cz

Abstract: The paper deals with the investigation of the ascorbic acid influence on the analytical results of polyphenol content in the samples of the spearmint, fennel, and thyme homogenates. The homogenates without and with addition of ascorbic acid and water were prepared and stabilized by high-pressure treatment. Their analysis was accomplished by high-performance liquid chromatography (HPLC) with DAD detection and by combination of liquid chromatography with mass spectrometry (LC-MS). Volatile terpenes were analyzed in all homogenates by combination of gas chromatography with mass spectrometry technique (GC-MS). The content of polyphenols of acidic nature, e.g., rosmarinic acid, revealed the highest difference between analytical results of the samples with and without ascorbic acid. Finally, prepared herb homogenates are suitable food supplements, which will find increasing application in various food preparations.

Keywords: herbs homogenates; *Mentha spicata*; *Foeniculum vulgare*; *Thymus vulgaris*; phenolic compounds; ascorbic acid addition

1. Introduction

Spearmint (*Mentha spicata*), fennel (*Foeniculum vulgare*), and thyme (*Thymus vulgaris*) are the most common herbs used not only in folk medicine, in the kitchen, and in the food industry, but the attention is increasingly being paid to them in modern pharmacy and in bio industry (e.g., products with insecticidal activity). The mentioned plants contain a number of phenolic biologically active substances e.g., diosmin, diosmetin, hesperidin, luteolin, apigenin, rosmarinic acid in spearmint [1–3], chlorogenic acid, miquelianin, 1,5-dicaffeoylquinic acid, kaempferol-3-glucuronide, kaempferol-3-arabinoside in fennel [4,5] and luteolin-7-glucuronide, apigenin-7-glucuronide, rosmarinic acid, and salvianolic acid derivatives in thyme [6,7]. The dominant terpenic substances include e.g., carvacrol, thymol, p-cymene in thyme [8], γ-terpinene, 4-allylanisole, anethole, limonene in fennel [9–11] and limonene, 1,8-cineole, carvone, and cymene in spearmint [12–15].

An overview of phenolic substances and their testing in pharmacy is provided by several reviews, e.g., on the effects of spearmint substances [16–18], thyme substances [18,19], and fennel substances [20,21].

High-pressure processing (HPP) is technology used for food processing and is denoted cold pasteurization. Products in suitable packaging are subjected to isostatic pressure 300–600 MPa. This process inactivates the most active forms of microorganisms and preserves the sensorial and nutritional quality of original raw materials [22].

Regarding the fact that the herb homogenates with a guaranteed content of active substances find increasing application in various preparations and food supplements, e.g., sage homogenate [23], it means that the homogenates must be standardized in some way. The technology of stabilizing homogenates has not yet been sufficiently studied, especially regarding the content of health-promoting substances. One way to preserve the content of biologically active substances could be acidification. Acidification is an effective process to minimize the risk of bacterial spore germination and subsequently to spoil fruit and vegetable juices treated with high pressure [24]. The goal of our work was, therefore, to determine the effect of the addition of ascorbic acid on the terpenic and phenolic substances content in the final spearmint, fennel, and thyme homogenates.

2. Materials and Methods

2.1. Plant Materials

Mentha spicata (spearmint), *Foeniculum vulgare* (fennel), and *Thymus vulgaris* (thyme) were grown on the grounds of the Research Institute of Plant Production in Prague and Olomouc.

2.2. Preparations of Homogenates

The aerial parts of fresh herb material were homogenized with or without the additions of water and ascorbic acid in different ratio, and the final pH value of the mixture was measured. The experimental setup and pH value are given in Table 1.

Table 1. Compositions and pH of the homogenates.

Sample	Sample Composition	pH Value
M-1	400 g M	6.17
M-2	400 g M + 400 g W	6.60
M-3	400 g M + 18 g AA	4.24
M-4	400 g M + 400 g W + 18 g AA	4.34
F-1	400 g F	5.71
F-2	400 g F + 200 g W	5.59
F-3	348 g F + 30 g W + 6 g AA	4.17
F-4	400 g F + 200 g W + 10 g AA	4.18
T-1	400 g T	6.20
T-2	200 g T + 300 g W	6.18
T-3	300 g T + 18 g AA	3.91
T-4	400 g T + 600 g W + 24 g AA	3.91

M. . .*Mentha spicata*, F. . .*Foeniculum vulgare*, T. . .*Thymus vulgaris*, W. . .water, AA. . .ascorbic acid.

Fresh herbs were collected, and a small part was spread on sieves and allowed to dry for determination of dry matter before being placed in a cold room. Subsequently, homogenization, i.e., mechanical disintegration of aerial parts of herbs, was performed using mixer Coupe R301 (Montceau-en-Bourgone, France) with individual amounts of water because plants differ in dry matter and toughness of aerial parts. The method is based on mechanical disintegration and must be adapted with respect to the individual state of plant tissue. Similarly, ascorbic acid dose was adapted with pH level and is individual for each plant. The dose was evaluated with respect to pH lowering to level 3.8–4.5. The procedure for homogenizing fresh herbs into a smooth paste that is useful for food preparation was developed in our laboratory and has not been previously published in international journals.

2.3. Microbial Analysis

A horizontal method for the enumeration of microorganisms by colony count at 30 °C by ISO 4833-1 [25] was used. ISO 21527-1:2008 [26] was followed for the number of yeast and molds determination. Plate count agar (Himedia, Maharashtra, India) was used for total count of microorganisms, and Yeast Glucose Chloramphenicol (YGC) Agar (Sigma-Aldrich, Prague, Czech Republic) was used for yeast and mold cultivation. Samples were cultivated at 30 °C, and colony-forming units (CFU) were enumerated.

2.4. High-Pressure Processing

Samples given in Table 1 represent homogenates that were filled into polyethylene terephthalate/aluminum/polyethylene (PET/Al/PE) containers, which were vacuum sealed and then treated with a high pressure of 500 MPa for 10 min, then cooled to 15 °C and stored in refrigerator between 5 and 8 °C. The pressurizing was performed by a high-pressure press CYX 6/0103 (ŽĎAS a. S., Žd'ár nad Sázavou, Czech Republic). High-pressure treatment was used as an antimicrobial intervention replacing pasteurization. Pressurized homogenates are microbially stabile for at least 2 years for food applications.

2.5. Extraction of Phenolic Compounds from Homogenate

The phenolic compounds were extracted from homogenates by methanol similarly as in [7], but at the higher temperature and for a shorter time. Then, 0.25 g of homogenate were extracted by 3 mL of 100% methanol, and the extraction was conducted at 50 °C for 1 h. After centrifugation (3500 rpm, 10 min), the sediment was washed twice with 1 mL of methanol. Supernatants were combined and the total volume was measured. Each sample was prepared in triplicate and stored at −18 °C. The extracts were analysed using HPLC and LC/MS.

2.6. Extraction of Terpenes from Homogenates

Samples of plant material homogenates (thyme, fennel, spearmint) were prepared identically in the following manner. Approximately 0.3 g of homogenate were weighed in triplicates from each sample (double the amount for water-diluted homogenates). The weighed amount was extracted three times with 2 mL of hexane. After the first addition of hexane, the mixture was shaken for 1 h. The hexane was then collected in a separate vial. Then, 2 mL of hexane were added again to the homogenate. This mixture was shaken for 0.5 h and then hexane was added to the first portion of hexane. Finally, 2 mL of hexane were added again and shaken for 0.5 h. The last portion of hexane was also added to the previous two. The hexane solution thus obtained was directly injected into the GC/MS. The hexane extract obtained from spearmint homogenates was diluted five times before measurement due to high concentration of the components.

2.7. Determination of Phenolic Compounds

Quantification of phenolic compounds by HPLC: The samples were analysed using an HPLC apparatus (Hewlett Packard 1050) (Hewlett-Packard, Palo Alto, CA, USA) with a diode array detector (DAD Agilent G1315B, Palo Alto, CA, USA) and column Phenomenex Luna C18(2) (3 μm, 2 × 150 mm) (Phenomenex, Torrance, CA, USA).

Mobile phase A: 5% acetonitrile + 0.1% *o*-phosphoric acid, mobile phase B: 80% acetonitrile + 0.1% *o*-phosphoric acid. Gradient for separation of fennel (35 °C): 9% to 33% of B during 35 min, 33% to 45% of B during 1 min, 45% to 80% of B during 2 min, 80% to 100% of B during 2 min. Gradient for separation of spearmint (25 °C): 2% to 42% of B during 40 min, 40% to 80% of B during 2 min, 80% of B during 3 min. Gradient for thyme (35 °C): 0% to 45% of B during 55 min, 45% to 80% of B during 5 min. The volume of the injected sample was 5 μL. Flow rate was 0.25 mL/min.

Standards (diosmin, diosmetin, hesperidin, rosmarinic acid, luteolin, apigenin, chlorogenic acid, miquelianin, 1,5-dicaffeoylquinic acid, luteolin-7-glucuronide) were purchased from Sigma-Aldrich, Praha, Czech Republic; methanol and acetonitrile from Merck, Praha, Czech Republic, *o*-phosphoric acid from Fluka, formic acid from Sigma-Aldrich, Praha, Czech Republic.

Identification of phenolic compounds was performed by LC/MS. For compound identification, we used atmospheric pressure chemical ionization (APCI-LC/MS) in positive and negative mode. The instrument (LCQ Accela Fleet, Thermo Fisher Scientific, San Jose, CA, USA) had the same column and used the same gradients as in HPLC, but mobile phases were acidified by 0.1% formic acid. Instrument conditions were the following: Vaporizer temperature 300 °C, sheath gas flow rate 58 L/min, auxiliary gas flow rate 10 L/min, discharge current 5 μA, capillary temperature 275 °C, and capillary voltage 2 V.

2.8. Determination of Volatile Terpenes

Terpenes from homogenates extracts were analysed on a Trace GC Ultra gas chromatograph (Thermo Fischer Scientific, San Jose, CA, USA) equipped with a Restek-fused silica capillary column, Rxi-5 ms, 30 m × 0.25 mm I.D. × 0.25 μm (Restek Corporation, Bellefonte, PA, USA), liner SKY, Splitless, 3 mm × 0.8 mm × 105 mm (Restek Corporation), and coupled to a mass selective detector ISQ (Thermo Fischer Scientific) working at 70 eV of ionization energy. Helium was used as a carrier gas at 1.0 mL/min with injection of 1 μL in splitless mode at 250 °C. Split flow after 1 min was 50 mL/min. The oven temperature was programmed as follows: 40 °C for 5 min, then increase to 150 °C at a rate of 3 °C/min, further increase to 250 °C at 10 °C/min, and finally increase to 290 °C at a rate of 25 °C/min. This temperature was then held for 2 min. Transfer line temperature was 250 °C, ion source temperature was set to 200 °C. Mass scanning was started at 7.00 min, masses were scanned in the full range 50–450 m/z. Qualitative analysis was performed using 36 purchased standards, and their list is given in the table in the Supplementary Materials.

2.9. Statistics

A Two-Way ANOVA [27] was conducted to determine to what extent H_2O and ascorbic acid have an influence on contents of specific compounds.

3. Results and Discussion

Plant homogenates were prepared as shown in Materials and Methods in Table 1.

Microbial stability of homogenates with water and ascorbic acid was suitable for food application during 21 days of storage (see Table 2). The number of microorganisms did not exceed 1.3×10^3 during storage in laboratory temperature. Yeast and mold were not detected in all samples during storage period. Microbial stability of homogenates with ascorbic acid can be supported by pH lower than 4.5 and antimicrobial impact of phytochemical components.

Table 2. The microbial stability of homogenates prepared with water and ascorbic acid.

Sample	Time of Storage (Day)	Total Count (CFU/g) at 5 °C	Total Count (CFU/g) at 20 °C
M-4	0	1.1×10^3	1.1×10^3
	7	1.3×10^3	1.1×10^3
	14	1.1×10^3	1.1×10^3
	21	9.1×10^2	9.3×10^2
F-4	0	8.2×10^1	8.2×10^1
	7	7.7×10^2	8.0×10^2
	14	3.7×10^2	3.3×10^2
	21	5.0×10^2	1.7×10^2
T-4	0	6.1×10^1	6.1×10^1
	7	3.8×10^2	1.4×10^2
	14	2.3×10^2	1.6×10^2
	21	2.5×10^2	1.7×10^2

The results of analyses of the phenolic compounds of interest (Table 3) showed that the differences in the content of most substances in different homogenate preparation were considerable. The greatest effect has the addition of ascorbic acid during the homogenization process on the measured content of phenolic acids, e.g., chlorogenic acid, rosmarinic acid, and their derivatives, while the measured number of phenolic acids is minimal without addition of ascorbic acid. In thyme homogenates with ascorbic acid addition, the measured rosmarinic acid content is 37,212 mg/kg d. m., while in thyme homogenates without ascorbic acid, the measured content of rosmarinic acid is only 196 mg/kg d. m., and in thyme homogenates without ascorbic acid but with water addition, its content is even a little less (101 mg/kg d. m.) (Table 3). It is obvious that measured content of phenolic acids is higher in the presence of ascorbic acid and results in lower pH value. Better stability of chlorogenic acid at low pH (pH 3) was described already in the literature [28], but extraction is influenced also by water addition, and statistically significant effect has water and ascorbic acid simultaneously. The higher content was also observed by the following flavonol glucuronides (quercetin-3-O-glucuronide (miquelianin), other quercetin derivative and kaempferol-3-O-glucuronide) and flavone glucuronides (luteolin-7-glucuronide and apigenin-7-glucuronide). All mentioned glucuronides have the highest content in the homogenates with added ascorbic acid whether water is added or not. This is due to the greater stability of glucuronides at lower pH [29]. Extraction of mentioned flavonols is also influenced by water addition, ascorbic acid presence, and their combination. Only quercetin derivatives in *Foeniculum vulgare* do not show a statistically significant difference for the combined use of water and ascorbic acid. Kaempferol-arabinoside also shows greater availability in the presence of ascorbic acid and water.

Regarding flavan hesperidine, its content is highest in homogenates with ascorbic acid addition but without the addition of water. Combination of other conditions also influences the yield of extraction. The other flavones (luteolin, diosmin, diosmetin) are influenced with water and ascorbic acid presence in homogenate, and with their combination. Apigenin behaves rather differently. There is an unclear dependence on the presence of ascorbic acid, but from a statistical point of view, only water addition has a significant effect (Table 3). In the opposite, the highest content of diosmetin was found in all homogenates without ascorbic acid addition. The content of diosmin is the highest in homogenate with water addition, but without ascorbic acid addition; in water-free homogenates with or without ascorbic acid addition, the content of diosmin is lower. In this context, it is clear that herb homogenate preparation and their analysis are very important steps, as our results show. The greatest changes are in the rosmarinic acid content. Pure rosmarinic acid is a stable substance, and its solution in ethanol is also stable at different temperatures (10–40 °C) and under different light exposure, as experimentally found [30].

Rosmarinic acid content in *Melissa officinalis* tinctures prepared from dry plant material was higher (2.96–22.18 mg/mL) than in the tinctures prepared from fresh ground-crushed material (less than 0.92 mg/mL) [31]. Olah et al. [32] found higher rosmarinic acid content in fresh *Rosmarinus officinalis* tinctures (0.35 mg/mL) than in the tinctures prepared from dried material (0.18 mg/mL), but it should be noted that the above-mentioned authors did not cut or crush fresh material. Thus, the enzymatic activity was not increasing. Six et al. [33] found that the amount of rosmarinic acid in a 50% ethanolic extract from the dried material of various Laminaceae decreases after 24 weeks by 14–27%, even by 41% in sage. Bodalska et al. [34] tested the stability of rosmarinic acid in commercial herbal medicinal products and found that in aqueous tinctures, rosmarinic acid is very unstable; stability is much better in water-ethanolic extracts. A very important step influencing the final content of the compounds of acidic nature is whether we will grind the samples or leave the whole fresh plant material intact [35]. In the fresh plant homogenates, the released enzymes act on chlorogenic and rosmarinic acids and their derivatives, and these compounds are, therefore, the subject of rapid degradation. The presence of ascorbic acid has at least dual positive effects on the content of the compounds. First, ascorbic acid decreases pH and low pH blocks enzymatic activity; and second, ascorbic acid suppresses ionization of acidic compounds in water. Therefore, they reveal higher yield during the extraction step. The content of rosmarinic acid in homogenates from *Mentha spicata* and *Thymus vulgaris* was influenced not only by ascorbic acid addition, but also with water addition. The combination of these two additions has also a statistically significant effect. In both homogenates, without ascorbic acid addition, rosmarinic acid content was very low. The same was observed also for the rosmarinic acid derivatives 1 and 2. For extraction of rosmarinic acid and their derivatives, the presence of water is also important; only rosmarinic acid derivative 1 was not influenced by water addition.

In acidic media, terpenes generally undergo various transformations. The complex mixtures of products obtained in these transformations are the major factors that hinder the proper interpretation of the analytical results, especially because essential oils (EOs) themselves are a very complex matrix. In addition, these processes can also be influenced by the phenolic substances present, as is the case, for example, with γ-terpinene. In the case of γ-terpinene, we can see the largest changes in the homogenate from *Thymus vulgaris* in Table 4. In the presence of ascorbic acid and water, the amount of γ-terpinene increases from 0.377 to 1.510 (measured by the relative peak area—here and further in the discussion). The opposite is true for 4-cymene. The content decreased in this homogenate from 3.3 to 0.593, as well as in *Foeniculum vulgare* homogenate from 0.570 to 0.110. In this homogenate, the largest decrease in α-pinene was also observed, from 0.717 to 0.257. In contrast, this decrease in β-pinene content was not observed in the acidic environment in the *Mentha spicata* homogenate.

It is clear from the literature [22] that for most plant enzymes, the response to pressure-induced inactivation is enzyme-specific and depends on the conditions applied with partial inactivation at most under commercially feasible HPP conditions. In general, enzymes are more resistant to inactivation than vegetative microorganisms, posing a challenge to the application of HPP for stabilization of fruit and vegetable products.

Table 3. The content of monitored phenolic compounds (mg/kg dry matter).

Mentha spicata	Water	Ascorbic Acid	Diosmin	Hesperidin	Rosmarinic Acid	Luteolin	Apigenin	Diosmetin
M-1	0	0	12,826 ± 524 [abc]	893 ± 46 [abc]	122 ± 27 [abc]	485 ± 26 [abc]	485 ± 17 [a]	922 ± 111 [abc]
M-2	water	0	22,364 ± 1039 [abc]	1201 ± 66 [abc]	95 ± 4 [abc]	652 ± 29 [abc]	571 ± 60 [a]	1888 ± 143 [abc]
M-3	0	ascorbic acid	13,862 ± 842 [abc]	2471 ± 144 [abc]	66,588 ± 2393 [abc]	495 ± 31 [abc]	495 ± 46 [a]	309 ± 44 [abc]
M-4	water	ascorbic acid	18,573 ± 571 [abc]	1281 ± 106 [abc]	33,271 ± 2013 [abc]	1000 ± 25 [abc]	599 ± 9 [a]	720 ± 26 [abc]
Foeniculum vulgare	**Water**	**Ascorbic Acid**	**Chlorogenic Acid**	**Miquelianin**	**Quercetin Derivative**	**1,5-dicaffeoylquinic Acid**	**Kaempferol-3-_O_-Glucuronide**	**Kaempferol-3-_O_-Arabinoside**
F-1	0	0	n.d.	527 ± 38 [abc]	372 ± 28 [ab]	n.d.	495 ± 26 [abc]	443 ± 17 [abc]
F-2	water	0	n.d.	555 ± 41 [abc]	309 ± 16 [ab]	81 ± 18 [abc]	481 ± 17 [abc]	420 ± 15 [abc]
F-3	0	ascorbic acid	1864 ± 46 [abc]	1933 ± 55 [abc]	889 ± 23 [ab]	716 ± 38 [abc]	798 ± 19 [abc]	579 ± 17 [abc]
F-4	water	ascorbic acid	2021 ± 40 [abc]	2416 ± 66 [abc]	853 ± 28 [ab]	617 ± 10 [abc]	1098 ± 35 [abc]	664 ± 28 [abc]
Thymus vulgaris	**Water**	**Ascorbic Acid**	**Luteolin-7-Glucuronide**	**Apigenin 7-Glucuronide**	**Rosmarinic Acid**	**Rosmarinic Acid Derivative 1**	**Rosmarinic Acid Derivative 2**	**Caffeoyl-Rosmarinic Acid**
T-1	0	0	563 ± 3 [abc]	1476 ± 32 [abc]	101 ± 8 [abc]	n.d.	n.d.	n.d.
T-2	water	0	954 ± 81 [abc]	2550 ± 280 [abc]	196 ± 63 [abc]	n.d.	n.d.	n.d.
T-3	0	ascorbic acid	7476 ± 39 [abc]	2608 ± 90 [abc]	37,212 ± 992 [abc]	3737 ± 212 [b]	4098 ± 127 [abc]	3061 ± 145 [abc]
T-4	water	ascorbic acid	8734 ± 84 [abc]	3011 ± 89 [abc]	33,145 ± 255 [abc]	3779 ± 174 [b]	4611 ± 97 [abc]	2775 ± 44 [abc]

a—influenced by water addition, b—influenced by ascorbic acid addition, c—influenced by both water and ascorbic acid addition, chlorogenic acid LOD 0.42 $\mu g \cdot mL^{-1}$; 1,5-dicaffeoylquinic acid LOD 0.55 $\mu g \cdot mL^{-1}$; rosmarinic acid LOD 0.079 $\mu g \cdot mL^{-1}$. n.d. not detected.

Table 4. The content of monitored terpenes (% of peaks from normalized peak area).

Mentha spicata	Water	Ascorbic Acid	β-Pinene	Myrcene	Limonene	Eucalyptol	*trans*-Caryophyllene	Piperitenone Oxide	
M-1	0	0	0.337 ± 0.017 ^{ac}	1.090 ± 0.029 ^{abc}	2.517 ± 0.168 ^{ab}	5.443 ± 0.076 ^{b}	1.883 ± 0.125 ^{ac}	83.010 ± 0.403 ^{ac}	
M-2	water	0	0.350 ± 0.008 ^{ac}	1.140 ± 0.041 ^{abc}	2.583 ± 0.078 ^{ab}	5.563 ± 0.181 ^{b}	1.537 ± 0.024 ^{ac}	82.500 ± 0.268 ^{ac}	
M-3	0	ascorbic acid	0.313 ± 0.005 ^{ac}	1.170 ± 0.037 ^{abc}	2.623 ± 0.063 ^{ab}	6.007 ± 0.046 ^{b}	1.800 ± 0.033 ^{ac}	83.303 ± 0.172 ^{ac}	
M-4	water	ascorbic acid	0.367 ± 0.012 ^{ac}	1.457 ± 0.026 ^{abc}	2.980 ± 0.128 ^{ab}	6.007 ± 0.118 ^{b}	1.720 ± 0.028 ^{ac}	81.890 ± 0.168 ^{ac}	
Foeniculum vulgare	**Water**	**Ascorbic Acid**	**α-Pinene**	**Phelandrene**	**4-Cymene**	**Limonene**	**Fenchone**	**Estragole**	***trans*-Anethole**
F-1	0	0	0.717 ± 0.005 ^{abc}	1.153 ± 0.005 ^{ac}	0.570 ± 0.008 ^{bc}	0.217 ± 0.005 ^{ab}	1.160 ± 0.033 ^{ab}	0.260 ± 0.014 ^{abc}	95.800 ± 0.059 ^{abc}
F-2	water	0	0.480 ± 0.054 ^{abc}	0.877 ± 0.052 ^{ac}	0.780 ± 0.078 ^{bc}	0.173 ± 0.017 ^{ab}	1.323 ± 0.068 ^{ab}	0.383 ± 0.009 ^{abc}	95.863 ± 0.130 ^{abc}
F-3	0	ascorbic acid	0.650 ± 0.008 ^{abc}	1.295 ± 0.004 ^{ac}	0.210 ± 0.008 ^{bc}	0.160 ± 0.000 ^{ab}	0.775 ± 0.029 ^{ab}	0.250 ± 0.024 ^{abc}	96.595 ± 0.061 ^{abc}
F-4	water	ascorbic acid	0.257 ± 0.012 ^{abc}	0.817 ± 0.024 ^{ac}	0.110 ± 0.008 ^{bc}	0.110 ± 0.008 ^{ab}	0.750 ± 0.024 ^{ab}	0.223 ± 0.005 ^{abc}	97.650 ± 0.029 ^{abc}
Thymus vulgaris	**Water**	**Ascorbic Acid**	**4-Cymene**	**γ-Terpinene**	**Linalool**	**Borneol**	**Terpinen-4-ol**	**Thymol**	**Carvacrol**
T-1	0	0	3.300 ± 0.062 ^{ab}	0.377 ± 0.031 ^{abc}	0.177 ± 0.046 ^{a}	0.230 ± 0.043 ^{a}	0.057 ± 0.017 ^{abc}	88.897 ± 0.893 ^{ac}	2.760 ± 0.228 ^{abc}
T-2	water	0	2.733 ± 0.070 ^{ab}	0.257 ± 0.012 ^{abc}	0.100 ± 0.000 ^{a}	0.120 ± 0.008 ^{a}	0.027 ± 0.005 ^{abc}	91.180 ± 0.204 ^{ac}	2.510 ± 0.070 ^{abc}
T-3	0	ascorbic acid	1.147 ± 0.021 ^{ab}	2.243 ± 0.108 ^{abc}	0.207 ± 0.029 ^{a}	0.220 ± 0.033 ^{a}	0.313 ± 0.025 ^{abc}	86.267 ± 1.319 ^{ac}	3.967 ± 0.180 ^{abc}
T-4	water	ascorbic acid	0.593 ± 0.056 ^{ab}	1.510 ± 0.104 ^{abc}	0.090 ± 0.000 ^{a}	0.100 ± 0.000 ^{a}	0.200 ± 0.000 ^{abc}	92.663 ± 0.081 ^{ac}	2.750 ± 0.273 ^{abc}

a—influenced by water addition; b—influenced by ascorbic acid addition; c—influenced by both water and ascorbic acid addition. "0" in the Table means that there is no addition of ascorbic acid or water.

By using herb homogenates as a food additive, a complex of substances having the same synergistic effect and the same biological effects as the original herb is introduced into the food, which is not ensured by adding only one major substance obtained from the herb by extraction. This was verified in the hops homogenate, where the complete homogenate had higher antimicrobial activity than individual alpha and beta bitter acids acting alone [36]. It is quite clear that plant extracts and herb homogenates in this case have shown a considerable promise in a range of applications in the food industry, which also results from the published literature, e.g., [37].

4. Conclusions

The differences in the content of most polyphenolic compounds in different homogenate preparation were considerable. The greatest effect has the addition of ascorbic acid during the homogenization process on the content of phenolic acids, e.g., rosmarinic acid and chlorogenic acid. In thyme homogenates, the rosmarinic acid content was almost 200 times higher compared to thyme homogenates without ascorbic acid. The content of flavonols glucuronides (quercetin-3-O-glucuronide, kaempferol-3-O-glucuronide) and flavone glucuronides (luteolin-7-glucuronide and apigenin-7-glucuronide) was also the highest in the presence of ascorbic acid. Flavones luteolin, apigenin, diosmin, and diosmetin behave rather differently; there is an unclear dependence on the presence of ascorbic acid. From our experiments, and from the literature, we can conclude that for the preservation of phenolic acids, flavonols, and flavone glucuronides in the fresh herb homogenates, it is very important to add ascorbic acid, which will block enzymatic activity. This step prevents rapid degradation of the compounds and suppresses ionization of acidic compounds in water. Therefore, compounds of acidic nature reveal a higher amount in the homogenates. The obtained results also show that in order to achieve the necessary accuracy in the analysis of natural substances containing phenolic compounds and terpenes, increased attention must be paid to setting and maintaining the optimal pH value.

Supplementary Materials: The following supporting information can be downloaded at: https://www.mdpi.com/article/10.3390/foods14132165/s1.

Author Contributions: Conceptualization, J.T. and N.V.; methodology, N.V.; validation and statistics, N.V., E.K. and P.N.; formal analysis, J.T.; high pressure treatment, J.S.; resources, J.T. and N.V.; data curation, N.V.; writing—original draft preparation, J.T.; writing—review and editing, M.H.; analyses of terpenes and data curation, J.B.; supervision, R.P.; project administration, R.P.; funding acquisition, R.P. All authors have read and agreed to the published version of the manuscript.

Funding: "This research was funded by Ministry of Agriculture of the Czech Republic (Medicinal plants in the food industry—a new direction for the prevention of civilization diseases, Project No. QL24010019), by the Ministry of Education, Youth and Sports of the Czech Republic (AdAgriF; CZ.02.01.01/00/22_008/0004635) and by Metrofood-CZ (Grant No: LM2023064).

Institutional Review Board Statement: Not applicable.

Informed Consent Statement: Not applicable.

Data Availability Statement: The original contributions presented in the study are included in the article, further inquiries can be directed to the corresponding author.

Conflicts of Interest: The authors declare no conflicts of interest.

References

1. Gökbulut, A.; Şarer, E. Simultaneous Determination of Phenolic Compounds in *Mentha spicata* L. subsp. Spicata by RP-HPLC. *Turk. J. Pharm. Sci.* **2010**, *7*, 249–254.
2. Bimakr, M.; Rahman, L.A.; Taip, F.S.; Ganjloo, A.; Salleh, L.M.; Selamat, J.; Hamid, A.; Zaidul, I.S.M. Comparison of different extraction methods for the extraction of major bioactive flavonoid compounds from spearmint (*Mentha spicata* L.) leaves. *Food Bioprod. Process.* **2011**, *89*, 67–72. [CrossRef]
3. Saeidi, I.; Hadjmohammadi, M.R.; Peyrovi, M.; Iranshahi, M.; Barfi, B.; Babaei, A.B.; Dust, A.M. HPLC determination of hesperidin, diosmin and eriocitrin in Iranian lime juice using polyamide as an adsorbent for solid phase extraction. *J. Pharm. Biomed. Anal.* **2011**, *56*, 419–422. [CrossRef]
4. Križman, M.; Baričevič, D.; Prošek, M. Determination of phenolic compounds in fennel by HPLC and HPLC–MS using a monolithic reversed-phase column. *J. Pharm. Biomed. Anal.* **2007**, *43*, 481–485. [CrossRef]
5. Roby, M.H.H.; Sarhan, M.A.; Selim, K.A.-H.; Khalel, I.K. Antioxidant and antimicrobial activities of essential oil and extracts of fennel (*Foeniculum vulgare* L.) and chamomile (*Matricaria chamomilla* L.). *Ind. Crops Prod.* **2013**, *44*, 437–445. [CrossRef]
6. Nagy, T.O.; Solar, S.; Sontag, G.; Koenig, J. Identification of phenolic components in dried spices and influence of irradiation. *Food Chem.* **2011**, *128*, 530–534. [CrossRef] [PubMed]
7. Roby, M.H.H.; Sarhan, M.A.; Selim, K.A.-H.; Khalel, I.K. Evaluation of antioxidant activity, total phenols and phenolic compounds in thyme (*Thymus vulgaris* L.), sage (*Salvia officinalis* L.), and marjoram (*Origanum majorana* L.) extracts. *Ind. Crops Prod.* **2013**, *43*, 827–831. [CrossRef]
8. Pavela, R.; Sedlák, P. Post-application temperature as a factor influencing the insecticidal activity of essential oil from Thymus vulgaris. *Ind. Crops Prod.* **2018**, *113*, 46–49. [CrossRef]
9. Odeh, A.; Allaf, A.W. Determination of polyphenol component fractions and integral antioxidant capacity of Syrian aniseed and fennel seed extracts using GC–MS, HPLC analysis, and photochemiluminescence assay. *Chem. Pap.* **2017**, *71*, 1731–1737. [CrossRef]
10. Agarwal, D.; Saxena, S.N.; Sharma, L.K.; Lal, G. Prevalence of Essential and Fatty Oil Constituents in Fennel (*Foeniculum vulgare* Mill) Genotypes Grown in Semi-Arid Regions of India. *J. Essent. Oil-Bear. Plants* **2018**, *21*, 40–51. [CrossRef]
11. Najdoska-Bogdanov, M.; Bogdanov, J.B.; Stefova, M. Simultaneous determination of essential oil components and fatty acids in fennel using gas chromatography with a polar capillary column. *Nat. Prod. Commun.* **2015**, *10*, 1619–1626. [CrossRef] [PubMed]
12. Bardaweel, S.K.; Bakchiche, B.; ALSalamat, H.M.; Rezzoug, M.; Gherib, A.; Flamini, G. Chemical composition, antioxidant, antimicrobial and Antiproliferative activities of essential oil of *Mentha spicata* L. (Lamiaceae) from Algerian Saharan atlas. *BMC Complement. Alter. Med.* **2018**, *18*, 201. [CrossRef]
13. Bishr, M.M.; Salama, O.M. Inter and intra GC-MS differential analysis of the essential oils of three *Mentha* species growing in Egypt. *Future J. Pharm. Sci.* **2018**, *4*, 53–56. [CrossRef]
14. Saba, I.; Anwar, F. Effect of harvesting regions on physico-chemical and biological attributes of supercritical fluid-extracted spearmint (*Mentha spicata* L.) leaves essential oil. *J. Essent. Oil Bear. Plants* **2018**, *21*, 400–419. [CrossRef]
15. Rodríguez-Solana, R.; Salgado, J.M.; Domínguez, J.M.; Cortés-Diéguez, S. Comparison of Soxhlet, Accelerated Solvent and Supercritical Fluid Extraction Techniques for Volatile (GC–MS and GC/FID) and Phenolic Compounds (HPLC–ESI/MS/MS) from *Lamiaceae* Species. *Phytochem. Anal.* **2015**, *26*, 61–71. [CrossRef] [PubMed]
16. Patel, K.; Gadewar, M.; Tahilyani, V.; Patel, D.K. A review on pharmacological and analytical aspects of diosmetin: A concise report. *Chin. J. Integr. Med.* **2013**, *19*, 792–800. [CrossRef]
17. Hostetler, G.L.; Ralston, R.A.; Schwartz, S.J. Flavones: Food Sources, Bioavailability, Metabolism, and Bioactivity. *Adv. Nutr.* **2017**, *8*, 423–435. [CrossRef]
18. Kim, G.-D.; Park, Y.S.; Jin, Y.-H.; Park, C.-S. Production and applications of rosmarinic acid and structurally related compounds. *Appl. Microbiol. Biotechnol.* **2015**, *99*, 2083–2092. [CrossRef]
19. Hosseinzadeh, S.; Kukhdan, A.J.; Hosseini, A.; Armand, R. The application of Thymus vulgaris in traditional and modern medicine: A review. *Glob. J. Pharmacol.* **2015**, *9*, 260–266.
20. Kooti, W.; Moradi, M.; Ali-Akbari, S.; Sharafi-Ahvazi, N.; Asadi-Samani, M.; Ashtary-Larky, D. Therapeutic and pharmacological potential of Foeniculum vulgare Mill: A review. *J. HerbMed. Pharmacol.* **2015**, *4*, 1–9.
21. Kushwah, P.; Patel, R.; Midda, A.; Kayande, N. Pharmacological review on *Foeniculum vulgare*. *Int. J. Adv. Sci. Res.* **2016**, *1*, 40–42.
22. Houška, M.; da Silva, V.M. (Eds.) *High Pressure Processing of Fruit and Vegetable Products*; CRC Press: Boca Raton, FL, USA, 2017.
23. Strohalm, J.; Houška, M.; Novotná, P. A Food Preparation Based on Hydrocolloids with the Addition of Mongolian Milkvetch, Chinese Knotweed and Red Sage. Utility Pattern No. 31446, 6 February 2018.
24. Houška, M.; Strohalm, J.; Kocurová, K.; Totušek, J.; Lefnerová, D.; Tříska, J.; Vrchotová, N.; Fiedrleová, V.; Holasová, M.; Gabrovská, D.; et al. High pressure and foods-fruit/vegetable juices. *J. Food Eng.* **2006**, *77*, 386–398. [CrossRef]

25. *ISO 4833-1:2013*; Microbiology of the Food Chain—Horizontal Method for the Enumeration of Microorganisms Part 1: Colony Count at 30 °C by the Pour plate Technique. Available online: https://www.iso.org/standard/53728.html (accessed on 1 May 2025).
26. *ISO 21527-1:2008*; Microbiology of Food and Animal Feeding Stuffs—Horizontal Method for the Enumeration of Yeasts and Moulds Part 1: Colony Count Technique in Products with Water Activity Greater than 0.95. Available online: https://www.iso.org/standard/38275.html (accessed on 1 May 2025).
27. Dunn, O.J.; Clark, V.A. *Applied Statistics: Analysis of Variance and Regression*; John Wiley & Sons, Inc.: Hoboken, NJ, USA, 1974.
28. Friedman, M.; Jürgens, H.S. Effect of pH on the stability of plant phenolic compounds. *J. Agric. Food Chem.* **2000**, *48*, 2101–2110. [CrossRef] [PubMed]
29. Patel, S.R. Bioanalytical challenges and strategies for accurately measuring acyl glucuronide metabolites in biological fluids. *Biomed. Chromatog.* **2020**, *34*, e4640. [CrossRef] [PubMed]
30. Zhang, Y.; Smuts, J.P.; Dodbiba, E.; Rangarajan, R.; Lang, J.C.; Armstrong, D.W. Degradation study of carnosic acid, carnosol, rosmarinic acid and rosemary extract (*Rosmarinus officinalis* L.) assessed using HPLC. *J. Agric. Food Chem.* **2012**, *60*, 9305–9314. [CrossRef]
31. Sanchez-Medina, A.; Etheridge, C.J.; Hawkes, G.E.; Hylands, P.J.; Pendry, B.A.; Hughes, M.J.; Corcoran, O. Comparison of rosmarinic acid content in commercial tinctures produced from fresh and dried lemon balm (*Melissa officinalis*). *J. Pharm. Pharm. Sci.* **2007**, *10*, 455–463. [CrossRef]
32. Olah, N.K.; Osser, G.; Campean, R.F.; Furtuna, F.R.; Benedec, D.; Filip, L.; Raita, O.; Hanganu, D. The study of polyphenolic compounds profile of some *Rosmarinus officinalis* L. extracts. *Pak. J. Pharm. Sci.* **2016**, *29* (Suppl. S6), 2355–2361.
33. Sik, B.; Lakatos, E.H.; Kapcsándi, V.; Székelyhidi, R.; Ajtony, Z. Investigation of the long-term stability of various tinctures belonging to the lamiaceae family by HPLC and spectrophotometry method. *Chem. Pap.* **2021**, *75*, 5781–5791. [CrossRef]
34. Bodalska, A.; Kowalczyk, A.; Fecka, I. Stability of rosmarinic acid and flavonoid glycosides in liquid forms of herbal medicinal products—A preliminary study. *Pharmaceuticals* **2021**, *14*, 1139. [CrossRef]
35. Horablaga, N.M.; Cozma, A.; Alexa, E.; Obistioiu, D.; Cocan, I.; Poiana, M.-A.; Lalescu, D.; Pop, G.; Imbrea, I.M.; Buzna, C. Influence of sample preparation/extraction method on the phytochemicalprofile and antimicrobial activities of 12 commonly consumed medicinal plants in Romania. *Appl. Sci.* **2023**, *13*, 2530. [CrossRef]
36. Čermák, P.; Palečková, V.; Houška, M.; Strohalm, J.; Novotná, P.; Mikyška, A.; Jurková, M.; Sikorová, M. Inhibitory effects of fresh hops on *Helicobacter pylori* strains. *Czech J. Food Sci.* **2015**, *33*, 302–307. [CrossRef]
37. Negi, P.S. Plant extracts for the control of bacterial growth: Efficacy, stability and safety issues for food application. *Int. J. Food Microbiol.* **2012**, *156*, 7–17. [CrossRef] [PubMed]

foods

MDPI

Article

Permeabilization of *Cryptosporidium* spp. Oocysts in Water, Apple and Carrot Juice by Pulsed Electric Field Technology

Alejandro Berzosa [1], Laura Garza-Moreno [2], Joaquín Quílez [2], Javier Raso [1], Ignacio Álvarez-Lanzarote [1,*] and Juan Manuel Martínez [1,*]

[1] Departamento de Producción Animal y Ciencia de Los Alimentos, Tecnología de Los Alimentos, Facultad de Veterinaria, Instituto Agroalimentario de Aragón (IA2), Universidad de Zaragoza, 50009 Zaragoza, Spain; aberzosa@unizar.es (A.B.); jraso@unizar.es (J.R.)

[2] Departamento de Patología Animal, Facultad de Veterinaria, Instituto Agroalimentario de Aragón (IA2), Universidad de Zaragoza, 50009 Zaragoza, Spain; lgarza@unizar.es (L.G.-M.); jquilez@unizar.es (J.Q.)

* Correspondence: ialvalan@unizar.es (I.Á.-L.); j.martinez@unizar.es (J.M.M.)

Abstract: *Cryptosporidium* spp. oocysts are highly resistant to conventional disinfection methods and have been associated with foodborne outbreaks linked to unpasteurized fruit and vegetable juices. This study aimed to evaluate the effectiveness of Pulsed Electric Fields (PEF) in permeabilizing *Cryptosporidium* oocysts in water, apple juice, and carrot juice. Oocysts were exposed to monopolar square-wave pulses (3 μs) at electric field strengths ranging from 15 to 35 kV/cm, with treatment times up to 180 μs, and application temperatures between 25 °C and 60 °C. Membrane permeabilization was assessed using propidium iodide uptake via fluorescence microscopy and flow cytometry. Results showed that oocyst permeabilization increased with electric field strength, treatment time, and temperature, with up to 90% permeabilization achieved at 35 kV/cm and 45 °C. Carrot juice treatments yielded higher permeabilization levels than apple juice, attributed to greater electrical conductivity and energy input. Temperatures below 60 °C alone had negligible effects, but synergistically enhanced PEF efficacy. These findings demonstrate that PEF, particularly when combined with mild heat, is a promising non-thermal technology for reducing *Cryptosporidium* viability in beverages, offering an effective alternative for improving the microbiological safety of minimally processed juices while preserving sensory and nutritional quality.

Keywords: *Cryptosporidium*; PEF; inactivation; electroporation; beverage; fruit; vegetable

1. Introduction

Cryptosporidium spp. is an intracellular protozoan parasite, of which 42 different species have been recognized. Among these, *C. hominis* and *C. parvum* are most notable, responsible for the majority of human infections [1]. *C. hominis* is primarily an anthroponotic species, although its presence has been sporadically reported in various animals. In contrast, *C. parvum* is a potentially zoonotic species, capable of infecting a wide variety of mammals, including ruminants and humans. It is worth noting that *Cryptosporidium* is a ubiquitous parasite with a low infectious dose (132 oocysts for *C. parvum* and 10–83 for *C. hominis*) and high resistance to disinfectants used for drinking water treatment, making it the most prevalent waterborne protozoan in developed countries [2].

Cryptosporidium spp. infects the epithelial cells of the gastrointestinal tract of vertebrates. Its biological cycle is monoxenous, meaning it develops entirely within the same host. Infection occurs orally through the ingestion of food or water contaminated with oocysts, and upon reaching the stomach, four sporozoites are released that invade the enterocytes [3]. Within the intestinal epithelial cells, the parasite undergoes two cycles of asexual reproduction that severely damage the epithelium, followed by a sexual phase that culminates in the formation of oocysts with sporozoites through intracellular sporogony. Most oocysts have a thick wall and are excreted in feces, potentially being ingested by another host, and the cycle begins anew.

The global prevalence of cryptosporidiosis in humans is 7.6%, increasing to 31.5% in developing countries (with an average of 10.4%) and ranging from 0.1% to 14.1% in developed countries, with an estimated average of 4.3% [4]. Cryptosporidiosis is a disease under surveillance in the European Union, although mandatory reporting varies between countries, being obligatory in some member states such as Germany, Spain, Ireland, and Sweden [5]. This disease is considered a global health emergency, especially for children and immunocompromised patients. According to the European Centre for Disease Prevention and Control (ECDC), there was an increase in cases in 2021 compared to previous years, with a total of 4476 cases reported by 24 EU/EEA countries. The most affected age group was children aged 0–4 years, with a notification rate of 6.4 per 100,000.

Although oocysts do not multiply in water, their small size allows them to evade conventional filtration treatments and they can survive for long periods when exposed to chlorine and other chemicals commonly used for drinking water purification [6,7]. Chlorine requires high doses and long exposure times to be effective, yet there is a threshold beyond which it has no additional impact [8]. Ozone, while effective, contaminates the air and is irritating to humans [9]. Distillation can remove up to 99.5% of microbes but is extremely energy-intensive, as it requires boiling water and condensing the vapor making it impractical for large-scale treatment [10]. In the absence of effective treatment, this resilient parasite can survive in drinking water and occasionally causes outbreaks that impact public health [11].

In addition to waterborne transmission, *Cryptosporidium* has gained prominence as a food safety concern, with an increasing number of foodborne outbreaks attributed to its presence. Transmitted via the fecal-oral route, its environmentally resilient oocysts can contaminate a wide range of food products, including fresh produce, unpasteurized juices, and raw milk. Notably, the infectious dose is extremely low, meaning that even minimal contamination poses a substantial health risk [12,13]. The first large-scale foodborne outbreak was reported in the United States in 1996 involving freshly prepared apple juice [13]. A 2005 outbreak in Denmark was traced to a contaminated salad bar, with ingredients like carrots and red peppers identified as the source [14]. More recently, an outbreak in Norway linked to self-pressed apple juice contaminated with *C. parvum* emphasized the microbiological risks associated with artisanal juice production [15]. Collectively, these outbreaks—along with additional cases reported in non-scientific sources—highlight that unpasteurized fruits and vegetables can consistently act as vehicles for *Cryptosporidium* transmission. This underscores the urgent need for effective and alternative safety measures in the production of minimally processed beverages. As the consumption of products such as juices, smoothies, purées, compotes, and fresh-cut fruits and vegetables continues to rise—largely due to their perceived nutritional benefits—the combination of *Cryptosporidium*'s low infectious dose and high environmental resilience presents a persistent and significant challenge to food safety within this category [16,17].

Traditional thermal treatments, such as flash pasteurization, have been shown to achieve significant reductions in oocyst viability. Specifically, heat treatments at 71.7 °C for 10–20 s have been shown to inactivate up to 99.999% [18]. However, reliance on heat treatment can negatively impact the sensory and nutritional qualities of fruit juices, altering flavor, aroma, and vitamin content, which is particularly undesirable in the production of fresh or premium juices marketed for their natural attributes [19]. Non-thermal methods are also under investigation for their potential to preserve nutritional and sensory qualities of juices while ensuring microbial safety. Ultraviolet (UV) irradiation has demonstrated potential to inactivate *C. parvum* in apple cider without altering taste or appearance, but its effectiveness can be significantly reduced in turbid or colored juices where suspended particles shield oocysts from UV exposure, leading to incomplete disinfection [20]. Similarly, chemical treatments using organic acids like malic and citric acid, or oxidizing agents such as hydrogen peroxide and ozone, have shown variable efficacy depending on concentration and exposure time, but they may also affect the organoleptic properties of the juice or leave residual compounds that could alter flavor, pH, or consumer acceptability [21,22]. High hydrostatic pressure (HHP), ultrasounds, and pulsed light technologies are emerging as promising alternatives, achieving significant oocyst reduction with minimal impact on juice quality, but these methods can involve high operational costs, specialized equipment, and may still require optimization for consistent efficacy across different juice matrices and contamination levels [23,24]. Continued research into these inactivation strategies is essential for developing safe, non-thermal juice processing methods that can reliably target *Cryptosporidium* and other protozoan pathogens.

The physical alternative proposed in this study to inactivate *Cryptosporidium* oocysts is Pulsed Electric Fields (PEF). PEF treatments involve subjecting a product placed between two electrodes, usually immersed in an aqueous solution, to high-intensity electric fields (between 0.5 and 30 kV/cm) by intermittently applying short-duration pulses (microseconds to milliseconds). Although the application of electric fields inherently introduces energy into the system, the total treatment time is extremely brief—usually less than one second in continuous flow systems. This short exposure time ensures that the temperature increase is minimal and transient. Consequently, thermal damage is effectively negligible, preserving the physicochemical and sensory qualities of the treated food. If the electric field intensity is sufficiently high, electroporation of the cytoplasmic membrane occurs, increasing its permeability to ions and macromolecules [25]. This phenomenon can be reversible or irreversible depending on the electrical parameters applied. In reversible electroporation, the membrane reseals after the treatment, whereas in irreversible electroporation, the structural integrity of the membrane is permanently compromised. This irreversible loss of selective permeability disrupts cellular homeostasis and can lead to cell death [26].

Although PEF is widely used for the non-thermal inactivation of bacteria in food [27], its application for parasite inactivation has been limited. Recent studies targeted the inactivation of nematode larvae as *Anisakis* spp. and *Trichinella* spp. [26,28], while there is limited information on the effect of PEF on infective forms of protozoa (oocysts), except for the work by Haas and Aturaliye [29]. However, 26 years have passed since this study, and PEF technology has greatly improved, offering much greater treatment possibilities today. PEF technology has been successfully scaled up for industrial use, with continuous-flow systems processing thousands of liters per hour. It is already applied in sectors like juice pasteurization and winemaking, demonstrating its feasibility for high-throughput applications. Despite these technological developments, no systematic investigation has been conducted to assess how PEF influences protozoan oocyst inactivation. Therefore, this study addresses a significant gap in the literature by being, to our knowledge, the first to evaluate PEF-mediated permeabilization of *Cryptosporidium* oocysts, under well-controlled experimental conditions, using modern and high-performance PEF equipment.

The objective of this work is to evaluate the potential of PEF treatments to inactivate *Cryptosporidium* spp. oocysts for the decontamination of water and food matrices such as fruit juices. The aim is to achieve irreversible electroporation of *Cryptosporidium* cell membranes, thereby compromising their physiology to the extent of causing inactivation and preventing water and food-borne transmission.

2. Materials and Methods

2.1. Isolation and Purification of Cryptosporidium Oocysts

Cryptosporidium spp. oocysts were isolated from fecal samples collected from 10-day-old calves naturally infected within a herd located in Aragón, Spain. Fecal samples were obtained directly from the rectum, placed in sterile containers and transported to the Faculty of Veterinary Medicine at the University of Zaragoza for further analysis. These fecal specimens were collected by the herd veterinarian for diagnostic analysis of parasitic protozoa.

Oocysts in fecal samples were concentrated and purified through successive centrifugations using a sodium chloride (NaCl) flotation method. Purified oocysts were suspended in microtubes at a final concentration of 3×10^6 oocysts/mL and stored at 4 °C until analysis.

2.2. PEF Treatment

The PEF equipment used in this study was a commercial device (Vitave, Prague, Czech Republic) capable of delivering pulses of up to 20 kV. The system applies monopolar square-wave pulses of variable width (500 ns–100 µs), with a maximum current intensity of 500 A, and allows operation at frequencies up to 50 kHz. The applied voltage was monitored using a high-voltage probe (P6015A, Tektronik, Wilsonville, OR, USA), while the current intensity was recorded with a current probe (HCT5514, Meatrol®Electrical, Shanghai, China). Both probes were connected to an oscilloscope (TDS 220, Tektronik Wilsonville, OR, USA).

Treatments at **room temperature**: a suspension of oocysts (0.44 mL, with a conductivity of 1 mS/cm) was introduced into a cylindrical static treatment chamber with parallel electrodes, featuring specific dimensions (gap: 0.25 cm; diameter: 1.5 cm), using a sterile 1 mL syringe (TERUMO, Leuven, Belgium) with a 20 G × 1″ (0.9 × 25 mm) needle (TERUMO, Leuven, Belgium). Samples for investigating the effect of PEF on oocysts were subjected to between 20 and 90 square-wave pulses of 3 µs at different electric field intensities (15, 25, and 35 kV/cm). These treatments corresponded to specific energies ranging from 14.65 to 320.27 kJ/kg. Control and treated oocyst experiments were performed in duplicate.

Temperature-controlled treatments: The batch parallel electrode treatment chamber used for the temperature-controlled treatments was based on a previous design of Heinz et al. [30] and later of [31]. The treatment chamber consisted of a cylindrical polypropylene tube closed with two polished stainless steel cylinders of 1.77 cm^2 surface and 4 cm length. The inner part of the electrodes was empty, and dielectric oil (conductivity: 1.4 µS/cm) tempered at different temperatures was recirculated through both electrodes to temper the treatment medium at different temperatures and to maintain a constant temperature in the medium during the PEF treatments. A small hole in the cylindrical plastic tube (chamber) was used to introduce and remove the microbial suspension. The distance between electrodes was 0.25 cm. Both the oocyst suspensions in water and the juices inoculated with oocysts introduced into the chamber for each treatment had a volume of 0.44 mL. Before and immediately after applying the PEF treatment, in order to know the temperature in different zones of the treatment chamber, the temperature was measured using an Infrared Thermal imaging camera (FLIR E6 PRO, Teledyne FLIR LLC, Wilsonville, OR, USA,) and a fiber-optic temperature probe (Fiso, Québec, QC, Canada).

2.3. Assessment of Thermal Effects on Oocyst Membrane Permeabilization in the Absence of PEF

To isolate the thermal effects from those induced by PEF and to better understand the heat resistance of oocysts, oocyst solutions were placed in the PEF treatment chamber pre-tempered at 40, 50, 60 and 70 °C for 1, 5, and 10 min, with the PEF system turned off during the thermal exposure. This approach allowed the assessment of membrane permeabilization attributable exclusively to temperature.

PEF-treated and thermal treated samples were processed immediately for microscopic observation and flow cytometry analysis in order to minimize the time lapse between treatment and evaluation.

2.4. Assessment of *Cryptosporidium* Oocyst Viability

2.4.1. Microscopy

To evaluate the viability of *Cryptosporidium* oocysts and potential viability loss following different treatments, a fluorescence microscopy-based method was employed, using the differential uptake of the vital dyes 4′,6-diamidino-2-phenylindole (DAPI) and propidium iodide (PI). DAPI is a membrane-permeant fluorochrome that binds to double-stranded DNA, staining the nuclei of all sporozoites of the oocysts regardless of viability. PI, in contrast, is a membrane-impermeant dye that only penetrates oocysts with compromised membranes; it intercalates with nucleic acids and emits red fluorescence. The uptake of PI by oocysts reflects membrane permeabilization, which serves as a functional indicator of compromised integrity and, consequently, loss of viability.

Aliquots containing 3×10^6 oocysts were transferred into 1.5 mL microtubes and pre-incubated in acidified Hanks' Balanced Salt Solution (HBSS, pH 2.75) for 1 h at 37 °C to facilitate dye uptake. This was followed by two washes with HBSS (pH 7.2) to restore physiological pH. Then, 10 µL of DAPI working solution (2 mg/mL in absolute methanol) and 10 µL of PI working solution (1 mg/mL in 0.1 M phosphate-buffered saline, pH 7.2) were added to each sample and incubated for 2 h at 37 °C in the dark. To prevent DAPI crystallization, samples were washed twice with HBSS adjusted to pH 4, following the protocol described by Bukhari et al. [32].

Oocysts were visualized using an epifluorescence microscope (Nikon, Mod. L-Kc, Nippon Kogaku KK, Tokyo, Japan) equipped with a UV excitation filter (350–365 nm) for DAPI and a green excitation filter (545–546 nm) for PI. A minimum of 100 oocysts were examined per sample. Oocysts showing only blue fluorescence (DAPI-positive, PI-negative) were considered viable, whereas those exhibiting both blue and red fluorescence (DAPI-positive, PI-positive) were classified as non-viable due to membrane permeabilization. This loss of selective membrane integrity reflects compromised viability and parallels the concept of electroporation-induced damage explored in later analyses.

The data of treated oocysts were normalized to the background permeability observed in the controls. The viability percentage of the negative control oocysts (*NC*) was adjusted to 100%, and the viability percentage of the treated oocysts (*TO*) was corrected by the same factor using the following formula:

$$Corrected\ TO = \left(\frac{Observed\ TO}{Observed\ NC}\right) \times 100 \quad (1)$$

The percentage of permeabilized oocysts was calculated as the difference between 100% and the viability percentage.

2.4.2. Flow Cytometry

To quantitatively assess membrane permeabilization in *Cryptosporidium* oocysts following PEF treatments, with greater sensitivity and resolution than microscopy, flow cytometry was employed as the principal analytical technique. A dual-fluorescence staining protocol was carried out using a Guava® easyCyte™ flow cytometer (Luminex®, Tokyo, Japan), which operates with a 488 nm blue laser for excitation and detects fluorescence through specific optical filters.

Cryptosporidium oocysts were identified using the Aqua-Glo™ G/C Direct Fluorescent Antibody reagent (Waterborne Inc., New Orleans, LA, USA), which contains fluorescein isothiocyanate (FITC)-labeled mouse monoclonal antibodies directed against cyst and oocyst outer wall antigenic sites (epitopes) of *Giardia lamblia* and *Cryptosporidium parvum*. This reagent is genus-specific and binds exclusively to the cysts and oocysts of these two parasites. Green fluorescence from FITC-labeled oocysts was detected using a 525/30 nm bandpass filter (GRN-B), enabling specific identification of antibody-bound oocysts. Membrane permeabilization—and consequently, loss of viability—was assessed using propidium iodide (PI), whose fluorescence was detected using a 695/50 nm bandpass filter (RED-B).

Following PEF treatment, oocysts (370 μL, 10^6 oocyst/mL) were pre-incubated at pH 2.75 for 1 h at 37 °C, washed twice with HBSS (pH 7.2), and incubated with 30 μL of PI solution (0.1 mg/mL) for 2 h at 37 °C in the dark. During the final 25 min of incubation, 4.5 μL of the antibody reagent was added to each tube. Samples were then analyzed in duplicate using the flow cytometer, with 8000 events recorded per sample at a flow rate of 0.59 μL/s. Events emitting green fluorescence above 3.0×10^3 a.u. were considered FITC-positive, while those emitting red fluorescence above 3.0×10^3 a.u. were considered PI-permeable. The percentage of permeabilization was calculated based on the proportion of oocysts exceeding this fluorescence threshold.

Data acquisition began by visualizing the total particle distribution via forward scatter (FSC) versus side scatter (SSC) (Figure A1a). To identify the oocyst population and exclude smaller contaminants with overlapping size distributions, events emitting green fluorescence (Figure A1d) were gated to isolate specifically labeled oocysts (FITC-positive) (Figure A1b). Within this gated oocyst population, red fluorescence from PI was analyzed to differentiate between permeabilized oocysts (PI-positive) and intact oocysts (PI-negative) (Figure A1e,g).

This gating strategy enabled accurate and selective quantification of PEF-induced membrane permeabilization by analyzing PI uptake specifically within the FITC-positive oocyst population, while excluding non-target particles and background debris.

2.5. Fruit Juices

The PEF treatments applied in the buffer were validated with the oocysts suspended in two fruit juice matrices. Commercial pasteurized 100% fruit juice made from Golden and Granny Smith apples was used. Carrot juice was prepared using the Juissen 2 cold press juicer (Sojamatic, Barcelona, Spain) and commercial carrots. The physicochemical properties of the juices are shown in Table 1. The juices were inoculated with 2×10^6 oocysts/mL and treated by PEF at 25 kV/cm (30 pulses of 3 μs) applied at different temperatures (20, 30, and 40 °C).

Table 1. Physicochemical properties of juices used in the validation of PEF treatments.

Property	Carrot Juice	Apple Juice
pH	5.94 ± 0.01	3.40 ± 0.01
°Brix	6.35 ± 0.05	11.9 ± 0.1
Conductivity	4.3 ± 0.01	1.64 ± 0.01

2.6. Experimental Design

Response surface methodology (RSM) was used to evaluate the effect of electric field strength (15–35 kV/cm), treatment time (60–180 μs), and input energy (14.65 to 320.27 kJ/kg) on the permeabilization of *Cryptosporidium* to PI. The data obtained were modeled with the following second-order polynomial equation:

$$Y = \beta_0 + \Sigma_{i=1}{}^{k} \beta_i X_i + \Sigma_{i=1}{}^{k} \beta_{ii} X_i{}^2 + \Sigma_{i<j}{}^{k} \beta_{ij} X_i X_j \quad (2)$$

In which Y is the response variable to be modeled, X_i and X_j are independent factors, β_0 is the intercept, β_i represents the linear coefficients, β_{ii} represents the quadratic coefficients, β_{ij} represents the cross-product coefficients, and k is the total number of independent factors. A backward regression procedure was used to determine the models' parameters. This procedure systematically removed the effects that were not significantly associated ($p > 0.05$) with the response until a model with a significant effect was obtained.

2.7. Statistical Analysis

Experiments were performed in triplicate, and the presented results are the mean ± 95% confidence interval. One-way analysis of variance (ANOVA) using Tukey's test was performed to evaluate the significance of differences between the mean values. The differences were considered significant at $p < 0.05$. A multiple regression analysis was conducted for fitting Equation (2) to the experimental data and significant terms of the model were determined by ANOVA. The central composite design and the corresponding data analysis were carried out by using the software package Design-Expert 13 (Stat-Ease Inc., Minneapolis, MN, USA).

3. Results and Discussion

3.1. Influence of the Electric Parameters on Cryptosporidium Oocysts Permeabilization to PI

3.1.1. Effect of Electric Field Strength and Treatment Time

The influence of electric field strength and treatment time (number of pulses × pulse width) on the permeabilization of *Cryptosporidium* to PI was investigated in order to find the optimal PEF treatment conditions that maximize inactivation. The results confirm that a certain level of permeabilization occurs after PEF treatments, with a clear direct effect of electric field strength and treatment duration. The range of permeabilized oocysts varied between 2.09% and 96.11%, depending on the treatment parameters, highlighting the importance of fine-tuning the electrical conditions to achieve effective membrane disruption.

3.1.2. Regression Modeling and Statistical Significance

In order to quantify the influence of electric field strength and treatment time on the permeabilization of *Cryptosporidium*, a multiple regression analysis was performed, fitting the experimental data to Equation (2). The analysis resulted in the following equation after removing non-significant terms ($p > 0.05$):

$$\%\ Permeabilized\ Cryptosporidium\ oocyst = 72.707 - 7.571E - 0.076t + 0.009Et + 0.194E^2 \quad (3)$$

Results of the analysis of variance (ANOVA) are shown in Table 1, including the statistical measures used to assess the adequacy of the generated models. The Model F-value of 101.82 implies the model is significant. There is only a 0.01% chance that an F-value this large could occur due to noise. p-values less than 0.05 indicate model terms are significant. In this case A, B, AB, and A^2 are significant model terms. The lack of fit F-value of 0.77 implies the lack of fit is not significant relative to the pure error. There is a 61.93%

chance that a lack of fit F-value this large could occur due to noise. Non-significant lack of fit is good; we want the model to fit.

The F-values of the model parameters displayed in Table 2 are indicators of the significance of the variables' effects. According to those F values, the electric field strength's linear term (F = 304.45), along with the electric field strength's quadratic (F = 39.64) were the two most significant variables, indicating that changes in these factors had the most influence on the oocysts permeabilization. The linear term of treatment time (F = 16.84) and the interaction between electric field and treatment time (F = 4.30) were also significant, but had lower F-values. The presence of that interaction term means that the effect of electric field strength on oocyst permeabilization slightly depends on the treatment time. The fit statistics of the model developed are detailed in Table 3.

Table 2. Coefficients and F-values of the mathematical equation to describe the influence of electric field strength and treatment time on the *Cryptosporidium* oocysts' permeabilization to PI after multiple regression modelling. Statistics to test the adequacy are also shown.

	Sum of Squares	df	Mean Square	F-Value	*p* Value
Model	40,577.82	4	10,144.46	101.82	<0.0001
A-Electric Field Strength	30,333.77	1	30,333.77	304.45	<0.0001
B-Treatment time	1677.53	1	1677.53	16.84	0.0002
AB	428.60	1	428.60	4.30	0.0437
$\mathbf{A^2}$	3949.28	1	3949.28	39.64	<0.0001
Residual	4583.20	46	99.63		
Lack of Fit	553.70	7	79.10	0.7656	0.6193
Pure Error	4029.50	39	103.32		

Table 3. Fit statistics of the model developed to describe the influence of electric field strength and treatment time on the *Cryptosporidium* oocysts permeabilization to PI.

Fit Statistics	
R^2	0.8985
Adjusted R^2	0.8897
Predicted R^2	0.8767
Adeq Precision	27.0390

The model showed a strong goodness-of-fit, with a coefficient of determination (R^2) of 0.899, indicating that approximately 89.9% of the variability in the response variable was explained by the fitted model. The adjusted R^2 (0.890) was close to the R^2 value, suggesting that the model was not overfitted and that the included predictors are relevant. Furthermore, the predicted R^2 (0.877) demonstrated a good agreement with both R^2 and adjusted R^2, confirming the model's predictive capability and its robustness in estimating responses for new observations.

3.1.3. Interpretation of the Model and Response Surface Visualization

In order to illustrate the influence of electric field strength and treatment time on the permeabilization of *Cryptosporidium* oocyst membrane, a graphical representation (Figure 1) has been generated using the regression model (Equation (3)) considering the responses within the range of experimental conditions assayed.

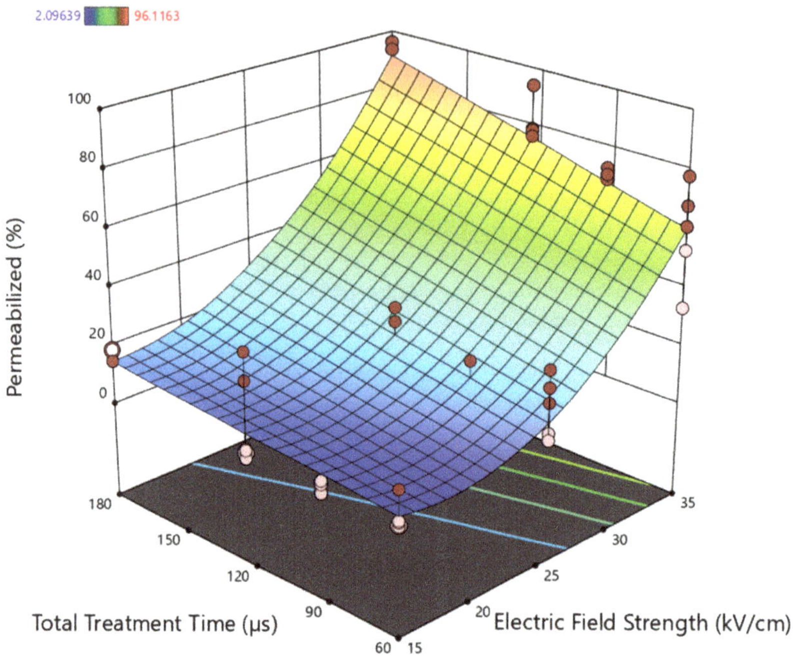

Figure 1. Three-dimensional surface plot illustrating the effects of electric field strength and treatment time on the PI permeabilization percentage of *Cryptosporidium* oocysts after Pulsed Electric Field (PEF) treatments. The surface represents the model prediction, with a color gradient indicating the magnitude of the response variable: blue for lower values and red for higher values. The color bar at the top-left corner corresponds to the predicted response values, ranging from 2.00689 (blue) to 96.1163 (red). Red dots represent actual experimental values that are higher than the model predictions, while pink dots indicate values that are lower than predicted.

As shown in the three-dimensional response surface, a progressive increase in permeabilization is observed with increasing both parameters. However, the contribution of each factor differs significantly. Electric field strength exhibits a non-linear, quadratic effect, evidenced by the pronounced curvature along the E-axis, suggesting a threshold behavior consistent with membrane electroporation theory [25]. At subcritical field levels < 25 kV/cm, the increase in membrane permeability is limited, but it escalates rapidly once the threshold is exceeded. A similar quadratic dependence of electric field strength has been reported in other parasites; for example, Abad et al. [28] described that the inactivation of *Anisakis* larvae followed a non-linear pattern with respect to electric field intensity, where moderate increases in the field resulted in significant higher lethality. Conversely, the effect of treatment time is linear and comparatively modest, indicating that prolonged exposure alone cannot compensate for subthreshold field intensities. This reinforces the principle that sufficient transmembrane potential, which is primarily governed by field strength, is essential for effective pore formation [33].

Furthermore, a synergistic interaction between electric field intensity and pulse duration is evident, as reflected by the positive coefficient of the E·t interaction term in the regression model. This synergy implies that simultaneous increases in both parameters produce a greater-than-additive effect on oocyst permeabilization. The presence of experimental data points both above and below the modeled surface highlights natural biological variability and potential experimental factors not captured by the model, such as oocyst age, aggregation, or microenvironmental conditions.

According to the model, electric field strengths higher than 30 kV/cm are required to achieve more than 50% permeabilization of *Cryptosporidium* oocysts. For instance, applying treatments at room temperature with 30 kV/cm and 180 µs resulted in 55.1% permeabilization of the oocysts. In contrast, more intense conditions (35 kV/cm and 180 µs) led to nearly 90% permeabilization of *Cryptosporidium* oocysts. The intensity of the PEF parameters (electric field and treatment time) required to achieve permeabilization highlights the strong resistance of *Cryptosporidium* oocysts to electroporation.

The only previous study of PEF on Cryptosporidium, conducted by Haas and Aturaliye [29], reported limited effectiveness of electroporation alone for oocyst inactivation, with only minimal effects observed even at energy levels up to 386 kJ/kg. However, the authors reported only that the maximum voltage used in the experiments was 3 kV; they did not directly report the electric field strength in kilovolts per centimeter (kV/cm). The electrode gap (distance between electrodes in the cuvette) is not explicitly stated in the paper, so we cannot calculate the electric field. If we assume they used a standard 0.2 cm (2 mm) cuvette, which is common in electroporation setups and coincides with the volume they used per sample, the maximum field strengths would be 3 kV/0.2 cm = 15 kV/cm. As demonstrated in the present study, the applied electric field used by Haas and Aturaliye [29] remained well below the threshold required to electroporate the resistant forms of Cryptosporidium. Their results indicated that electroporation was insufficient as a standalone inactivation method but still enhanced the efficacy of chemical disinfectants such as combined chlorine and hydrogen peroxide through synergistic interactions. These discrepancies in the efficacy of PEF may be attributed to differences in the technological development of PEF equipment and treatment parameters (e.g., electric field strength, pulse duration). While Haas and Aturaliye [29] applied up to 20 pulses from 0 to 99 µs, they did not specify the exact treatment time. The present study implements a comprehensive and well-defined experimental design, applying 20 to 60 pulses of 3 µs each, with precise control over electric field intensity (15–35 kV/cm) and application temperature. This setup enables accurate monitoring of the actual voltage and current applied, thanks to the probes connected to the equipment. Furthermore, viability assessment in the current work was supported by flow cytometry using genus-specific antibodies, providing higher resolution than the excystation-based methods used in 1999. Collectively, these findings suggest that advancements in PEF technology and methodology have significantly improved its standalone potential for Cryptosporidium inactivation.

3.2. Influence of the PEF Treatment Temperature on the Cryptosporidium Oocysts Permeabilization to PI

3.2.1. Evaluation of Thermal Resistance of Oocysts in the Absence of PEF

After demonstrating that PEF could be an effective procedure for *Cryptosporidium* oocyst permeabilization, but considering their high resistance and the intense treatment parameters required, the effect of applying the treatments at temperatures above room temperature was investigated to determine whether it could enhance the effectiveness of PEF and facilitate its implementation.

Before selecting the temperatures for the application of PEF treatments, a study was conducted to evaluate the thermal resistance of isolated *Cryptosporidium* oocysts. The aim was to rule out a purely thermal effect and to identify moderate temperatures that could exert a synergistic effect with PEF treatments without affecting oocyst viability on their own. Figure 2 shows the percentage of *Cryptosporidium* oocysts permeabilized to PI after different exposure times (1, 5, and 10 min) at four temperatures (40, 50, 60, and 70 °C). Permeabilization was negligible in all cases, except at the highest temperature (70 °C) for all the times assayed and after the longest exposure time (10 min) at 60 °C. Therefore, it can be said that temperatures below 60 °C are ineffective for inactivating *Cryptosporidium* oocysts, even after prolonged exposure times (5 min).

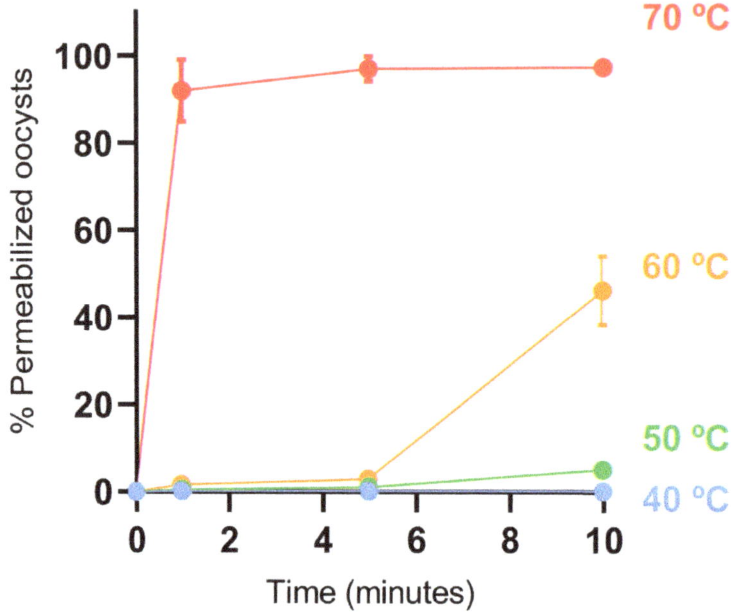

Figure 2. Thermal resistance of *Cryptosporidium* oocysts in buffer: percentage of permeabilized oocysts after exposure to 40 (blue), 50 (green), 60 °C (orange) and 70 °C (red) for 1, 5, and 10 min.

Studies on the permeabilization of *Cryptosporidium* oocysts to PI after heat treatment were commonly conducted to assess post-treatment viability. PI is a fluorochrome that does not penetrate intact cell membranes, so its uptake by oocysts indicates damage to the membrane, presumably associated with a loss of viability. This staining gained popularity due to its simplicity and rapid results [34–36]. Campbell et al. [34] reported a highly significant correlation (r = 0.997) between a dual-staining assay using DAPI and PI and in vitro excystation, suggesting that PI staining can reliably indicate oocyst viability under certain conditions. However, comparisons with other methods suggest that PI staining tends to overestimate viability compared to infectivity assays because some non-infectious oocysts may still exclude the dye and be misclassified as viable. Additionally, SYTO-17 staining, a fluorogenic nucleic acid dye method, has demonstrated high sensitivity (~96%) and specificity in distinguishing heat-killed from viable oocysts. Like other dye-based methods, it still relies on the assumption that membrane integrity equates to infectivity [37]. More recent approaches include RT-qPCR, PMA-qPCR, and RNA FISH, which aim to detect metabolic activity or intact nucleic acids associated with viable oocysts. For instance, PMA-qPCR selectively amplifies DNA only from oocysts with intact membranes, improving specificity. Meanwhile, RNA FISH targets ribosomal RNA to identify viable oocysts but its reliability can be affected by RNA degradation in environmental matrices [36]. All of these molecular methods focus on the integrity of a specific part of the oocyst, such as the membrane, or on certain metabolic activities, but they do not directly measure its infective capacity, which—aside from cell culture assays—would require confirmation using animal models. However, as animal models (e.g., mice) are increasingly avoided due to ethical concerns, and loss of membrane integrity correlates well with infectivity, PI staining remains, in our view, a valid and practical approach for preliminary infectivity assessments.

The results obtained in the present study are consistent with previous publications on the inactivation of *Cryptosporidium* oocysts that include thermal treatments (temperature-time combinations) within the temperature range studied. In a pivotal study by Fayer [38], oocysts suspended in water were subjected to controlled heating and infectivity was evaluated through bioassays in neonatal BALB/c mice, followed by histological examination of intestinal tissues. The results demonstrated that oocysts remained infectious when water temperatures reached up to 67.5 °C within 1 min. However, complete inactivation was achieved when oocysts were exposed to temperatures of 72.4 °C. These findings in animal models are consistent with our results with PI uptake and reinforce the notion that moderate heat treatments below 60 °C are ineffective. Importantly, this study, as in our case, eliminated confounding factors such as food matrices or organic matter, thereby providing robust data on heat inactivation in clean aqueous suspensions. Nevertheless, the authors noted that results might differ in more complex matrices like milk, emphasizing the need for matrix-specific validation. Similarly, the study of Jenkins et al. [35] on the thermal inactivation of *C. parvum* oocysts using differential uptake of DAPI and PI demonstrated that exposure to increasing temperatures accelerated the shift in oocyst membrane permeability from $DAPI^-PI^-$ (impermeable, viable) to $DAPI^+PI^+$ (permeable, non-viable). Nine minutes at 70 °C was required to achieve 75% $DAPI^+PI^+$ oocysts, while 25% were still considered potentially viable. In contrast, 2 min of exposure at 80 °C resulted in complete transition to $DAPI^+PI^+$, indicating total inactivation. In our study, we used a fluorescein-labeled antibody that specifically binds to the wall of *Cryptosporidium* oocysts. Therefore, after parameter selection, gaining, and gating, we ensured that the events analyzed by flow cytometry corresponded to oocysts, making our assessment of PI uptake potentially more accurate than in older studies where DAPI and microscopy was used. Temesgen et al. (2021) [39] employed a novel approach to assess the viability of *C. parvum* oocysts following thermal inactivation treatments by developing a reverse transcription quantitative PCR (RT-qPCR) method based on the expression of genes induced by oxidative stress. This technique is grounded in the principle that only metabolically active (i.e., viable) oocysts respond to oxidative stress by upregulating specific genes, such as thioredoxin and COWP7. Their results indicated that oocysts exposed to 60 °C for 2 min still exhibited measurable expression of these stress-responsive genes, suggesting a lack of inactivation, whereas heating at 80 °C for 3 min was required to achieve complete inactivation. This study, which evaluates metabolic viability rather than PI uptake, is also consistent with the findings of the present work. Therefore, it is confirmed that temperatures below 60 °C should not have an effect on *Cryptosporidium* inactivation on their own, and thus temperatures below this threshold (25–60 °C) were selected for the experiments evaluating the influence of application temperature during PEF treatments.

3.2.2. Synergistic Effect of Temperature and PEF

The PEF treatment conditions selected for this experiment (25 kV/cm, 90 µs) were chosen based on the results obtained from the response surface model described earlier. Specifically, the model revealed a threshold electric field around 25 kV/cm, below which oocyst permeabilization was minimal. To remain at this threshold level and avoid inducing high baseline permeabilization, which could mask potential synergistic effects between PEF and temperature, 25 kV/cm was selected as the field intensity. Additionally, as the model indicated that treatment time had a relatively minor effect compared to field strength, a moderate duration of 90 µs was employed. This exposure was sufficient to induce approximately 20–30% membrane permeabilization at room temperature, thereby creating a dynamic window to evaluate potential enhancements when combined with sublethal thermal stress.

The influence of the temperature of application of PEF treatment (25 kV/cm, 90 μs) on the *Cryptosporidium* oocysts permeabilization to PI is shown in Figure 3. A significant temperature-dependent increase in oocyst permeabilization is observed when applying PEF treatments. At 25 °C, the permeabilization percentage is relatively low (~30%) with high variability. A progressive increase in temperature from 30 °C to 45 °C results in a steady increase in permeabilization, reaching approximately 85% at 45 °C. From 50 °C onwards, the effect appears to plateau, with values consistently around or above 90%, suggesting that a maximum permeabilization effect is achieved, or even that the oocysts are so severely damaged that they release their genetic material, preventing PI from binding and producing a signal. Error bars indicate that variability decreases as temperature increases, especially beyond 45 °C. These results indicate that increasing the temperature of PEF application significantly enhances oocyst permeabilization, with temperatures ≥ 40 °C maximizing the effect. This suggests a synergistic interaction between thermal energy and electric pulses, improving the efficacy of PEF treatment on the oocysts' inactivation.

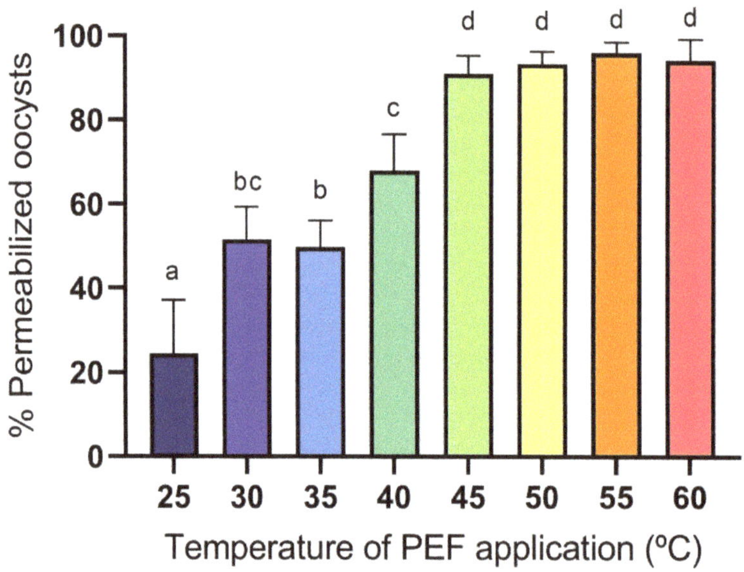

Figure 3. Influence of the temperature of application of PEF treatment (25 kV/cm, 90 μs) on the *Cryptosporidium* oocysts' permeabilization to PI. Different letters indicate significant differences ($p \leq 0.05$).

Consistent with our findings, multiple studies have examined the effects of PEF treatments on microbial inactivation—including yeasts and bacteria—at different temperatures, emphasizing the synergistic benefits of combining PEF with moderate heat. Montanari et al. [40] explored the inactivation of the yeast *Saccharomyces* (*S.*) *cerevisiae* using PEF combined with mild heat (50 °C). Preheating the yeast suspension enhanced the efficacy of PEF treatments, leading to greater microbial reductions. The combination of heat and PEF was more effective than either treatment alone. Yan et al. [41] examined the inactivation of *S. cerevisiae, Escherichia coli*, and *Bacillus* (*B.*) *velezensis* under different initial temperatures (24 °C, 30 °C, 40 °C, and 50 °C). The results indicated that increasing the temperature reduced the critical electric field intensity required for microbial inactivation. This suggests that temperature enhances the electroporation process, leading to more effective microbial inactivation. Notably, *B. velezensis*, known for its spore-forming capability, showed increased resistance, but the combination of higher temperatures with PEF treatments significantly improved inactivation rates. In this sense, similarly to bacterial spores, the oocysts of *Cryptosporidium* exhibit remarkable structural adaptations that confer resistance to environmental stresses, ensuring their survival in adverse conditions. Bacterial endospores, such as those formed by *Bacillus* and *Clostridium* species, are characterized by a multilayered structure comprising an exosporium, spore coat, cortex, and core. The spore coat, rich in proteins, provides chemical and enzymatic resistance, while the cortex, composed of peptidoglycan, aids in maintaining dehydration of the core. The core contains DNA stabilized by small acid-soluble spore proteins (SASPs) and high concentrations of calcium dipicolinate, contributing to resistance against heat, radiation, and desiccation [42]. Similarly, *Cryptosporidium* oocysts possess a robust wall structure with inner and outer layers composed of proteins and carbohydrates. This wall confers resistance to chemical disinfectants and environmental pressures. The oocyst wall's unique composition, including acid-fast lipids, contributes to its durability [43]. Both structures (bacterial spores and parasitic oocysts) are metabolically dormant, enabling them to withstand unfavorable conditions and persist in the environment. Their resilience poses significant challenges in public health, particularly in water treatment and food sterilization processes, necessitating advanced methods for effective inactivation.

As mentioned in the introduction, some recent research has explored the application of PEF for inactivating parasites such as *Anisakis* spp. and *Trichinella* spp. [26,28]. However, to the best of our knowledge, the influence of application temperature during PEF treatments on parasite inactivation had not been studied to date. Previous studies about permeabilization of bacteria and yeast, together with the results of the present investigation, underscore the importance of combination with mild temperature in enhancing the effectiveness of PEF treatments for microbial inactivation.

3.2.3. Technological Implications of Thermally-Assisted PEF Treatments

Our study demonstrated that PEF treatments (25 kV/cm, 90 μs) applied at 45 °C induce substantial membrane permeabilization in *Cryptosporidium* oocysts, achieving up to 90% PI uptake. This high level of permeabilization is indicative of severe membrane compromise and likely loss of viability. Importantly, these effects were achieved at a field intensity notably lower than what would be required to reach similar permeabilization levels at room temperature (35 kV/cm, 180 μs). This highlights a critical benefit of thermally assisted PEF: by leveraging moderate thermal energy, it is possible to reduce the electric field intensity necessary to disrupt oocyst membranes effectively.

From a technological and industrial perspective, this finding is highly relevant. Operating at lower field intensities reduces the demand for high-voltage equipment, making PEF systems more energy-efficient, economically viable, and scalable for large-volume applications. High-voltage generators required to produce fields above 30–35 kV/cm are often expensive and difficult to implement in continuous industrial processes, whereas fields below 25 kV/cm can be achieved more easily with commercially available PEF systems, facilitating integration into existing processing lines. Therefore, combining PEF with moderate heating presents a practical strategy to enhance oocyst inactivation while minimizing infrastructure complexity.

In this context, it is worth comparing PEF with other non-thermal or minimally thermal technologies investigated for *Cryptosporidium* inactivation. UV irradiation, particularly using medium-pressure lamps, has been investigated for the inactivation of *C. parvum*. Clancy et al. [44] demonstrated that a dose of 3 mJ/cm^2 achieved over a 3-log reduction in infectivity when applied to thin liquid layers, as assessed using neonatal mouse models. Similarly, Craik et al. [45] confirmed that UV doses of 10–25 mJ/cm^2 could achieve 2–3 log reductions. More recent studies using UV-LEDs [46] indicated that wavelengths around

284–289 nm were particularly effective, achieving 2-log reductions in less than 40 min of exposure using in vitro assays. However, although these results may suggest that UV could be an effective disinfection strategy, its practical application is severely limited by matrix turbidity and poor penetration depth, both of which invalidate its efficacy under real-world conditions. Similarly, high hydrostatic pressure (HHP) has been assayed for inactivating resilient *C. parvum* oocysts. Slifko et al. [24] reported that HHP at 550 MPa for 60 s eliminated *C. parvum* infectivity in fruit juices, achieving over 3.4-log reductions. Similarly, Collins et al. [47] observed up to a 93.3% reduction in infectivity in oyster tissues at 550 MPa for 180 s using neonatal mouse bioassays. While highly effective, the requirement for specialized high-pressure equipment—and particularly the significant challenges related to designing systems capable of continuous treatment—prevents its practical implementation, making it another unfeasible alternative to liquid treatments. Pulsed light (PL) has also been studied, with the aim of inactivating *Cryptosporidium* oocysts in both water and food matrices through both laboratory and field-scale setups. Huffman et al. [48] reported that a point-of-use PL device achieved >4 $\log_{10}$ inactivation of *C. parvum* in water using cell culture and animal infectivity assays, highlighting its potential for household water purification systems with very low flow. On produce, PL has shown variable efficacy depending on the commodity. Craighead et al. [23] achieved up to 4.3 $\log_{10}$ reduction in oocyst infectivity on mesclun lettuce and between 2.2–2.5 $\log_{10}$ on spinach and tomatoes after 45–90 s of treatment with minimal impact on produce quality. However, despite its proven antimicrobial efficacy on surfaces, PL is not scalable for industrial food and water treatment applications because of its limited penetration depth, which, similarly to UV-irradiation, restricts inactivation to surface-level contaminants. Anecdotally, cold atmospheric plasma (CP) has shown some positive results in inactivating *Cryptosporidium* oocysts on fresh produce, such as cilantro, achieving up to a 2.03-$\log_{10}$ reduction after 180 s of exposure. However, treatment durations longer than 30 s caused visible damage to the cilantro leaves, such as drying and darkening, which may compromise sensory and nutritional quality, and standardization of the treatments was impossible due to different factors [49]. Other studies have explored the use of ultrasounds as a non-thermal method to inactivate *C. parvum* oocysts in water. Olvera et al. [50] investigated the effect of a 1 MHz ultrasound on *C. parvum* oocysts and found that continuous irradiation for 20 min led to significant inactivation, with an estimated 102.7 oocysts killed per second and minimal energy consumption. Similarly, Oyane et al. [51] utilized three different sonicators employing the squeeze-film effect and observed that a cylindrical sonicator operating at 26.6 kHz and 30 W achieved a 97% inactivation rate of oocysts at a flow rate of 33 mL/min, corresponding to a residence time of approximately 5.2 min. These findings suggested that ultrasound treatment can effectively reduce *C. parvum* viability, with efficacy influenced by factors such as frequency, power, and exposure time. However, these ultrasound-based disinfection methods require long treatment times, are very energy-demanding, particularly at higher frequencies where cavitation intensity is reduced, and most importantly are nowadays impossible to scale-up for continuous treatment.

In contrast, PEF technology offers a combination of effectiveness, energy efficiency, and scalability. Treatments are extremely rapid, compatible with continuous-flow systems, and—when combined with moderate heating—can achieve substantial inactivation of resilient pathogens like *Cryptosporidium* using lower electric fields. This positions PEF as a competitive and industrially viable alternative among current emerging disinfection technologies.

3.3. Validation of PEF Treatments Applied at Different Temperatures for the Inactivation of Cryptosporidium Oocysts in Carrot Juice and Apple Juice

3.3.1. Effect of PEF Temperature on Oocysts Permeabilization in Juices

Several documented outbreaks of *Cryptosporidium* have been linked to the consumption of contaminated apple juice. One notable incident occurred in 1999 in the United States, where unpasteurized apple cider caused an outbreak affecting dozens of individuals, including children. Investigations revealed that the juice had been contaminated with *Cryptosporidium* oocysts, likely due to poor hygienic practices during production or the use of contaminated water [15,52] Similarly, *Cryptosporidium* oocysts have been detected on a variety of fresh produce, including carrots, lettuce, spinach, and other leafy greens, raising public health concerns due to their potential for foodborne transmission [53]. Contamination typically occurs through contact with fecally contaminated irrigation water, manure-based fertilizers, or during handling and processing. Carrots, in particular, may be at risk due to their close contact with soil, especially when consumed raw or minimally processed. The robust wall structure of the oocysts allows them to adhere to irregular surfaces and resist standard washing procedures [17] Furthermore, consumer demand for non- or minimally processed fruit juices is rising due to health awareness, preference for clean-label products, and interest in natural nutrient retention. Market projections show strong growth, with sales expected to nearly double [54]. The growing demand for minimally processed juices, combined with past outbreaks, raises significant safety concerns and highlights the need for effective non-thermal disinfection technologies. This is particularly critical given that *Cryptosporidium* oocysts are highly resistant to conventional chlorination of water used to wash fruits and vegetables and can remain viable in the acidic and refrigerated conditions typical of fruit juices.

In both apple and carrot juice, heat treatment alone up to 40 °C did not lead to statistically significant differences in oocyst permeabilization compared to the untreated control, even when the incubation at 40 °C was extended beyond one hour. The influence of the temperature of a PEF treatment application (25 kV/cm, 90 µs) in apple juice and carrot juice on the *Cryptosporidium* oocysts permeabilization is detailed in Figure 4. As shown in Figure 4a, when the PEF treatments were performed in apple juice, the percentage of permeabilized oocysts increased significantly with temperature. At 20 °C, permeabilization was minimal (<5%), while at 30 °C it reached approximately 15%. A marked increase was observed at 40 °C, with over 60% of oocysts permeabilized. In agreement with the results obtained with oocysts suspended in the buffer, these results suggest that higher application temperatures enhance the efficacy of PEF in disrupting oocyst membranes, likely due to synergistic effects between electric pulses and thermal stress. However, the lower permeabilization observed compared to buffer treatments suggests that the organic matter in the juice may exert a protective effect on the oocysts. On the other hand, Figure 4b shows the influence of PEF application temperature in carrot juice. The percentage of permeabilized oocysts remained high across all tested temperatures, with values around 75% at 20 °C and 30 °C, and slightly increasing to approximately 85% at 40 °C. These results indicate that PEF treatments were much more effective at permeabilizing oocyst membranes in this matrix in comparison to apple juice. As considerable permeabilization had already occurred at 20 and 30 °C, the improvement in permeabilization at the higher temperature was more modest.

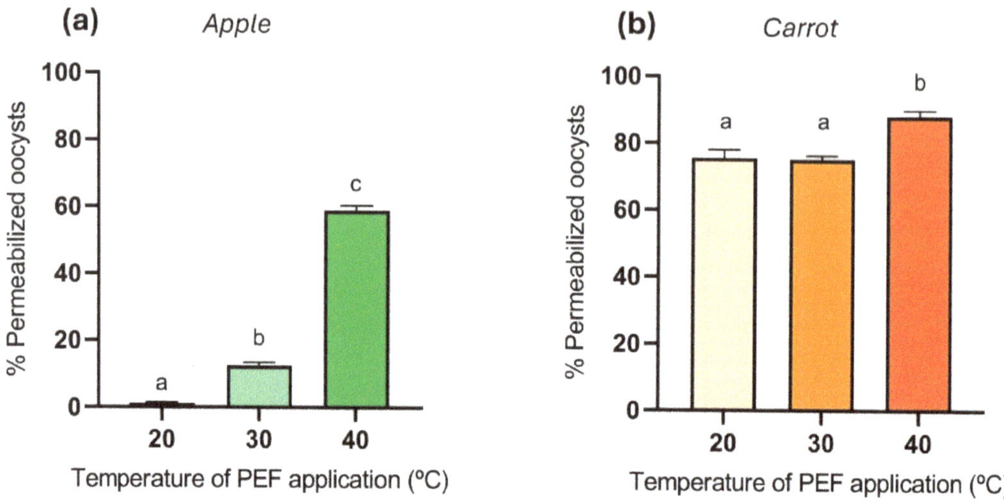

Figure 4. Influence of the temperature of application of PEF treatment (25 kV/cm, 90 µs) on the *Cryptosporidium* oocysts permeabilization to PI in (**a**) apple and (**b**) carrot juices. Different letters indicate significant differences ($p \leq 0.05$).

3.3.2. Influence of Matrix Properties on PEF Efficacy

Although the PEF treatments applied to apple and carrot juices were equivalent in terms of electric field strength, pulse number, and pulse width, the energy delivered to the system differed substantially due to the variation in electrical conductivity between the two juices. Carrot juice showed higher conductivity due to its natural content of minerals, organic acids, amino acids, and vitamins, which increase ionic strength and facilitate greater current flow during PEF treatment, leading to higher energy input under the same electric field conditions. The conductivity of apple juice was 1.6 mS/cm, whereas carrot juice exhibited a higher conductivity of 4.3 mS/cm. Consequently, the specific energy inputs for apple juice were 106, 143, and 164 kJ/kg at 20, 30, and 40 °C, respectively, while those for carrot juice were higher, reaching 235, 248, and 262 kJ/kg at the same temperatures. The greater energy input during the treatments applied to carrot juice likely accounted for the higher level of oocyst permeabilization observed in this matrix compared to apple juice. The energy input applied during PEF treatments plays a critical role in the inactivation of microorganisms. Higher energy inputs generally lead to increased membrane permeabilization, which enhances microbial inactivation by facilitating irreversible electroporation and loss of cellular integrity. However, the extent of inactivation also depends on factors such as microbial type, medium conductivity, and temperature. Several studies have demonstrated that increasing the specific energy input improves the effectiveness of PEF in reducing microbial loads in liquid foods [55,56]. Specifically in parasites, Abad et al. [28] demonstrated that increasing the specific energy input significantly enhanced the efficacy of PEF treatments in inactivating *Anisakis* larvae.

Contrary to expectations, the treatments were more effective in the neutral-pH juice (carrot juice, pH 5.94) compared to the acidic juice (apple juice, pH 3.40). This may be explained by the higher energy applied during the PEF treatments in carrot juice, which could have masked any potential synergistic effect of acidic pH on oocyst permeabilization in the apple juice matrix. In the case of bacterial and yeast inactivation, the pH of the treatment medium significantly influences the efficacy of PEF treatments. Several studies have demonstrated that acidic environments increase microbial susceptibility to PEF. For example, Jin et al. [57] showed that reducing the pH from 7.0 to 4.0 during PEF treatment led to a substantial decrease in *E. coli* counts in liquid whole egg, indicating a synergistic interaction between low pH and PEF. Likewise, Li et al. [58] reported that *S. cerevisiae* exhibited higher inactivation rates at pH 4.0 than at pH 7.0 under PEF, suggesting that acidic conditions may weaken membrane integrity and enhance electroporation. These findings highlight the relevance of pH in optimizing PEF-based microbial inactivation strategies. However, since the treatments in our study differ significantly in terms of energy input, no conclusions can be drawn regarding the specific effect of matrix pH on the permeabilization of oocysts following PEF treatments at different application temperatures.

In addition to the pH of the medium, the efficacy of inactivation treatments for *Cryptosporidium* oocysts is strongly influenced by the surrounding matrix. Studies have shown that organic-rich environments—such as food matrices or manure—can exert protective effects on oocysts, thereby reducing the effectiveness of treatments compared to simpler systems like buffer solutions. For instance, Petersen et al. [59] observed that the inactivation rate of *C. parvum* oocysts was notably lower in cattle slurry compared to distilled water when treated with ammonia, suggesting that components in the slurry may shield oocysts from chemical agents. Similarly, Temesgen et al. [39] highlighted that standard inactivation methods effective in water might not yield the same results in food matrices, emphasizing the need to consider matrix composition in treatment strategies. Natural juice components—such as total polyphenols, polysaccharides, and vitamins—may influence the effectiveness of PEF treatment by either protecting or enhancing the inactivation of *Cryptosporidium* oocysts. Our results, in agreement with the literature, underscore the importance of tailoring inactivation approaches to specific environmental contexts and food matrices to ensure the effective control of *Cryptosporidium* oocysts.

3.3.3. Industrial Feasibility, Regulatory Considerations and Limitations

Several studies have demonstrated effective microbial inactivation, particularly of bacteria, in both apple and carrot juices using PEF, with no significant alterations in color, flavor, or nutritional composition. The minimal impact on organoleptic and nutritional properties positions PEF as a promising non-thermal alternative to conventional pasteurization, especially in industrial settings where product quality and shelf-life extension are critical [60,61]. However, despite its strong potential, the industrial adoption of PEF for juice pasteurization remains limited. This is largely due to the high capital investment required for equipment, the complexity of process validation for different juice matrices, and the absence of standardized regulatory guidelines for microbial safety. Consequently, PEF is currently applied primarily in niche markets focused on premium juice products, where quality preservation justifies the associated costs.

Regulatory guidelines for microbial safety in foods typically focus on bacterial pathogens, while protozoan parasites such as *Cryptosporidium* remain less specifically addressed, especially in fruit and vegetable juices. Although the oocysts are highly resistant to conventional disinfection methods, including chlorination, and have been associated with foodborne outbreaks related to fresh produce and beverages, current regulations—such as the FDA's HACCP guidance—acknowledge *C. parvum* as a relevant pathogen but do not establish explicit microbial limits for its presence in juices [62]. Similarly, no specific regulatory thresholds have been defined by other major authorities as the European Food Safety Authority [63]. This regulatory gap poses a significant challenge for the juice industry, as oocysts can persist in minimally processed products without altering their organoleptic properties, making contamination difficult to detect and consequently, posing a risk for public health. In this context, non-thermal processing technologies such as PEF offer a promising industrial strategy for achieving oocyst inactivation while preserving the sensory and nutritional quality of the final product.

However, despite the potential of PEF, some limitations of the present study should be acknowledged. The assessment of oocyst inactivation was based on PI uptake as a proxy for membrane permeabilization, which does not directly confirm the loss of infectivity. In vivo assays should be performed to validate this outcome. Furthermore, the experiments were conducted under controlled laboratory conditions using model matrices, and although apple and carrot juices are relevant systems, the variability in industrial formulations and processing environments may affect treatment reproducibility. Future studies should address these limitations to strengthen the robustness and scalability of PEF as a disinfection strategy.

4. Conclusions

This study demonstrates that Pulsed Electric Field (PEF) technology can effectively permeabilize *Cryptosporidium* oocysts suspended in water, apple juice, and carrot juice, particularly when treatments are applied at moderate temperatures (>40 °C). The findings highlight the critical role of electric field strength, treatment time, and application temperature in achieving high levels of oocyst membrane disruption, with permeabilization exceeding 90% under optimized conditions. Notably, the surrounding matrix influenced treatment efficacy, with higher inactivation observed in carrot juice—likely due to greater conductivity and thus energy input—compared to apple juice. These results support the potential of PEF as a non-thermal alternative for enhancing the microbiological safety of minimally processed fruit juices, with minimal impact on organoleptic quality. Given the growing demand for fresh-like beverages and the resilience of *Cryptosporidium* to conventional sanitizers, PEF offers a scalable, energy-efficient solution for industrial applications targeting protozoan pathogens. Further studies incorporating infectivity assays are recommended to fully validate oocyst inactivation and guide regulatory acceptance.

Author Contributions: Conception and design of the study: A.B., L.G.-M., J.Q., J.R., I.Á.-L. and J.M.M. Acquisition of data: A.B., L.G.-M. and J.M.M. Analysis and interpretation of the data: A.B., L.G.-M. and J.M.M. Drafting of the article: A.B., L.G.-M. and J.M.M. Critical revision of the article: J.Q., J.R., I.Á.-L. and J.M.M. Funding acquisition: I.Á.-L. All authors have read and agreed to the published version of the manuscript.

Funding: This research was supported by Departamento de Ciencia, Universidad y Sociedad del Conocimiento and Fondo Social Europeo-Gobierno de Aragón (ZeroPARASITOS PROY_A16_24, A03_23R).

Data Availability Statement: The original contributions presented in the study are included in the article, further inquiries can be directed to the corresponding authors.

Acknowledgments: A.B. is thankful for the financial assistance granted (Gran Number: FPU20/02527) by the Ministerio de Universidades (Spain), which supported his academic pursuits.

Conflicts of Interest: The authors declare no conflicts of interest.

Appendix A

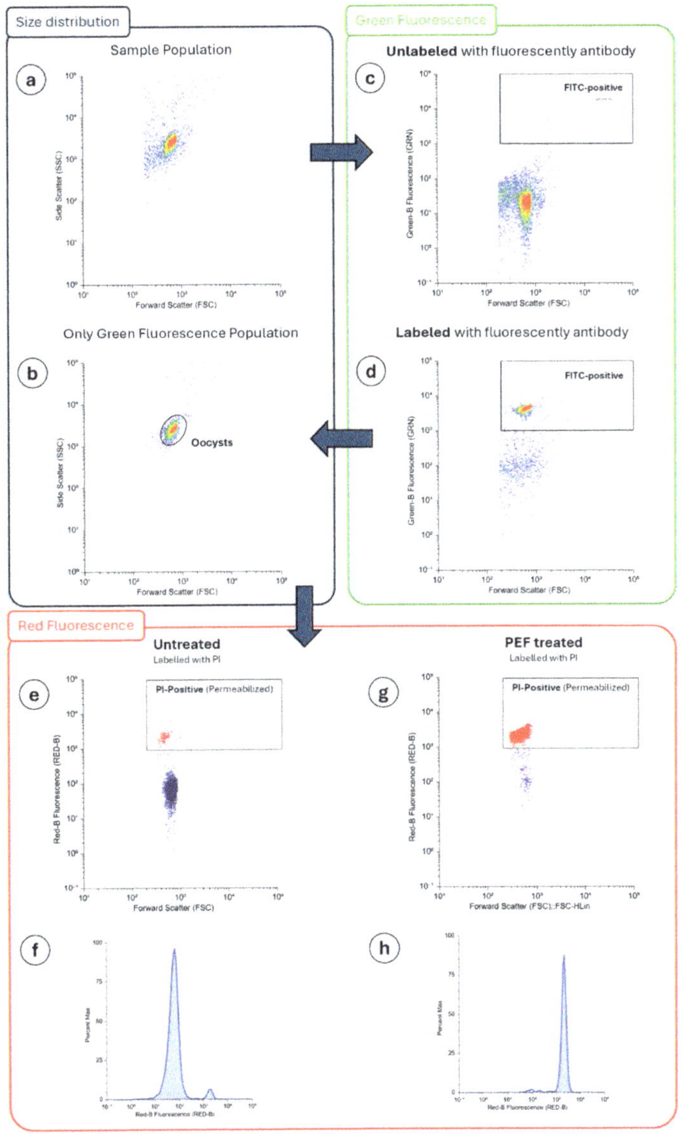

Figure A1. Flow cytometry gating strategy used to evaluate membrane permeabilization of *Cryptosporidium* oocysts after PEF treatment. (**a**) Initial FSC vs. SSC plot showing the total particle population in the samples. (**b**) Sub-gating of the oocyst population based on size and internal complexity (FSC vs. SSC), selecting only events exhibiting green fluorescence (FITC-positive) to ensure specific identification of antibody-labeled oocysts. (**c**,**d**) Green fluorescence (GRN-B) from samples without and with FITC-labeled antibody, respectively; only labeled samples exhibit a clear FITC-positive signal used to identify oocysts. (**e**,**g**) PI uptake in untreated and PEF-treated oocysts, respectively, measured in the red fluorescence channel (RED-B); PI-positive oocysts appear in red. (**f**,**h**) Histogram overlays representing red fluorescence intensity (RED-B) in untreated and PEF-treated oocysts, respectively, showing an increase in PI uptake after PEF treatment. Data correspond to 8000 events acquired per sample.

References

1. Zahedi, A.; Ryan, U. *Cryptosporidium*–An Update with an Emphasis on Foodborne and Waterborne Transmission. *Res. Vet. Sci.* **2020**, *132*, 500–512. [CrossRef] [PubMed]
2. Costa, D.; Razakandrainibe, R.; Valot, S.; Vannier, M.; Sautour, M.; Basmaciyan, L.; Gargala, G.; Viller, V.; Lemeteil, D.; Ballet, J.J.; et al. Epidemiology of Cryptosporidiosis in France from 2017 to 2019. *Microorganisms* **2020**, *8*, 1358. [CrossRef]
3. De Felice, L.A.; Basset, C.; Unzaga, J.M. Chapter 13: *Cryptosporidium* spp. In *Protozoos Parásitos de Importancia Sanitaria: Un Abordaje Transdisciplinar*; Editorial de la Universidad Nacional de La Plata (EDULP): Plata, Argentina, 2023. [CrossRef]
4. Garcia-R, J.C.; Hayman, D.T.S. A Review and Analysis of Cryptosporidiosis Outbreaks in New Zealand. *Parasitology* **2023**, *150*, 606–611. [CrossRef] [PubMed]
5. European Centre for Disease Prevention and Control (ECDC). *Cryptosporidiosis–Annual Epidemiological Report 2021*; European Centre for Disease Prevention and Control: Stockholm, Sweden, 2024.
6. Augendre, L.; Costa, D.; Escotte-Binet, S.; Aubert, D.; Villena, I.; Dumètre, A.; La Carbona, S. Surrogates of Foodborne and Waterborne Protozoan Parasites: A Review. *Food Waterborne Parasitol.* **2023**, *33*, e00212. [CrossRef]
7. Nasser, A.M. Removal of *Cryptosporidium* by Wastewater Treatment Processes: A Review. *J. Water Health* **2016**, *14*, 1–13. [CrossRef]
8. Adeyemo, F.E.; Singh, G.; Reddy, P.; Bux, F.; Stenström, T.A. Efficiency of Chlorine and UV in the Inactivation of *Cryptosporidium* and *Giardia* in Wastewater. *PLoS ONE* **2019**, *14*, e0216040. [CrossRef] [PubMed]
9. Korich, D.G.; Mead, J.R.; Madore, M.S.; Sinclair, N.A.; Sterling, C.R. Effects of Ozone, Chlorine Dioxide, Chlorine, and Monochloramine on Cryptosporidium Parvum Oocyst Viability. *Appl. Environ. Microbiol.* **1990**, *56*, 1423–1428. [CrossRef]
10. Kristanti, R.A.; Hadibarata, T.; Syafrudin, M.; Yılmaz, M.; Abdullah, S. Microbiological Contaminants in Drinking Water: Current Status and Challenges. *Water Air Soil Pollut.* **2022**, *233*, 299. [CrossRef]
11. Kunz, J.M.; Lawinger, H.; Miko, S.; Gerdes, M.; Thuneibat, M.; Hannapel, E.; Roberts, V.A. Surveillance of Waterborne Disease Outbreaks Associated with Drinking Water-United States, 2015–2020. *Morb. Mortal. Wkly. Rep.* **2024**, *73*, 1–23. [CrossRef]
12. Duffy, G.; Moriarty, E.M. *Cryptosporidium* and Its Potential as a Food-Borne Pathogen. *Anim. Health Res. Rev.* **2003**, *4*, 95–107. [CrossRef]
13. Donnelly, J.K.; Stentiford, E.I. The Cryptosporidium Problem in Water and Food Supplies. *LWT-Food Sci. Technol.* **1997**, *30*, 111–120. [CrossRef]
14. Ethelberg, S.; Lisby, M.; Vestergaard, L.S.; Enemark, H.I.; Olsen, K.E.P.; Stensvold, C.R.; Nielsen, H.V.; Porsbo, L.J.; Plesner, A.A.M.; Mølbak, K. A Foodborne Outbreak of *Cryptosporidium Hominis* Infection. *Epidemiol. Infect.* **2009**, *137*, 348–356. [CrossRef] [PubMed]
15. Robertson, L.J.; Temesgen, T.T.; Tysnes, K.R.; Eikås, J.E. An Apple a Day: An Outbreak of Cryptosporidiosis in Norway Associated with Self-Pressed Apple Juice. *Epidemiol. Infect.* **2019**, *147*, e139. [CrossRef] [PubMed]
16. Ali, M.; Ji, Y.; Xu, C.; Hina, Q.; Javed, U.; Li, K. Food and Waterborne Cryptosporidiosis from a One Health Perspective: A Comprehensive Review. *Animals* **2024**, *14*, 3287. [CrossRef]
17. Gharpure, R.; Perez, A.; Miller, A.D.; Wikswo, M.E.; Silver, R.; Hlavsa, M.C. Cryptosporidiosis Outbreaks-United States, 2009–2017. *Morb. Mortal. Wkly. Rep.* **2019**, *68*, 568–572. [CrossRef]
18. Deng, M.Q.; Cliver, D.O. Inactivation of *Cryptosporidium parvum* Oocysts in Cider by Flash Pasteurization. *J. Food Prot.* **2001**, *64*, 523–527. [CrossRef]
19. Barrett, D.M.; Beaulieu, J.C.; Shewfelt, R. Color, Flavor, Texture, and Nutritional Quality of Fresh-Cut Fruits and Vegetables: Desirable Levels, Instrumental and Sensory Measurement, and the Effects of Processing. *Crit. Rev. Food Sci. Nutr.* **2010**, *50*, 369–389. [CrossRef]
20. Hanes, D.E.; Worobo, R.W.; Orlandi, P.A.; Burr, D.H.; Miliotis, M.D.; Robl, M.G.; Bier, J.W.; Arrowood, M.J.; Churey, J.J.; Jackson, G.J. Inactivation of *Cryptosporidium parvum* Oocysts in Fresh Apple Cider by UV Irradiation. *Appl. Environ. Microbiol.* **2002**, *68*, 4168–4172. [CrossRef]
21. Kniel, K.E. Evaluation of Chemical Treatments and Ozone on the Viability of *Cryptosporidium parvum* Oocysts in Fruit Juices. Ph.D. Dissertation, Faculty of the Virginia Polytechnic Institute and State University, Blacksburg, VA, USA, 2002.
22. Kniel, K.E.; Sumner, S.S.; Lindsay, D.S.; Hackney, C.R.; Pierson, M.D.; Zajac, A.M.; Golden, D.A.; Fayer, R. Effect of Organic Acids and Hydrogen Peroxide on *Cryptosporidium parvum* Viability in Fruit Juices. *J. Food Prot.* **2003**, *66*, 1650–1657. [CrossRef]
23. Craighead, S.; Huang, R.; Chen, H.; Kniel, K.E. The Use of Pulsed Light to Inactivate *Cryptosporidium parvum* Oocysts on High-Risk Commodities (Cilantro, Mesclun Lettuce, Spinach, and Tomatoes). *Food Control* **2021**, *126*, 107965. [CrossRef]
24. Slifko, T.R.; Raghubeer, E.; Rose, J.B. Effect of High Hydrostatic Pressure on *Cryptosporidium parvum* Infectivity. *J. Food Prot.* **2000**, *63*, 1262–1267. [CrossRef] [PubMed]
25. Kotnik, T.; Kramar, P.; Pucihar, G.; Miklavčič, D.; Tarek, M. Cell Membrane Electroporation-Part 1: The Phenomenon. *IEEE Electr. Insul. Mag.* **2012**, *28*, 14–23. [CrossRef]

26. Martínez, J.M.; Abad, V.; Quílez, J.; Reina, D.; Pérez-Martin, J.E.; Raso, J.; Cebrián, G.; Álvarez-Lanzarote, I. Inactivation of Trichinella Spp. in Naturally Infected Boar Meat after Pulsed Electric Field (PEF) Treatments. *Food Control* **2024**, *163*, 110482. [CrossRef]
27. Delso, C.; Martínez, J.M.; Cebrián, G.; Condón, S.; Raso, J.; Álvarez, I. Microbial Inactivation by Pulsed Electric Fields. In *Pulsed Electric Fields Technology for the Food Industry: Fundamentals and Applications*; Raso, J., Heinz, V., Alvarez, I., Toepfl, S., Eds.; Springer International Publishing: Cham, Switzerland, 2022; pp. 169–207. ISBN 978-3-030-70586-2.
28. Abad, V.; Alejandre, M.; Hernández-Fernández, E.; Raso, J.; Cebrián, G.; Álvarez-Lanzarote, I. Evaluation of Pulsed Electric Fields (PEF) Parameters in the Inactivation of *Anisakis* Larvae in Saline Solution and Hake Meat. *Foods* **2023**, *12*, 264. [CrossRef]
29. Haas, C.N.; Aturaliye, D. Semi-Quantitative Characterization of Electroporation-Assisted Disinfection Processes for Inactivation of *Giardia* and *Cryptosporidium*. *J. Appl. Microbiol.* **1999**, *86*, 899–905. [CrossRef]
30. Heinz, V.; Phillips, S.T.; Zenker, M.; Knorr, D. Inactivation of Bacillus Subtilis by High Intensity Pulsed Electric Fields under Close to Isothermal Conditions. *Food Biotechnol.* **1999**, *13*, 155–168. [CrossRef]
31. Saldaña, G.; Puértolas, E.; Álvarez, I.; Meneses, N.; Knorr, D.; Raso, J. Evaluation of a Static Treatment Chamber to Investigate Kinetics of Microbial Inactivation by Pulsed Electric Fields at Different Temperatures at Quasi-Isothermal Conditions. *J. Food Eng.* **2010**, *100*, 349–356. [CrossRef]
32. Bukhari, Z.; Glen, W.G.; Clancy, J.L. Effects of pH on a fluorogenic vital dyes assay (4′, 6′-diamidino-2-phenyl-indole and propidium iodide) for *Cryptosporidium* sp. oocysts. *Water Res.* **1999**, *33*, 3037–3039. [CrossRef]
33. Weaver, J.C.; Chizmadzhev, Y.A. Theory of Electroporation: A Review. *Bioelectrochem. Bioenerg.* **1996**, *41*, 135–160. [CrossRef]
34. Campbell, A.T.; Robertson, L.J.; Smith, H.V. Viability of Cryptosporidium Parvum Oocysts: Correlation of In Vitro Excystation with Inclusion or Exclusion of Fluorogenic Vital Dyes. *Appl. Environ. Microbiol.* **1992**, *58*, 3488–3493. [CrossRef]
35. Jenkins, M.B.; Anguish, L.J.; Bowman, D.D.; Walker, M.J.; Ghiorse, W.C. Assessment of a Dye Permeability Assay for Determination of Inactivation Rates of *Cryptosporidium parvum* Oocysts. *Appl. Environ. Microbiol.* **1997**, *63*, 3844–3850. [CrossRef] [PubMed]
36. Vande Burgt, N.H.; Auer, A.; Zintl, A. Comparison of in Vitro Viability Methods for *Cryptosporidium* Oocysts. *Exp. Parasitol.* **2018**, *187*, 30–36. [CrossRef] [PubMed]
37. Tomonaga, T.; Kumar Rai, S.; Uga, S. Differentiation between Viable and Dead Cryptosporidium Oocysts Using Fluorochrome Staining. *Kobe J. Med. Sci.* **2015**, *61*, E138–E143.
38. Fayer, R. Effect of High Temperature on Infectivity of *Cryptosporidium parvum* Oocysts in Water. *Appl. Environ. Microbiol.* **1994**, *60*, 2732–2735. [CrossRef]
39. Temesgen, T.T.; Tysnes, K.R.; Robertson, L.J. Use of Oxidative Stress Responses to Determine the Efficacy of Inactivation Treatments on Cryptosporidium Oocysts. *Microorganisms* **2021**, *9*, 1463. [CrossRef]
40. Montanari, C.; Tylewicz, U.; Tabanelli, G.; Berardinelli, A.; Rocculi, P.; Ragni, L.; Gardini, F. Heat-Assisted Pulsed Electric Field Treatment for the Inactivation of Saccharomyces Cerevisiae: Effects of the Presence of Citral. *Front. Microbiol.* **2019**, *10*, 1737. [CrossRef]
41. Yan, Z.; Yin, L.; Hao, C.; Liu, K.; Qiu, J. Synergistic Effect of Pulsed Electric Fields and Temperature on the Inactivation of Microorganisms. *AMB Express* **2021**, *11*, 47. [CrossRef]
42. Nicholson, W.L.; Munakata, N.; Horneck, G.; Melosh, H.J.; Setlow, P. Resistance of Bacillus Endospores to Extreme Terrestrial and Extraterrestrial Environments. *Microbiol. Mol. Biol. Rev.* **2000**, *64*, 548–572. [CrossRef] [PubMed]
43. Bushkin, G.G.; Motari, E.; Carpentieri, A.; Dubey, J.P.; Costello, C.E.; Robbins, P.W.; Samuelson, J. Evidence for a Structural Role for Acid-Fast Lipids in Oocyst Walls of *Cryptosporidium*, *Toxoplasma*, and *Eimeria*. *mBio* **2013**, *4*, e00387-13. [CrossRef]
44. Clancy, J.L.; Bukhari, Z.; Hargy, T.M.; Bolton, J.R.; Dussert, B.W.; Marshall, M.M. Using UV to Inactivate Cryptosporidium. *J. Am. Water Works Assoc.* **2000**, *92*, 97–104. [CrossRef]
45. Craik, S.A.; Weldon, D.; Finch, G.R.; Bolton, J.R.; Belosevic, M. Inactivation of *Cryptosporidium parvum* oocysts using medium-and low-pressure ultraviolet radiation. *Water Res.* **2001**, *35*, 1387–1398. [CrossRef] [PubMed]
46. Matsubayashi, M.; Teramoto, I.; Urakami, I.; Naohara, J.; Sasai, K.; Kido, Y.; Kaneko, A. Evaluation of *Cryptosporidium parvum* Oocyst Inactivation Following Exposure to Ultraviolet Light-Emitting Diodes by in Vitro Excystation and Dye Staining Assays. *Parasitol. Int.* **2022**, *88*, 102557. [CrossRef] [PubMed]
47. Collins, M.V.; Flick, G.J.; Smith, S.A.; Fayer, R.; Croonenberghs, R.; O'keefe, S.; Lindsay, D.S. The Effect of High-Pressure Processing on Infectivity of *Cryptosporidium parvum* Oocysts Recovered from Experimentally Exposed Eastern Oysters (*Crassostrea virginica*). *J. Eukaryot. Microbiol.* **2005**, *52*, 500–504. [CrossRef]
48. Huffman, D.E.; Slifko, T.R.; Salisbury, K.; Rose, J.B. Inactivation of bacteria, virus and *Cryptosporidium* by a point-of-use device using pulsed broad spectrum white light. *Water Res.* **1999**, *34*, 2491–2498. [CrossRef]
49. Craighead, S.; Hertrich, S.; Boyd, G.; Sites, J.; Niemira, B.A.; Kniel, K.E. Cold Atmospheric Plasma Jet Inactivates *Cryptosporidium parvum* Oocysts on Cilantro. *J. Food Prot.* **2020**, *83*, 794–800. [CrossRef]

50. Olvera, M.; Eguía, A.; Rodríguez, O.; Chong, E.; Pillai, S.D.; Ilangovan, K. Inactivation of *Cryptosporidium parvum* Oocysts in Water Using Ultrasonic Treatment. *Bioresour. Technol.* **2008**, *99*, 2046–2049. [CrossRef] [PubMed]
51. Oyane, I.; Furuta, M.; Stavarache, C.E.; Hashiba, K.; Mukai, S.; Nakanishi, M.; Kimata, I.; Maeda, Y. Inactivation of *Cryptosporidium parvum* by Ultrasonic Irradiation. *Environ. Sci. Technol.* **2005**, *39*, 7294–7298. [CrossRef]
52. Centers for Disease Control and Prevention Outbreaks of Escherichia Coli O157:H7 Infection and Cryptosporidiosis Associated With Drinking Unpasteurized Apple Cider—Connecticut and New York, October 1996. *JAMA* **1997**, *277*, 781–782. [CrossRef]
53. Robertson, L.J.; Gjerde, B. Occurrence of Parasites on Fruits and Vegetables in Norway. *J. Food Prot.* **2001**, *64*, 1793–1798. [CrossRef]
54. Grand View Research. *Cold Press Juice Market Size, Share & Trends Analysis Report by Type (Fruit, Vegetable), by Category (Conventional, Organic), by Packaging, by Distribution Channel, by Region, and Segment Forecasts, 2024–2030*; Grand View Research: San Francisco, CA, USA, 2024.
55. Heinz, V.; Toepfl, S.; Knorr, D. Impact of Temperature on Lethality and Energy Efficiency of Apple Juice Pasteurization by Pulsed Electric Fields Treatment. *Innov. Food Sci. Emerg. Technol.* **2003**, *4*, 167–175. [CrossRef]
56. Toepfl, S.; Mathys, A.; Heinz, V.; Knorr, D. Review: Potential of High Hydrostatic Pressure and Pulsed Electric Fields for Energy Efficient and Environmentally Friendly Food Processing. *Food Rev. Int.* **2006**, *22*, 405–423. [CrossRef]
57. Jin, T.; Zhang, H.; Hermawan, N.; Dantzer, W. Effects of PH and Temperature on Inactivation of Salmonella Typhimurium DT104 in Liquid Whole Egg by Pulsed Electric Fields. *Int. J. Food Sci. Technol.* **2009**, *44*, 367–372. [CrossRef]
58. Li, L.; Yang, R.; Zhao, W. The Effect of Pulsed Electric Fields (PEF) Combined with Temperature and Natural Preservatives on the Quality and Microbiological Shelf-Life of Cantaloupe Juice. *Foods* **2021**, *10*, 2606. [CrossRef]
59. Petersen, H.H.; Dalsgaard, A.; Vinneras, B.; Jensen, L.S.; Le, T.T.A.; Petersen, M.A.; Enemark, H.L.; Forslund, A. Inactivation of *Cryptosporidium parvum* Oocysts and Faecal Indicator Bacteria in Cattle Slurry by Addition of Ammonia. *J. Appl. Microbiol.* **2021**, *130*, 1745–1757. [CrossRef] [PubMed]
60. Barsotti, L.; Cheftel, J.C. Food Processing by Pulsed Electric Fields. II. Biological Aspects. *Food Rev. Int.* **1999**, *15*, 181–213. [CrossRef]
61. Charles-Rodríguez, A.V.; Nevárez-Moorillón, G.V.; Zhang, Q.H.; Ortega-Rivas, E. Comparison of Thermal Processing and Pulsed Electric Fields Treatment in Pasteurization of Apple Juice. *Food Bioprod. Process.* **2007**, *85*, 93–97. [CrossRef]
62. Food and Drug Administratio. *Guidance for Industry: Juice Hazard Critical Control Point Hazards and Controls Guidance*, 1st ed.; Food and Drug Administratio: Silver Spring, MD, USA, 2004.
63. Robertson, L.J.; Chalmers, R.M. Foodborne Cryptosporidiosis: Is There Really More in Nordic Countries? *Trends Parasitol.* **2013**, *29*, 3–9. [CrossRef]

foods

MDPI

Article

Modulating Yogurt Fermentation Through Pulsed Electric Fields and Influence of Milk Fat Content

Graciela A. Miranda-Mejía [1], Anaberta Cardador-Martínez [2], Viridiana Tejada-Ortigoza [1], Mariana Morales-de la Peña [3,*] and Olga Martín-Belloso [3,4,*]

[1] Tecnologico de Monterrey, Escuela de Ingeniería y Ciencias, Ave. Eugenio Garza Sada 2501, Monterrey 64849, NL, Mexico; a01206309@tec.mx (G.A.M.-M.); viri.tejada@tec.mx (V.T.-O.)
[2] Tecnologico de Monterrey, Escuela de Ingeniería y Ciencias, Epigmenio González 500, Querétaro 76130, Mexico; mcardador@tec.mx
[3] Department of Food Technology, Engineering and Science, Universitat de Lleida, Rovira Roure 191, 25198 Lleida, Spain
[4] Agrotecnio Center, Rovira Roure 191, 25198 Lleida, Spain
* Correspondence: mariana.morales@udl.cat (M.M.-d.l.P.); olga.martin@udl.cat (O.M.-B.)

Abstract: Yogurt is a highly consumed dairy product valued for its nutritional and probiotic properties. Its production involves the use of lactic acid bacteria, which drive biochemical transformations during fermentation. Optimizing fermentation time without compromising yogurt quality is essential for improving processing efficiency. Pulsed electric fields (PEFs) constitute a promising technology that stimulates microbial activity. In this study, a yogurt starter inoculum suspended in milk (IM) with different fat content (0.5–2.8%) was treated with low-intensity PEFs (1 kV/cm, 800–1600 µs) to enhance fermentation kinetics. pH, soluble solids, lactose, lactic acid, and riboflavin concentrations were monitored during 6 h, comparing PEF-treated IM (PEF-IM) and untreated IM (C-IM). PEF-treatments applied to IM reduced the fermentation time of inoculated milk by 4.3–20.4 min compared to C-IM. The lowest fermentation time (5.1 $\pm$ 0.16 h) was observed in milk added with PEF-IM (2.8% fat) treated at 1 kV/cm for 1600 µs. Milk inoculated with PEF-IM exhibited enhanced lactose consumption (1.6–3.1%) and higher lactic acid production (7.2%) than milk with C-IM. Riboflavin concentration (0.9–7%) decreased between 2 and 4 h, but it stabilized at the end of fermentation. Obtained results suggest that PEFs promote reversible electroporation in microbial cells, facilitating nutrient uptake and acidification, making it a promising assisted-fermentation approach to improve yogurt production.

Keywords: pulsed electric fields; yogurt; fermentation; lactic acid bacteria

1. Introduction

Among dairy products obtained through fermentation, yogurt is highly demanded for its nutritional, functional, and sensory characteristics [1,2]. Yogurt is produced by inoculating milk with lactic acid bacteria (LAB), specifically *Streptococcus thermophilus* (*S. thermophilus*) and *Lactobacillus delbrueckii* subsp. *bulgaricus* (*L. bulgaricus*), which exhibit a symbiotic relationship. *S. thermophilus* initiates the fermentation by producing formic acid and CO_2, which stimulate *L. bulgaricus* growth. In turn, *L. bulgaricus* hydrolyzes milk proteins and releases peptides and amino acids, promoting the growth of *S. thermophilus* and contributing also to yogurt flavor and texture [3–5]. These microorganisms metabolize lactose into lactic acid, which causes a drop in pH and generates protein coagulation. Yogurt is considered ready once the inoculated milk achieves a pH range between 4.5 and 4.7 (cut-off pH), ensuring its desired consistency, flavor, and microbial stability [3–5].

In addition to acidification, LAB also influences the nutritional profile of yogurt. Some strains can synthesize B-group vitamins such as riboflavin (vitamin B_2), while others may uptake available riboflavin from the milk during fermentation, depending on their genetic characteristics and physiological state, thereby modulating the final vitamin content in yogurt [6,7].

Milk composition, particularly its fat content, also plays a key role in the fermentation performance. Higher fat concentration is associated with slower acidification rates due to the buffering capacity of fat globules and the reduced availability of aqueous-phase nutrients. Nevertheless, milk fat improves texture and sensory attributes and may also contribute to LAB stability under processing conditions [8].

Fermentation is one of the most energy-demanding and time-consuming stages in yogurt production, usually requiring between 4 and 6 h under standard industrial conditions. Therefore, finding alternative approaches able to improve fermentation efficiency without compromising product quality, represents a great challenge for the food industry [9,10]. In this context, novel technologies have gained attention as promising processes to assist fermentation, showing numerous advantages. Among them, pulsed electric fields (PEFs) are recognized as a promising non-thermal process characterized by their short processing time, from micro to milliseconds, and high potential for enhancing microbial performance when applied at low intensity. Reversible electroporation induced by low-intensity PEFs (0.1–3 kV/cm) has been reported to transiently increase microbial membrane permeability. This facilitates greater nutrient uptake and accelerates metabolic activity, promoting faster fermentation without compromising cell viability [11–13]. Regarding milk fermentation, electroporation may enhance the metabolic activity of *S. thermophilus* and *L. bulgaricus* by improving their uptake of lactose and micronutrients, which are essential for acid production and symbiotic growth [14].

Within the last decade, PEF processing (0.2–3.67 kV/cm) of yogurt starter cultures has been studied to improve fermentation kinetics without compromising product quality. PEF treatment of *S. thermophilus* and *L. bulgaricus* has been shown to enhance acidification kinetics in reconstituted skimmed milk, reducing the pH lag phase by 12 min [15]. Similarly, improvements in the fermentation rate, texture, and reduction of syneresis have also been observed when mixed LAB cultures were suspended in peptone water and exposed to PEFs [16]. Recently, applying PEFs to LAB inoculated into whole milk significantly shortened total fermentation time by 31.2 min without compromising the physicochemical or sensory quality of the obtained yogurt after 14 days of refrigerated storage [11].

On the other hand, it has been reported that PEF effects on microorganisms are highly influenced by the composition of the medium ant its physicochemical properties, such as conductivity [17–19]. Despite this, based on the current available literature, no studies have addressed how the treatment of the inoculum by PEFs and medium characteristics affect LAB performance during milk fermentation to obtain yogurt.

Hence, the aim of this study was to assess the effect of applying low-intensity PEFs at different treatment times on inoculum suspended in milk (IM) with 0.5 and 2.8% fat content, by monitoring the fermentation time and analyzing changes in the lactose, lactic acid, and riboflavin content of milk supplemented with the IM throughout the fermentation process.

2. Materials and Methods

2.1. Yogurt Manufacturing and Determination of Fermentation Endpoint

The procedure to obtain natural yogurt was conducted according to the method previously established in Miranda-Mejía et al. [11]. Initially, the inoculum was made under sterile conditions in a laminar flow hood, using 800 mL of commercial UHT (ultra-high temperature) treated milk with 0.5 or 2.8% fat content (Alpura, Querétaro, Mexico) and 200 g of commercial thermophilic culture composed of *S. thermophilus* and *L. bulgaricus* (Novonesis, Bagsvaerd, Capital region, Denmark; culture YF-L904, batch No: 3751231, MEX), reaching an initial microbial load of 2.4×10^9 CFU/mL. The mixture of starter culture and milk with different fat content (IM-0.5 and IM-2.8) was blended at 150–170 rpm for 20 min to ensure homogeneity. Subsequently, 330 μL of IM-0.5 or IM-2.8 was transferred into 330 mL of milk with 0.5 or 2.8% fat content, respectively, gently mixed, and incubated at 42 °C for 6 h [20,21]. The pH was monitored every hour until reaching the cut-off pH (4.7), which marked the end of the fermentation and yogurt formation.

An acidification curve was plotted using pH values vs. time obtained during 6 h of yogurt fermentation. Then, the fermentation time was calculated by solving x with Wolfram Alpha (Wolfram Research, Champaign, IL, USA) from Equation (1):

$$y = x^2 - x + a \quad (1)$$

where y corresponds to the cut-off pH (4.7), a is a constant parameter, and x denotes the fermentation endpoint.

2.2. Pulsed Electric Field Treatment

PEF treatment was carried out using an EPULSUS®-LPM1A-10 system (Energy Pulse Systems, Lisbon, Portugal) with a maximum output of 10 kV, 200 A, and 3 kW. The equipment featured a batch parallel treatment chamber comprising two stainless steel electrodes separated by a 10 cm gap and insulated with a nylon dielectric material, operating in monopolar mode. The IM-0.5 or IM-2.8 were subjected to PEF treatment before adding them to the milk, when microorganisms were in the lag phase. IM (80 mL) with electrical conductivity values of 22.9 mS/cm (0.5% fat content) and 18.5 mS/cm (2.8% fat content) were located inside the treatment chamber and PEF processed during different treatment times (t). t was determined by multiplying the pulse number by the pulse width (μs). The temperature of IM was taken before and right after PEF treatment using a digital thermometer (PECULA TP101, Shenzhen, China). Under all tested conditions, the temperature remained below 24 ± 0.5 °C. To independently assess the effect of the PEF, untreated IM (C-IM) was used as a control and subjected to the same handling procedures without PEF application, ensuring all other fermentation conditions were identical. PEF-treated IM (PEF-IM) and C-IM were added to the milk and incubated under the same conditions to monitor fermentation kinetics.

2.3. Experimental Design

A 2^3 factorial experiment was set up to investigate the influence of milk fat content (0.5 and 2.8%) and PEF processing at an electric field strength (E) of 1 kV/cm, pulsed with (τ) of 8 μs, and a frequency (f) of 10 Hz during different t (800, 1200, and 1600 μs) on microbial performance through fermentation. The selection of these conditions was based on prior research and preliminary experiments [11]. A total of six PEF treatments were applied to IM as indicated in Table 1, having the milk with C-IM as reference.

The evaluation of lactose, lactic acid, and riboflavin concentration changes during fermentation was performed on the milk added with PEF-IM, achieving the lowest fermentation time (PEF-IM$_{OPT}$), comparing results with milk containing C-IM. This evaluation allowed a comparative analysis of fermentation biochemical changes between milks with PEF-IM at optimum conditions and C-IM.

Table 1. PEF treatments (1 kV/cm, pulse width 8 μs at 10 Hz) applied to inoculated milk (IM) with different fat content (0.5 and 2.8%).

Treatment	Fat (%)	*t* (μs)
C-IM0.5	0.5	0
PEF-IM0.5-800		800
PEF-IM0.5-1200		1200
PEF-IM0.5-1600		1600
C-IM2.8	2.8	0
PEF-IM2.8-800		800
PEF-IM2.8-1200		1200
PEF-IM2.8-1600		1600

Inoculum size: 200 g of starter culture in 800 mL of milk.

2.4. pH, Total Soluble Solids, and Conductivity Measurements

Measurements of total soluble solids (°Brix) and pH were performed on 20 mL of IM with a refractometer (HANNA HI 96813, HANNA Instruments, Woonsocket, RI, USA) and a potentiometer (OAKTON pH 510, OAKTON Instruments, Vernon Hills, IL, USA), respectively. Conductivity was measured (80 mL) using a conductivity meter (OAKTON 6+, OAKTON Instruments, Vernon Hills, IL, USA).

2.5. Microbial Count

Bacterial count was performed according to Kang et al. (2019) with some modifications [22]. The culture medium was prepared by dissolving 70 g of MRS (de Man–Rogosa–Sharpe) (DIFCO, Becton, Dickinson, and Company, Sparks, MD, USA) agar in 1 L of distilled water, sterilized (121 °C/15 min at 15 psi), and poured into Petri dishes. The plates were left to solidify at room temperature for 40 min, covered with parafilm, and placed in storage upside down under refrigerated conditions (4 ± 1 °C) until use. IM were stirred at 160 rpm for 1 h to ensure a homogeneous mixture. Six serial dilutions were prepared, inoculated in petri dishes, and incubated 24 h at 42 ± 0.5 °C. Bacterial colonies were enumerated with a digital colony counter, and the number of colony-forming units per milliliter (CFU/mL) was calculated with Equation (2).

$$\frac{CFU}{mL} = \frac{\text{Averages colony count}}{\text{Dilution factor} \times \text{volume plated}} \quad (2)$$

2.6. Lactose and Organic Acids Extraction and Quantification

Lactose and organic acids extractions were performed following the procedure of Leclercq-Perlat (1999) with some modifications [23]. Milks with PEF-IM$_{OPT}$ and C-IM (750 μL for lactose and 1000 μL for organic acids) were blended with 1 mL of deionized water and placed in a water bath (50 °C) for 1 h. After incubation, the mixtures were vortexed at 160,000 rpm for 2 min and immediately cooled using an ice bath to approximately reach 25 °C. For lactose extraction, 1250 μL each of Carrez I and Carrez II reagents, along with 250 μL of sodium hydroxide, were incorporated into the mixture. In the case of organic acid extraction, 1 mL of a 240 g/L trichloroacetic acid solution (SIGMA-Aldrich, St. Louis, MO, USA) and an additional 1 mL of distilled water were added. Both extracts were subsequently vortexed again for homogenization and incubated for 1 h at 25 °C. The resulting solutions were filtered through 0.45 μm PTFE membranes (Agilent Technologies, Santa Clara, CA, USA) and transferred into 2 mL Eppendorf tubes. Lactose (L_e) and Organic acid (OA_e) extracts were preserved at 4 °C and analyzed within 15 days to avoid degradation. To minimize analyte degradation and ensure consistency across all extracts, the procedures were carried out using freshly prepared reagents. Extractions were performed under chilled conditions, and the contact time with Carrez and TCA solutions was standardized to reduce variability caused by exposure duration.

For HPLC analysis, L_e and OA_e were transferred to HPLC vials and identified following the procedure described by Picque et al. (1993) with some modifications [24]. Vials were injected in an Agilent 1200 series system (Agilent Technologies, Santa Clara, CA, USA) equipped with a refractive index detector (RID) with an optical temperature of 30 °C for lactose (Lac), and a diode array detector (DAD) for (LA), propionic (PA), and butyric (BA) acids. A Zorbax Eclipse XBD-C18 column (4.6 × 150 mm, 5 μm, Agilent, Santa Clara, CA, USA) maintained at 50 °C was used to separate the compounds with an isocratic flow rate of 1 mL/min of the mobile phase (5 mM sulfuric acid) for 15 min. A constant injection volume of 10 μL was used for all analytes and standards. Identification of Lactose (Lac) was based on the retention time and RID spectral comparison with 10 g/L of standard (SIGMA-Aldrich, St. Louis, MO, USA). Lactic acid (LA), propionic acid (PA), and butyric acid (BA) were identified by comparing their UV-Vis spectra and retention times with certified reference standards: LA (40 g/L), PA (20 g/L), and BA (20 g/L), all from SIGMA-Aldrich (St. Louis, MO, USA). The quantification of Lac was based on the peak integration of RID signals, whereas organic acids were quantified by integrating the peak area at 210 nm. Calibration curves were prepared for each compound obtaining R^2 values of 0.9997 for Lac, 0.9952 for LA, 0.9962 for PA, and 0.9762 for BA. Results were presented as the percentage of lactose and organic acids in 100 mL of milk added with IM.

2.7. *Riboflavin Extraction and Quantification*

Riboflavin (vitamin B_2) extraction and quantification were performed according to the procedure of Albalá-Hurtado et al. (1997) and Gliszczyńska-Świglo and Koziolowa (2000) with some modifications [25,26]. Milks with PEF-IMOPT and C-IM (1050 μL) were transferred to 50 mL centrifuge tubes and blended with 0.1 g of trichloroacetic acid (TCA, SIGMA-Aldrich, St. Louis, MO, USA). The mixture was agitated on a magnetic stirring plate (CIMAREC) for 10 min and then centrifuged at 1250× *g* for 10 min (MULTIFUGE X1R, Thermo Scientific, Waltham, MA, USA). After collecting the supernatant, the remaining pellet was re-extracted adding 300 μL of 4% TCA, stirred (10 min), and centrifuged under the same conditions. The solid phase was discarded, and both supernatants were combined in a 2 mL amber Eppendorf tube, adjusting the volume with 4% TCA. The extract was passed through a 0.45 μm PTFE filter (Agilent Technologies, GER, Santa Clara, CA, USA) and kept in amber vials (4 °C) until it was analyzed. Throughout the extraction and filtration steps, ambient light exposure was minimized by covering the centrifuge tubes with aluminum foil.

Riboflavin analysis was conducted in an Agilent 1200 series HPLC system (Agilent Technologies, Santa Clara, CA, USA) equipped with a fluorescence detector (FLD) set at 35 °C. Separation of riboflavin was performed by injecting 20 μL of riboflavin extract into the system, using a Tracer Spherisorb ODS2-C18 column (4.6 × 250 mm, 5 μm, Agilent, Santa Clara, CA, USA) set at room temperature (25 °C), with an isocratic elution of 0.7 mL/min of the mobile phase during a 20 min runtime. The mobile phase was prepared by first dissolving 1.8 g of octane sulfonic acid (OSA, SIGMA-Aldrich, St. Louis, MO, USA) in 800 mL of double-distilled water. Then, 24 mL of glacial acetic acid (MEYER, Mexico City, Mexico) and 5 mL of triethylamine (SIGMA-Aldrich, St. Louis, MO, USA) were added to the solution, which was thoroughly mixed. The pH was adjusted to 3.6 ± 0.1 using either glacial acetic acid to lower the pH or triethylamine to raise it, depending on the initial value. Once the target pH was achieved, 150 mL of LC-grade methanol (SIGMA-Aldrich, St. Louis, MO, USA) was added. The solution was mixed again and volumetrically filled to one liter with double-distilled water. Finally, it was filtered using a 0.45 μm membrane.

Riboflavin identification was conducted by comparing the retention time and fluorescence signal with those of the standard solution (100 mg/L riboflavin, SIGMA-Aldrich, GER, St. Louis, MO, USA), which was prepared in aqueous acetic acid (2.4%, *v*/*v*). Quantification was performed by integrating the areas of the obtained peaks at 450 nm (excitation wavelength) and 525 nm (emission wavelength). Data were compared to a calibration curve, which showed a correlation coefficient of $R^2 = 0.9811$. The riboflavin content was reported in percentage per 100 mL of milk containing IM.

2.8. Statistical Analysis

PEF experiments were carried out in duplicate, and each treatment was analyzed in triplicate. Statistical differences between group means were evaluated using Tukey's post-hoc test and an independent Student's *t*-test. A significance threshold of $p < 0.05$ was applied to determine significant differences between treatments. All statistical evaluations were conducted using Minitab software (Minitab, LLC, State College, PA, USA, Version 19.2020.10).

3. Results and Discussion

3.1. Effect of PEF Treatment on Fermentation Process

3.1.1. Fermentation Time

A progressive pH decline from 6.63 ± 0.01 (0 h) to 4.48 ± 0.01 was observed during the fermentation process of milk with C-IM. The cut-off pH (4.7) was reached at 5.9 ± 0.07 h in C-IM0.5 and at 5.4 ± 0.18 h for C-IM2.8. These fermentation time are within the typical yogurt fermentation time range (5.0s–6.5 h) reported by other authors [11,27]. The observed changes in pH are associated with the consumption of fermentable carbohydrates, mainly lactose, by the LAB during fermentation [24].

As shown in Figure 1, milks with PEF-IM (0.5 and 2.8% fat content) at different treatment times exhibited a reduction in fermentation time compared to milk with C-IM. A longer PEF treatment time consistently resulted in a shorter fermentation time, regardless of milk fat content. According to Miranda-Mejía et al. (2024) and Chanos et al. (2020), this effect might be attributed to the reversible electroporation in LAB induced by the PEF processing at low electric field intensity (1 kV/cm) for long treatment times, enhancing cell membrane permeability [11,15]. This mechanism may facilitate the lactose uptake of microorganisms from the media, converting it to lactic acid and accelerating LAB metabolism [28]. Although post-treatment microbial viability was not directly quantified, the accelerated acidification and reduction in fermentation time observed in milks with PEF-IM suggest that the LAB remained functionally active and that PEFs may have enhanced their metabolic performance. These results are in agreement with previous studies reporting that low-intensity PEFs (1–3 kV/cm) can stimulate LAB through reversible electroporation, increasing nutrient transport and promoting lactic acid production without compromising cell viability [16,29]. In addition to electroporation, a sublethal cellular stress response might be triggered during PEF processing, stimulating the metabolic activity of LAB, shortening the lag phase, and enhancing acidification. Therefore, the combination of membrane permeabilization and controlled stress induction at 1 kV/cm appeared to enhance lactose metabolism and lactic acid production by the LAB during fermentation [15,30].

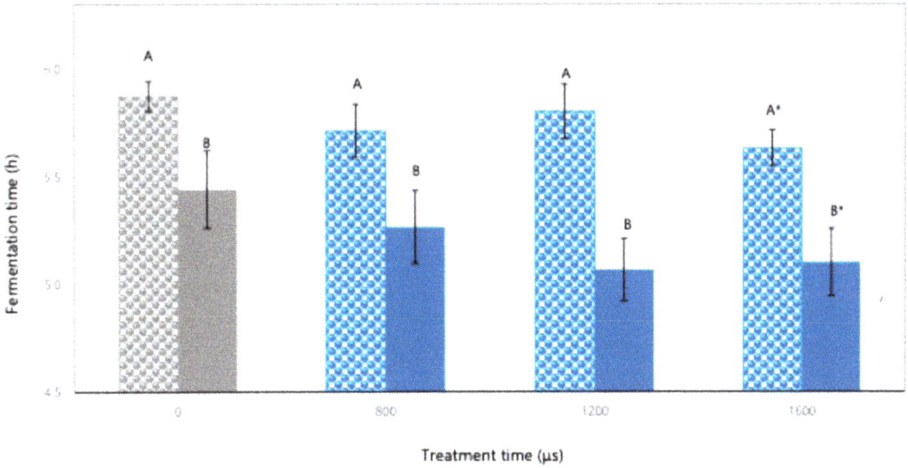

Figure 1. Fermentation time (h) of milk with PEF treated inoculum (PEF-IM) at different treatment times (800, 1200, and 1600 µs) having milk with untreated inoculum (C-IM) as control (0 µs, gray bars). Dotted-patterned bars correspond to milk with 0.5% fat content, and solid bars to milk with 2.8% fat content. A, B indicate significant differences ($p < 0.05$) between matrices at the same treatment time. * indicates significant differences ($p < 0.05$) within the same matrix at different treatment times. Treatment codes correspond to those listed in Table 1.

Interestingly, milk with PEF-IM0.5 resulted in a longer fermentation time compared to milk with PEF-IM2.8, irrespective of the processing time. This trend could be explained by the higher concentration of phospholipids, lipophilic vitamins, and other bioactive lipids in milk with a higher fat content (2.8%), which might protect LAB and contribute to membrane stabilization, improving microbial tolerance to PEF-induced stress [31,32]. In contrast, milk with a lower fat content (0.5%) lacks these protective compounds, which may limit bacterial adaptation, particularly under longer treatments. Despite these compositional differences, both inoculated milk matrices were added with the same concentration of starter culture, achieving an initial bacterial population of ~2.4×10^9 CFU/mL. This indicates that milk fat content did not influence the initial microbial viability before PEF application. This is consistent with previous studies reporting that variations in milk fat content do not significantly alter the initial survival counts of starter cultures when inoculated at fixed proportions, especially in UHT milk systems [5,33,34]. Therefore, the differences observed during the fermentation of milk with PEF-IM could be attributed to the matrix-dependent response to electroporation and conductivity properties, rather than the initial microbial load. Notably, electrical conductivity differed significantly between matrices. IM with 2.8% fat content exhibited a moderate conductivity (~18.5 mS/cm), whereas IM with 0.5% fat content showed higher values (~22.9 mS/cm). While moderate conductivity enhances electric field propagation and electroporation efficiency, excessively high conductivity may lead to energy dispersion and localized overheating, potentially compromising microbial performance [28].

Among all evaluated conditions, IM with 2.8% fat content, PEF-treated with pulses of 8 µs at 1 kV/cm during 1600 µs and 10 Hz (PEF-IM2.8-1600) was the most effective treatment, resulting in the greatest fermentation time reduction (20.4 min). These findings confirm that the interaction between PEF parameters, milk fat content, and conductivity plays a critical role in enhancing fermentation kinetics through synergistic effects on LAB metabolic performance.

3.1.2. Changes in Soluble Solids, Lactose, and Organic Acids During Fermentation as Affected by PEF_{OPT}

Soluble Solids

Changes in soluble solids (Brix) during the fermentation of milk with $PEF\text{-}IM_{OPT}$ and its corresponding C-IM are presented in Figure 2. During the first 2 h, soluble solids remained constant at approximately 27 °Brix in both milks with $PEF\text{-}IM_{OPT}$ and C-IM, which is the initial soluble solid content of whole milk (2.8% fat content). Between hours 2 and 4, a clear decrease in soluble solids was observed, especially in milk with $PEF\text{-}IM_{OPT}$. In this period, the soluble solids dropped from 27.0 to 23.0 in milk with $PEF\text{-}IM_{OPT}$, while in the milk with C-IM, they decreased slightly, from 27.1 to 26.4. These differences suggest that PEF enhanced the activity of LAB, possibly by improving membrane permeability and promoting faster sugar uptake [11,13]. From hour 4 onward, the values stabilized around 21 °Brix in both milks with $PEF\text{-}IM_{OPT}$ and C-IM. This result indicates that PEFs influenced the rate at which sugars were consumed, particularly during the early stages of fermentation, without altering the final soluble solids level [16,29].

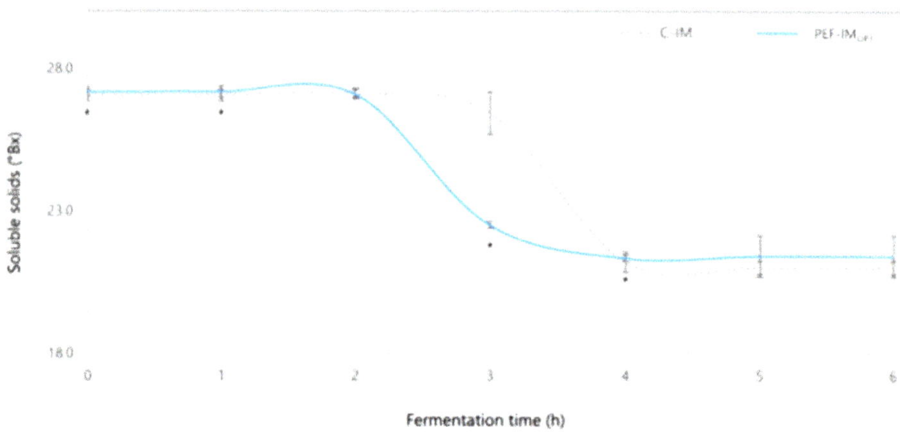

Figure 2. Soluble solids (°Brix) values during fermentation of milk with inoculum treated with PEF at optimal conditions ($PEF\text{-}IM_{OPT}$) and untreated inoculum (C-IM). * corresponds to a statistical difference ($p < 0.05$) between milk with C-IM and $PEF\text{-}IM_{OPT}$.

Lactose and Organic Acids

The lactose concentration showed a gradual decline during fermentation in milk supplemented with $PEF\text{-}IM_{OPT}$, with the most pronounced reduction occurring during the first 2 h (Figure 3A). At both 2 and 4 h, lactose content in milk with $PEF\text{-}IM_{OPT}$ was significantly lower by approximately 1.6 to 3.1% than in the milk with C-IM, indicating a faster substrate consumption of LAB during the early phase of fermentation. This suggests that PEF treatment to IM before the fermentation stage enhanced the metabolic activity of LAB, likely by increasing membrane permeability and stimulating nutrient transport systems, as previously reported [13]. From hour 4 onwards, lactose concentrations stabilized in both milks, with $PEF\text{-}IM_{OPT}$ and C-IM, indicating that most of the lactose transformation reactions occurred during the early fermentation phase. These results support the hypothesis that low-intensity PEFs accelerate the rate but not the extent of carbohydrate utilization, promoting a faster transition through the exponential growth phase of LAB. The accelerated lactose depletion observed in milk with PEF-IM could be linked to PEF effects on enzymatic activity. Although the precise molecular mechanisms remain under investigation, it has been proposed that PEFs may enhance the function of lactose permeases (LacS) and β-galactosidase (LacZ), facilitating more efficient substrate uptake and hydrolysis during early fermentation [35]. Consistently, previous studies have shown that PEF treatment enhanced enzymatic activity and acid production in *L. bulgaricus* without compromising cell viability [29].

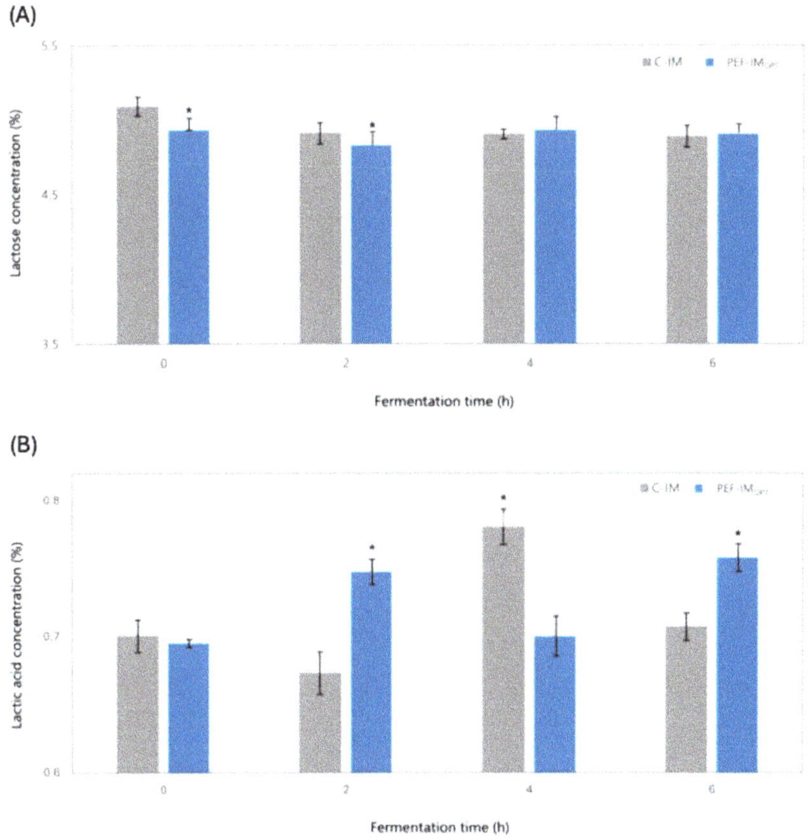

Figure 3. Changes in lactose (**A**) and lactic acid (**B**) concentration of milk with PEF treated IM at optimal conditions (PEF-IM$_{OPT}$) and untreated inoculum (C-IM) during fermentation. * corresponds to a statistical difference ($p < 0.05$) between the CIM and PEF-IM$_{OPT}$.

As expected, LA concentrations increased progressively throughout fermentation in milk supplemented with PEF-IM$_{OPT}$ and C-IM (Figure 3B). At 2 and 6 h of fermentation, milk with PEF-IM$_{OPT}$ exhibited significantly higher LA content than milk supplemented with C-IM, which was 11% higher at 2 h and 7.2% higher at 6 h ($p < 0.05$), indicating that low-intensity PEF treatment enhanced early metabolic activity of LAB and sustained acid production until the end of fermentation. These results align with the reduction observed in total soluble solids and pH, suggesting an accelerated onset of carbohydrate metabolism [11]. Interestingly, at the 4th hour of fermentation, LA concentration in milk with C-IM was 10.3% higher than in the milk supplemented with PEF-IM$_{OPT}$ ($p < 0.05$). This transient drop in LA accumulation in the milk with PEF-IM may reflect a brief physiological adjustment of the LAB following PEF exposure. Previous studies have shown that reversible electroporation induced by low-intensity PEF can temporarily alter membrane permeability, potentially disrupting intracellular processes such as acid synthesis [29,36]. However, the increased LA content in milk with PEF-IM$_{OPT}$ by 6 h suggests that the LAB successfully adapted to this transient stress and re-established their acidification capacity. A comparable recovery pattern has been observed, indicating that the metabolic activity of LAB can normalize after an initial lag period post-PEF [16]. Likewise, previous studies have reported increased acidification rates and enzymatic activity in *L. bulgaricus* after PEF exposure, along with a faster pH drop and greater LA accumulation in yogurt fermented with PEF-treated cultures [11,29].

Moreover, no traces of propionic or butyric acids were detected in either milk with PEF-IM_{OPT} or milk with C-IM throughout fermentation. This indicates that the starter cultures exhibited selective metabolic activity favoring lactic acid production, with no signs of contamination by undesirable bacteria. The lack of these organic acids is considered favorable, as they are often associated with off-flavors such as rancid, bitter, or pungent notes in dairy products [37–39]. These findings suggest that PEFs enhance lactic acid production without altering the profile of secondary organic acids, as supported by previous reports [11].

Riboflavin

Riboflavin (vitamin B_2) was evaluated during fermentation in milks with PEF-IM_{OPT} and C-IM to monitor how riboflavin concentration changed as a potential response of LAB to PEFs. Figure 4 presents the concentration of riboflavin in both milks with PEF-IM_{OPT} and C-IM at 0, 2, 4, and 6 h of fermentation. A noticeable decrease in riboflavin content was observed between 2 and 4 h of fermentation in both milks, being 17% lower in milk with PEF-IM_{OPT} compared to milk with C-IM at hour 2. After this drop, riboflavin concentrations stabilized by hour 6, with no significant differences between the two inoculated milks. These findings suggest that PEFs did not affect the final riboflavin content in the obtained yogurt but may have transiently influenced its concentration during the early fermentation phase [40,41].

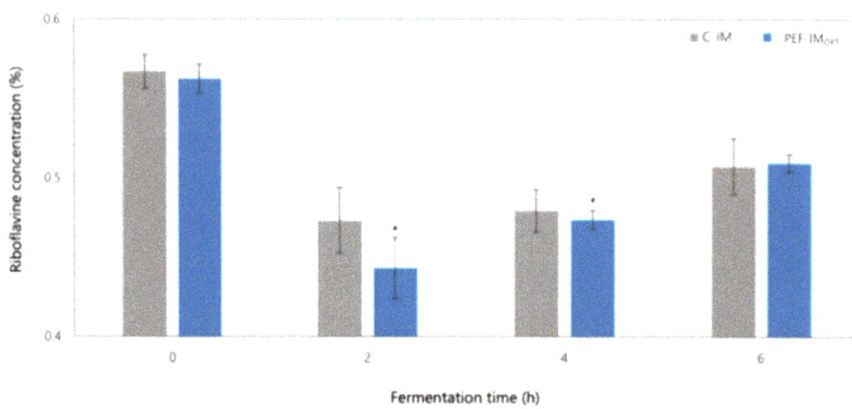

Figure 4. Changes in riboflavin concentration in milk with PEF-treated inoculum at optimal conditions (PEF-IM_{OPT}) and untreated inoculum (C-IM) during fermentation. * corresponds to a statistical difference ($p < 0.05$) between the C-IM and PEF-IM_{OPT}.

The temporary decrease in riboflavin content on milk supplemented with PEF-IM_{OPT} may be explained by the reversible electroporation effect induced by low-intensity PEFs, which increases membrane permeability in LAB, potentially enhancing the uptake of extracellular compounds such as riboflavin [40,41]. Additionally, some studies have shown that the metabolic change in riboflavin depends on the strain-specific behavior of LAB. During fermentation, some strains can synthesize riboflavin de novo, while others can uptake it directly from the media depending on their genetic makeup and physiological state [7,42]. Given the observed reduction in riboflavin concentration in milk with PEF-IM_{OPT} at the 2nd hour of fermentation, it is possible that the LAB strains from the starter culture used it as a nutrient for fermentation reactions, especially under PEF-induced stress. By the end of fermentation, the stabilization of riboflavin content in both milks with PEF-treated or untreated IM confirmed that PEF application did not negatively compromise the vitamin's final concentration in the obtained yogurt. Rather than altering the nutritional quality of the product, PEFs may transiently affect vitamin availability during early microbial adaptation, highlighting the complexity of microbial responses to non-thermal technologies such as PEFs [16,40,41].

Although this study provides insights into the effects of PEFs on vitamin dynamics during fermentation, comprehensive evaluations of the combined impact of PEFs and milk composition remain limited and, to the best of the authors' knowledge, there is not related information available in the literature. Hence, PEF effects on microbial vitamin uptake mechanisms and its interaction with matrix components in dairy systems warrant further investigation.

4. Conclusions

This study demonstrates that PEF treatment applied for a longer time to IM with 2.8% fat content significantly reduced the fermentation time of inoculated milk compared to a conventional process. Adding PEF-IM treated at 1 kV/cm for 1600 μs to the milk showed the greatest effect, reaching the cut-off pH (4.7) to obtain yogurt up to 20.4 min earlier than milk with untreated IM. The reduction in fermentation time on milk with PEF-IM was associated with enhanced LAB metabolic activity, likely due to induced membrane permeability leading to a faster lactose depletion and lactic acid accumulation. Furthermore, moderate conductivity values (~18.5 mS/cm) of treated media could promote a uniform electric field propagation during PEF processing, improving the electroporation effect. Although riboflavin levels showed a temporary decline during early fermentation in milk with PEF-IM, final concentrations remained stable, confirming that PEFs preserved the nutritional quality of the obtained yogurt. Overall, low-intensity PEFs demonstrated potential as a viable non-thermal strategy to improve yogurt process efficiency by accelerating milk fermentation without adversely affecting the physicochemical characteristics and vitamin content of the obtained yogurt. Further studies are needed to deepen our understanding of the microbial response to PEFs and to evaluate its scalability for industrial applications.

Author Contributions: Conceptualization, G.A.M.-M. and M.M.-d.l.P.; methodology, G.A.M.-M. and M.M.-d.l.P.; software, G.A.M.-M.; validation, G.A.M.-M., M.M.-d.l.P., V.T.-O. and O.M.-B.; formal analysis, G.A.M.-M. and M.M.-d.l.P.; investigation, G.A.M.-M.; resources, M.M.-d.l.P., V.T.-O. and A.C.-M.; data curation, G.A.M.-M.; writing—original draft preparation, G.A.M.-M.; writing—review and editing, G.A.M.-M., M.M.-d.l.P., V.T.-O. and O.M.-B.; visualization, G.A.M.-M. and M.M.-d.l.P.; supervision, M.M.-d.l.P., V.T.-O. and A.C.-M.; project administration, M.M.-d.l.P.; funding acquisition, M.M.-d.l.P. and O.M.-B. All authors have read and agreed to the published version of the manuscript.

Funding: This research received no external funding.

Institutional Review Board Statement: Not applicable.

Informed Consent Statement: Not applicable.

Data Availability Statement: The original contributions presented in the study are included in the article. Further inquiries can be directed to the corresponding authors.

Acknowledgments: Graciela A. Miranda-Mejía thanks Tecnológico de Monterrey and the Secretaría de Ciencia, Humanidades, Tecnología e Innovación (SECIHTI, CVU: 1078348) for the predoctoral grant. The authors also thank the BioFoods Research Lab and Novonesis company for their kind support through the provision of materials, reagents, and microorganisms for this study. During the preparation of this manuscript, the authors used ChatGPT (OpenAI, GPT-4, 2025) to refine wording, improve clarity, and ensure coherence in the structure of the scientific content. The authors have reviewed and edited the output and take full responsibility for the content of this publication.

Conflicts of Interest: The authors declare no conflicts of interest.

Abbreviations

The following abbreviations are used in this manuscript:

PEF	pulsed electric fields
IM	inoculum suspended in milk
PEF-IM	PEF-treated IM
C-IM	untreated-IM, as control
LAB	lactic acid bacteria
IM-0.5	inoculum suspended in milk with 0.5% fat content
IM-2.8	inoculum suspended in milk with 2.8% fat content
PEF-IMOPT	PEF-IM at optimum conditions
MRS	de Man–Rogosa–Sharpe
t	treatment times
CFU/mL	colony forming units per milliliter
Le	lactose extracts
OAe	organic acid extracts
RID	refractive index detector
Lac	lactose
DAD	diode array detector
LA	lactic acid
PA	propionic acid
BA	butyric acid
HPLC	high-performance liquid chromatography
TCA	trichloroacetic acid
FLD	fluorescence detector
OSA	octane sulfonic acid
C-IM0.5	untreated-IM, as control with 0.5% fat content
C-IM2.8	untreated-IM, as control with 2.8% fat content

References

1. Drouin-Chartier, J.P.; Brassard, D.; Tessier-Grenier, M.; Côté, J.A.; Labonté, M.É.; Desroches, S.; Couture, P.; Lamarche, B. Systematic Review of the Association between Dairy Product Consumption and Risk of Cardiovascular-Related Clinical Outcomes. *Adv. Nutr.* **2016**, *7*, 1026–1040. [CrossRef] [PubMed]
2. Fernández, E.F.; Hernández, J.A.M.; Suárez, V.M.; Villares, J.M.M.; Yurrita, L.R.C.; Cabria, M.H.; Rey, F.J.M. Documento de Consenso: Importancia Nutricional y Metabólica de La Leche. *Nutr. Hosp.* **2015**, *31*, 92–101. [CrossRef]
3. Bazán, D.L.; del Río, P.G.; Domínguez, J.M.; Cortés-Diéguez, S.; Mejuto, J.C.; Pérez-Guerra, N. The Chemical, Microbiological and Volatile Composition of Kefir-like Beverages Produced from Red Table Grape Juice in Repeated 24-h Fed-Batch Subcultures. *Foods* **2022**, *11*, 3117. [CrossRef]
4. Doleyres, Y.; Schaub, L.; Lacroix, C. Comparison of the Functionality of Exopolysaccharides Produced in Situ or Added as Bioingredients on Yogurt Properties. *J. Dairy Sci.* **2005**, *88*, 4146–4156. [CrossRef] [PubMed]
5. Soukoulis, C.; Panagiotidis, P.; Koureli, R.; Tzia, C. Industrial Yogurt Manufacture: Monitoring of Fermentation Process and Improvement of Final Product Quality. *J. Dairy Sci.* **2007**, *90*, 2641–2654. [CrossRef]
6. Leblanc, J.G.; Laiño, J.E.; del Valle, M.J.; Vannini, V.; van Sinderen, D.; Taranto, M.P.; de Valdez, G.F.; de Giori, G.S.; Sesma, F. B-Group Vitamin Production by Lactic Acid Bacteria—Current Knowledge and Potential Applications. *J. Appl. Microbiol.* **2011**, *111*, 1297–1309. [CrossRef] [PubMed]

7. Juarez del Valle, M.; Laiño, J.E.; Savoy de Giori, G.; LeBlanc, J.G. Riboflavin Producing Lactic Acid Bacteria as a Biotechnological Strategy to Obtain Bio-Enriched Soymilk. *Food Res. Int.* **2014**, *62*, 1015–1019. [CrossRef]
8. Sodini, I.; Montella, J.; Tong, P.S. Physical Properties of Yogurt Fortified with Various Commercial Whey Protein Concentrates. *J. Sci. Food Agric.* **2005**, *85*, 853–859. [CrossRef]
9. El Darra, N.; Turk, M.F.; Ducasse, M.A.; Grimi, N.; Maroun, R.G.; Louka, N.; Vorobiev, E. Changes in Polyphenol Profiles and Color Composition of Freshly Fermented Model Wine Due to Pulsed Electric Field, Enzymes and Thermovinification Pretreatments. *Food Chem.* **2016**, *194*, 944–950. [CrossRef] [PubMed]
10. Morales-de la Peña, M.; Miranda-Mejía, G.A.; Martín-Belloso, O. Recent Trends in Fermented Beverages Processing: The Use of Emerging Technologies. *Beverages* **2023**, *9*, 51. [CrossRef]
11. Miranda-Mejía, G.A.; Martín del Campo-Barba, S.T.; Arredondo-Ochoa, T.; Tejada-Ortigoza, V.; la Peña, M.M. de Low-Intensity Pulsed Electric Fields Pre-Treatment on Yogurt Starter Culture: Effects on Fermentation Time and Quality Attributes. *Innov. Food Sci. Emerg. Technol.* **2024**, *95*, 103708. [CrossRef]
12. Vazquez-Cabral, D.; Valdez-Fragoso, A.; Rocha-Guzman, N.E.; Moreno-Jimenez, M.R.; Gonzalez-Laredo, R.F.; Morales-Martinez, P.S.; Rojas-Contreras, J.A.; Mujica-Paz, H.; Gallegos-Infante, J.A. Effect of Pulsed Electric Field (PEF)-Treated Kombucha Analogues from Quercus Obtusata Infusions on Bioactives and Microorganisms. *Innov. Food Sci. Emerg. Technol.* **2016**, *34*, 171–179. [CrossRef]
13. El Kantar, S.; Koubaa, M. Pulsed Electric Field Treatment for the Stimulation of Microorganisms: Applications in Food Production. *Res. Agric. Eng.* **2022**, *68*, 80–92. [CrossRef]
14. Ulmer, A.; Erdemann, F.; Mueller, S.; Loesch, M.; Wildt, S.; Jensen, M.L.; Gaspar, P.; Zeidan, A.A.; Takors, R. Differential Amino Acid Uptake and Depletion in Mono-Cultures and Co-Cultures of Streptococcus Thermophilus and Lactobacillus Delbrueckii Subsp. Bulgaricus in a Novel Semi-Synthetic Medium. *Microorganisms* **2022**, *10*, 1771. [CrossRef]
15. Chanos, P.; Warncke, M.C.; Ehrmann, M.A.; Hertel, C. Application of Mild Pulsed Electric Fields on Starter Culture Accelerates Yogurt Fermentation. *Eur. Food Res. Technol.* **2020**, *246*, 621–630. [CrossRef]
16. Stühmeier-Niehe, C.; Lass, L.; Brocksieper, M.; Chanos, P.; Hertel, C. Pre-Treatment of Starter Cultures with Mild Pulsed Electric Fields Influences the Characteristics of Set Yogurt. *Foods* **2023**, *12*, 442. [CrossRef] [PubMed]
17. Tremín, J. Inactivación Microbiana Por Pulsos Eléctricos de Alto Voltaje. 2014, pp. 1–45. Available online: https://zaguan.unizar.es/record/16002/files/TAZ-TFG-2014-1503.pdf (accessed on 15 April 2025).
18. Bombón-Pilliza, L.A. Influencia de La Aplicación de Pulsos Eléctricos de Alta Intensidad (Peai) Sobre La Carga Microbiana Del Néctar de Fresa. 2012. Available online: https://repositorio.uta.edu.ec/items/68f896b9-4c8b-4920-8f9b-027d46de448e (accessed on 15 April 2025).
19. Pineda-Posadas, J.A.; Ortiz-Rodríguez, E.; Ortiz-Rodríguez, L. Campos Eléctricos Pulsados. *Handbook Tecnologías Emergentes Aplicadas en Alimentos*. 2022, pp. 53–61. Available online: https://www.ecorfan.org/handbooks/Handbooks_Tecnologias_Emergentes_Aplicadas_en_Alimentos_TI/Handbooks_Tecnologias_Emergentes_Aplicadas_en_Alimentos_TI_6.pdf (accessed on 26 May 2025).
20. *Novonesis Internal Technical Protocol for Thermophilic Culture YF-L904*; Document provided by the company uopon request; Novonesis: Gabsvaerd, Capital Region, Denmark, 2020.
21. Turengano-Roldan, C. NORMA Oficial Mexicana NOM-181-SCFI-2010, Yogurt-Denominación, Especificaciones Fisicoquímicas y Mic. Available online: https://www.dof.gob.mx/normasOficiales/4209/seeco/seeco.htm (accessed on 18 August 2020).
22. Kang, S.S.; Kim, M.K.; Kim, Y.J. Comprehensive Evaluation of Microbiological and Physicochemical Properties of Commercial Drinking Yogurts in Korea. *Food Sci. Anim. Resour.* **2019**, *39*, 820–830. [CrossRef]
23. Leclercq-Perlat, M.-N.; Oumer, A.; Bergere, J.-L.; Spinnler, H.; Corrieu, G. Growth of Debaryomyces Hansenii on a Bacterial Surface-Ripened Soft Cheese. *J. Dairy Res.* **1999**, *66*, 271–281. [CrossRef]
24. Picque, D.; Lefier, D.; Grappin, R.; Corrieu, G. Monitoring of Fermentation by Infrared Spectrometry: Alcoholic and Lactic Fermentations. *Anal. Chim. Acta* **1993**, *279*, 67–72. [CrossRef]
25. Albalá-Hurtado, S.; Veciana-Nogués, M.T.; Izquierdo-Pulido, M.; Mariné-Font, A. Determination of Water-Soluble Vitamins in Infant Milk by High-Performance Liquid Chromatography. *J. Chromatogr. A* **1997**, *778*, 247–253. [CrossRef]
26. Gliszczyńska-Świglo, A.; Koziolowa, A. Chromatographic Determination of Riboflavin and Its Derivatives in Food. *J. Chromatogr. A* **2000**, *881*, 285–297. [CrossRef]
27. Wherry, B.; Barbano, D.M.; Drake, M.A. Use of Acid Whey Protein Concentrate as an Ingredient in Nonfat Cup Set-Style Yogurt. *J. Dairy Sci.* **2019**, *102*, 8768–8784. [CrossRef]
28. Feng, Y.; Yang, T.; Zhang, Y.; Zhang, A.; Gai, L.; Niu, D. Potential Applications of Pulsed Electric Field in the Fermented Wine Industry. *Front. Nutr.* **2022**, *9*, 1048632. [CrossRef]
29. Peng, K.; Koubaa, M.; Bals, O.; Vorobiev, E. Effect of Pulsed Electric Fields on the Growth and Acidification Kinetics of Lactobacillus Delbrueckii Subsp. Bulgaricus. *Foods* **2020**, *9*, 1146. [CrossRef] [PubMed]
30. Vivanco, D.; Ardiles, P.; Castillo, D.; Puente, L. Emerging Technology: Pulsed Electric Fields (Pef) for Food Treatment and Its Effect on Antioxidant Content. *Rev. Chil. Nutr.* **2021**, *48*, 609–619. [CrossRef]

31. Lanciotti, R.; Gianotti, A.; Patrignani, F.; Belletti, N.; Guerzoni, M.E.; Gardini, F. Use of Natural Aroma Compounds to Improve Shelf-Life and Safety of Minimally Processed Fruits. *Trends Food Sci. Technol.* **2004**, *15*, 201–208. [CrossRef]
32. Papadimitriou, K.; Alegría, Á.; Bron, P.A.; de Angelis, M.; Gobbetti, M.; Kleerebezem, M.; Lemos, J.A.; Linares, D.M.; Ross, P.; Stanton, C.; et al. Stress Physiology of Lactic Acid Bacteria. *Microbiol. Mol. Biol. Rev.* **2016**, *80*, 837–890. [CrossRef]
33. Bintsis, T.; Papademas, P. The Evolution of Fermented Milks, from Artisanal to Industrial Products: A Critical Review. *Fermentation* **2022**, *8*, 679. [CrossRef]
34. Iftikhar, M.A.; Pasha, T.N.; Inayat, S.; Javed, K.; Ullah, R. Effect of Different Milk Composition on Physico-Chemical Characteristics of Set Type Yoghurt. *Pak. J. Zool.* **2022**, *56*, 1297–1304. [CrossRef]
35. Zheng, H.; Liu, E.; Shi, T.; Ye, L.; Konno, T.; Oda, M.; Ji, Z.S. Strand-Specific RNA-Seq Analysis of the Lactobacillus Delbrueckii Subsp. Bulgaricus Transcriptome. *Mol. Biosyst.* **2016**, *12*, 508–519. [CrossRef]
36. Emanuel, E.; Dubrovin, I.; Pogreb, R.; Pinhasi, G.A.; Cahan, R. Resuscitation of Pulsed Electric Field-Treated Staphylococcus Aureus and Pseudomonas Putida in a Rich Nutrient Medium. *Foods* **2021**, *10*, 660. [CrossRef] [PubMed]
37. Bokulich, N.A.; Bamforth, W. The Microbiology of Malting and Brewing. *Microbiol. Mol. Biol. Rev.* **2013**, *77*, 157–172. [CrossRef] [PubMed]
38. Bücher, C.; Burtscher, J.; Domig, K.J. Propionic Acid Bacteria in the Food Industry: An Update on Essential Traits and Detection Methods. *Compr. Rev. Food Sci. Food Saf.* **2021**, *20*, 4299–4323. [CrossRef] [PubMed]
39. Peruzy, M.F.; Blaiotta, G.; Aponte, M.; De Sena, M.; Murru, N. Late Blowing Defect in Grottone Cheese: Detection of Clostridia and Control Strategies. *Ital. J. Food Saf.* **2022**, *11*, 96–102. [CrossRef]
40. Thakur, K.; Tomar, S.K.; De, S. Lactic Acid Bacteria as a Cell Factory for Riboflavin Production. *Microb. Biotechnol.* **2016**, *9*, 441–451. [CrossRef]
41. Djukić-Vuković, A.; Meglič, S.H.; Flisar, K.; Mojović, L.; Miklavčič, D. Pulsed Electric Field Treatment of Lacticaseibacillus Rhamnosus and Lacticaseibacillus Paracasei, Bacteria with Probiotic Potential. *LWT* **2021**, *152*, 112304. [CrossRef]
42. Averianova, L.A.; Balabanova, L.A.; Son, O.M.; Podvolotskaya, A.B.; Tekutyeva, L.A. Production of Vitamin B2 (Riboflavin) by Microorganisms: An Overview. *Front. Bioeng. Biotechnol.* **2020**, *8*, 570828. [CrossRef]

foods

MDPI

Communication

Enhanced Peelability and Quality of Whiteleg Shrimp (*Litopenaeus vannamei*) Using Pulsed Electric Field (PEF) Treatment

Gyeong-Seo Park [1], Hyeon Seo [1], Han-Baek Lee [1], Ji-Won Lee [1], Hafiz Muhammad Shahbaz [2], Se-Ho Jeong [1] and Dong-Un Lee [1,*]

[1] Department of Food Science and Technology, Chung-Ang University, Anseong 17546, Republic of Korea; yuni2070@naver.com (G.-S.P.); clear0330@naver.com (H.S.); hanbaek98@naver.com (H.-B.L.); melona7237@naver.com (J.-W.L.); calvin0223@naver.com (S.-H.J.)
[2] Department of Nutrition and Health, College of Medicine and Health Sciences, United Arab Emirates University, Al Ain 15551, United Arab Emirates; shahbaz@uaeu.ac.ae
* Correspondence: dong-un.lee@cau.ac.kr; Tel.: +82-31-670-3034

Abstract: This study investigated the effects of pulsed electric field (PEF) treatment on the peeling efficiency and textural properties of whiteleg shrimp (*Litopenaeus vannamei*). Shrimp samples were treated at field strengths of 0, 1.0, 1.5, and 2.0 kV/cm to assess PEF impact on peeling force, incomplete peeling percentage, and texture profile. The results showed that PEF treatment significantly reduced the peeling force from 50.88 N in controls to 42.99 N at 2.0 kV/cm, while the percentage of incompletely peeled shrimp decreased from 27.5% to 15.9%. Texture profile analysis indicated that PEF treatment had no impact on the key properties of hardness and chewiness (no significant difference), with a reduction in springiness observed at higher field strengths. Improvements in peelability are attributed to electroporation, which disrupts collagen in the connective tissue between the shrimp shell and muscle. These findings indicate that PEF treatment is an efficient, non-thermal method for enhancing shrimp peeling processes while preserving textural integrity. PEF technology offers a promising alternative to traditional mechanical and thermal methods in the seafood processing industry.

Keywords: pulsed electric field; shrimp peeling; texture analysis; electroporation; non-thermal processing

1. Introduction

The whiteleg shrimp (*Litopenaeus vannamei*) is one of the most widely consumed seafoods globally, valued for its high content of amino acids, polyunsaturated fatty acids, and other essential nutrients [1]. The Global Shrimp Aquaculture Production Survey projected a 4.8 percent growth in world farmed shrimp production in 2024, reaching approximately 5.88 million metric tons [2]. Growing consumer demand for convenient and ready-to-eat seafood products has driven rapid expansion in the shrimp processing industry, which now represents approximately 45% of the global seafood processing market [3]. A significant portion of this market is linked with peeled shrimp, a product that requires efficient and effective processing methods [4].

Shrimp peeling presents several challenges due to the tight attachment of the shell to the muscle by connective fibers, making shell removal a labor-intensive process. Traditionally, shrimp are peeled either manually or mechanically, both of which can lead to substantial shrimp meat loss and high labor costs [5]. To improve efficiency and reduce waste, the seafood industry has explored the innovative techniques of high-pressure processing (HPP), ultrasound, and enzymatic treatments [6,7], with each method aiming to loosen the shell while preserving the texture and quality of the shrimp meat.

HPP has been effective in loosening shrimp shells through changes in the structure of collagen in the shrimp epidermis. This leads to easier peeling but may affect the textural properties of the meat [6,8]. Ultrasound technology uses high-frequency sound waves to induce cavitation to loosen the shrimp shell by creating small pores and facilitating mass transfer between the muscle and shell; however, while effective, ultrasound may also impact the integrity of the shrimp meat if not properly controlled [7]. Enzymatic treatments break down proteins in the attachment fibers, allowing the shell to be removed with minimal mechanical force [9]; however, the widespread use of enzymes in the seafood industry is limited by cost and a potential impact on the sensory qualities of shrimp.

In recent years, pulsed electric field (PEF) technology has emerged as a promising non-thermal method for food processing [10–12]. The PEF method applies short bursts of a high-intensity electric field to food products, causing electroporation, which creates pores in cell membranes and facilitates mass transfer without significantly affecting the enzymatic or structural integrity of the food [13]. PEF technology has been successfully applied to fruits and vegetables, where it aids in drying, extraction, and peeling [11]. For example, PEF technology has been used to enhance the peeling of tomatoes and peaches while reducing mechanical damage and improving processing efficiency [10]. However, its application in seafood, particularly in shrimp processing, remains relatively unexplored. Most studies have focused on plant-based foods, and no significant investigations have been conducted on the application of the PEF method in shrimp peeling [14]. The present study aims to fill this gap by exploring the effects of the PEF method on the peeling and quality characteristics of whiteleg shrimp.

2. Materials and Methods

2.1. Shrimp Preparation

Live whiteleg shrimp (*Litopenaeus vannamei*) measuring 16–18 cm in length and weighing 40–60 shrimp per kg were sourced from an indoor aquaculture facility, Freshrim (Ansan-Si, Republic of Korea). Upon harvesting, shrimp were immediately packed in ice and transported to the laboratory within 1 h to ensure freshness. Upon arrival, shrimp were thoroughly washed with cold water to remove impurities, and excess water was drained. Shrimp were prepared for experiment within 1 h of arrival by removing legs and excising two segments from the first abdominal section. This preparation step was performed to ensure uniformity across all samples used for PEF treatment and subsequent analysis.

2.2. PEF Treatment

PEF treatment was administered using a 5 kW pulse generator (HVP-5, DIL, Quakenbruck, Germany) equipped with a batch chamber comprising two parallel electrodes separated by an 80 mm gap. Four shrimp abdominal sections were placed in the chamber with 200 mL of tap water as the processing medium. Shrimp samples were subjected to PEF treatment under the following electric field strengths: 0 kV/cm (control); 1.0 kV/cm (PEF_1.0); 1.5 kV/cm (PEF_1.5); and 2.0 kV/cm (PEF_2.0). The pulse width was maintained at 20 μs with a pulse frequency of 100 Hz, and a total of 500 pulses were applied for each treatment. A total of 20 shrimp were utilized for each treatment condition, resulting in 40 abdominal sections (two sections per shrimp). This ensured sufficient replicates for statistical analysis while maintaining consistency across experimental groups.

2.3. Measurement of Peelability After PEF Treatment

The peelability of treated shrimp was evaluated using a texture analyzer (TAHDi/500, TAHD, London, UK) following the methodology described by Dang et al. [7]. A total of 40 shrimp samples were used for this analysis. Briefly, each sample of two abdominal segments was weighed, affixed to a pin in the probe, and clipped into the texture analyzer to perform tension tests (Video S1). These tests were designed to quantify the force required to peel the shrimp shell from the meat. Peeling work was defined as the product of the force applied (N) and the distance traveled (mm) to remove the shell divided by the weight of the two segments (g) using Equation (1). The percentage of incompletely peeled shrimp was calculated as the ratio of the total number of shrimps with shells remaining attached to the total number of shrimps tested, as shown in Equation (2).

$$W = \frac{Fd}{m_i} \quad (1)$$

where W (mJ/g) is the peeling work, F (N) is the force required to peel the shrimp shell, d (mm) is the distance required to remove the shell, and mi is the weight of two shrimp segments.

$$Incompletely\ peeled\ (\%) = \frac{Number\ of\ shrimp\ having\ any\ shell\ attached}{Total\ number\ of\ shrimp\ used\ for\ TA\ peeling} \times 100 \quad (2)$$

2.4. Measurement of Texture Properties of Treated Shrimp

We performed a texture profile analysis (TPA) to determine the texture properties of the shrimp samples, using a texture analyzer (TAHDi/500, TAHD, UK) as outlined by Lin et al. [15]. A total of 40 abdominal sections per treatment group, as described in Section 2.2, were prepared for TPA and cutting force tests. For these tests, 14 sections with nearly identical sizes were selected to ensure consistency in the cross-sectional area and minimize variability. Following the peeling process, shrimp abdominal sections were subjected to TPA using a P/75 probe under controlled conditions with a pre-, test, and post-test speed of 1.0 mm/s, a compression ratio of 50%, and a trigger force of 0.02 N. TPA measured key texture attributes, including hardness, springiness, and cohesiveness. In addition to TPA, cutting force was measured using a Warner–Bratzler flat blade attached to the texture analyzer. The cutting force required to cut through shrimp samples horizontally was recorded, based on the methodology used by Boonsumrej et al. [16] with minor modification. The maximum force (N) was measured on 14 shrimp abdominal sections with a pre-test speed of 5.0 mm/s, a test and post-test speed of 2.0 mm/s, and a trigger force of 4.9 N.

2.5. Statistical Analysis

All experimental data were analyzed using a one-way analysis of variance (ANOVA). A significance level of $p < 0.05$ was applied. Post hoc comparisons were performed using Duncan's multiple range test to identify significant differences between treatment groups. Statistical analyses were conducted using IBM Statistical Package for the Social Science (SPSS) software version 20 (IBM Corp., Armonk, NY, USA). Results were reported as means ± standard deviations.

3. Results and Discussion

3.1. Effect of PEF on Shrimp Peelability

The effect of PEF treatment on the peelability of shrimp is shown in Figure 1. As the intensity of the PEF treatment increased, both the peeling force and the percentage of incompletely peeled shrimp decreased. The peeling force was reduced from 50.88 ± 11.7 N in the control group to 42.99 ± 11.4 N at a PEF strength of 2.0 kV/cm. Similarly, the percentage of incompletely peeled shrimp decreased from 27.5% in the control group to 15.9% for the PEF 2.0 treatment. Furthermore, PEF treatment had no discernible effect on shrimp appearance before or after peeling (Figure 2).

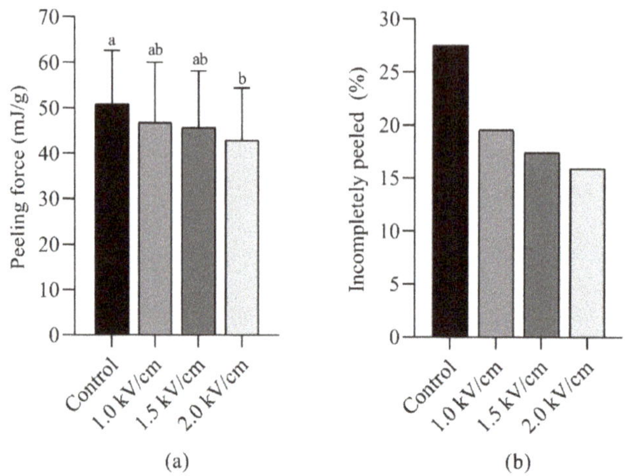

Figure 1. Peeling force (**a**) and incompletely peeled percentage (**b**) for shrimp after PEF treatment with different field strengths (n = 40). The error bar for the peeling force values (**a**) represents standard deviation, and letters (**a**,**b**) indicate significant differences for each sample ($p < 0.05$).

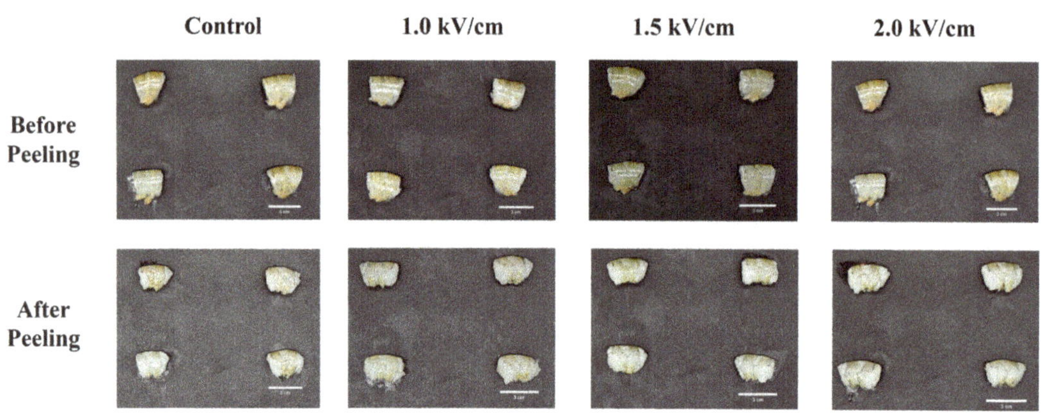

Figure 2. Appearance of PEF-treated shrimp before and after peeling. The scale bar of the images is 3 cm.

The improvement in peeling efficiency observed in this study is consistent with earlier findings that non-thermal HPP and ultrasound treatments can modify the structural integrity of the connective tissue between the shrimp shell and muscle [4,5]. Collagen, a major component of this connective tissue, is susceptible to degradation under physical treatments. In the case of PEF treatment, electroporation likely disrupts collagen fibers and facilitates easier shell detachment, as reported by Gómez et al. (2019) for other food matrices undergoing PEF treatment. This mechanism parallels that of ultrasound-induced vapor cavitation, which generates bubbles within the connective tissue to increase peeling efficiency [7].

Notably, results from this study support findings from HPP treatment studies on shrimp, whereby collagen denaturation at higher pressures significantly enhanced peelability [4]. Similarly, previous work by Xu et al. [17] on ice water pretreatment of shrimp found that the mechanical disruption of muscle fibers under controlled conditions improved shell removal without compromising product quality. These findings suggest that PEF treatment has a comparable effect by causing minor disruptions in shrimp epidermal collagen, allowing for easier separation of the shell from the muscle.

3.2. Textural Properties of PEF-Treated Shrimp

Table 1 summarizes the texture profile analysis (TPA) and cutting force results for shrimp after PEF treatment. No significant differences were found between the control and PEF-treated samples for hardness, cohesiveness, adhesiveness, gumminess, chewiness, resilience, or cutting force. However, a notable decrease in springiness was observed as the intensity of PEF treatment increased. Springiness values decreased by 9.9% at PEF 1.0, 8.7% at PEF 1.5, and 17.9% at PEF 2.0 compared to controls.

Table 1. Texture profile analysis of shrimp for different PEF strengths.

Field Strength (kV/cm)	Texture Properties							
	Hardness (N)	Adhesiveness	Springiness	Cohesiveness	Gumminess	Chewiness (N)	Resilience	Cutting Force (N)
Control	46.49 ± 8.19^{a}	-33.88 ± 12.12^{a}	0.62 ± 0.11^{a}	0.57 ± 0.12^{a}	26.81 ± 7.23^{a}	17.39 ± 8.85^{a}	0.37 ± 0.07^{a}	19.54 ± 2.84^{a}
PEF_1.0	47.37 ± 5.63^{a}	-28.22 ± 6.42^{a}	0.56 ± 0.04^{ab}	0.50 ± 0.03^{a}	23.81 ± 3.95^{a}	13.44 ± 2.88^{a}	0.35 ± 0.32^{a}	17.70 ± 1.75^{a}
PEF_1.5	48.95 ± 7.32^{a}	-33.74 ± 17.25^{a}	0.56 ± 0.13^{ab}	0.54 ± 0.13^{a}	27.35 ± 11.17^{a}	16.88 ± 13.58^{a}	0.35 ± 0.06^{a}	18.84 ± 2.13^{a}
PEF_2.0	45.23 ± 7.98^{a}	-26.31 ± 5.64^{a}	0.51 ± 0.05^{b}	0.51 ± 0.03^{a}	23.28 ± 5.25^{a}	12.08 ± 3.41^{a}	0.36 ± 0.03^{a}	20.00 ± 3.30^{a}

All values are expressed as the mean ± standard deviation (n = 14). Different letters (a, b) indicate significant differences in the same column ($p < 0.05$).

The reduction in springiness observed in this study may be linked to structural changes in shrimp muscle fibers caused by electro-permeabilization during PEF treatment. Similar effects were reported by Gómez et al. [18], who observed that PEF treatment can loosen muscle fibers and reduce springiness in meat tissues. The extent of electro-permeabilization depends on the strength of the electric field applied, with higher intensities causing greater disruption to muscle fibers. This could explain the progressive decrease in springiness with increasing PEF intensity in this study. However, studies on the PEF treatment of abalone and turkey breast meat by Arroyo et al. [19] and Luo et al. [20] showed that the protein structure of muscle fibers was not significantly affected, suggesting that the effect of PEF treatment on textural properties may vary depending on the particular food matrix. While springiness was affected by PEF treatment, the lack of significant changes in the textural parameters of hardness and chewiness is noteworthy for maintaining the sensory quality of shrimp. Textural properties such as hardness and chewiness are critical factors in consumer acceptance [17]. Additionally, the cutting force results indicate that PEF-treated shrimp retained structural integrity, which is beneficial for processing and handling.

The observed phenomena can be attributed to disparities in the chemical composition of the muscle and shell. The moisture and protein content of shrimp are higher in the muscle than in the shell, resulting in higher electrical conductivity [21,22]. The lower electrical conductivity of the shell and its connecting fibers may lead to localized intensification of the electric field, enhancing the electroporation effect in these regions. This phenomenon could explain the selective disruption of connective tissue near the shell while the structural integrity of the deeper muscle layers is preserved. The PEF effect is contingent on the electrical conductivity of tissues, and lower conductivity increases the electric field strength under the same energy. In this study, the peel part was positioned horizontally with the electrode, enabling the pulse energy to act on the peel and subsequently transfer to the deeper muscle layers. Consequently, there might be a discrepancy in the PEF effect between the peel and muscle, which could explain the observed ease of peeling. However, the effect's limitations in properties like hardness and cutting force should be noted.

The retention of textural properties such as hardness and chewiness alongside the improved peeling efficiency ensures that PEF-treated shrimp meet both consumer expectations for quality and industry requirements for processing efficiency. These findings highlight the efficacy of PEF treatment in enhancing processing efficiency while preserving critical quality attributes, positioning it as a promising approach for use in the seafood industry. Future studies should systematically investigate the role of electrical conductivity and tissue composition in influencing the selective effects of PEF treatment, particularly to optimize its application across different seafood matrices.

3.3. Mechanism and Industrial Implications

This study demonstrates that PEF treatment enhances shrimp peeling efficiency while maintaining key textural properties, offering a promising, non-thermal alternative to labor-intensive conventional methods [5]. Its potential for minimizing meat loss and serving as a complementary tool to existing peeling methods underscores its value for the shrimp processing industry by providing energy-efficient and cost-effective solutions.

In comparison with enzymatic and ultrasonic methods, PEF treatment offers the advantage of being non-thermal, thus preserving the freshness and quality of shrimp. As reported by Koch et al. [11], non-thermal processing technologies are gaining traction in the food industry due to their ability to enhance food quality while reducing energy consumption. For example, PEF treatment has been shown to improve peeling efficiency in tomatoes and peaches [10]. Its application in seafood offers potential for improving processing efficiency in this sector.

Additionally, the potential of PEF treatment to preserve the textural properties of shrimp could address consumer demand for high-quality, minimally processed seafood products. This is particularly relevant in the context of the growing market for convenience seafood, where ready-to-eat and easy-to-peel shrimp are highly desired [4]. Furthermore, the non-invasive nature of PEF treatment could complement existing processing technologies, such as HPP or enzymatic treatments, offering a more versatile approach to seafood processing.

Although the current study demonstrates the benefits of PEF treatment in shrimp processing, further research is needed to optimize PEF conditions for different shrimp sizes and species. Future studies could investigate synergies between PEF and other non-thermal technologies, such as ultrasound or cold plasma, to further enhance processing efficiency. Moreover, investigations into the long-term effects of PEF treatment on shrimp quality during storage would provide valuable insights for applications in industrial settings.

4. Conclusions

PEF treatment significantly improved the peeling efficiency of whiteleg shrimp by reducing the percentage of incompletely peeled shrimp and the peeling force required. Additionally, PEF treatment maintained key textural properties and ensured shrimp quality. These findings indicate that PEF is an effective, non-thermal method for enhancing shrimp processing efficiency while maintaining product integrity; however, further investigations are required to better understand the impact of PEF treatment on shrimp connective tissue and develop optimized protocols for various shrimp species.

Supplementary Materials: The following supporting information can be downloaded at: https://www.mdpi.com/article/10.3390/foods14020148/s1, Video S1: Supplementary_Video_Shrimp_Peeling process.mp4.

Author Contributions: Conceptualization, G.-S.P. and D.-U.L.; methodology, G.-S.P., H.S., H.-B.L., and J.-W.L.; software, H.S.; validation, H.-B.L. and J.-W.L.; formal analysis, G.-S.P.; investigation, G.-S.P., H.S., H.-B.L., and J.-W.L.; resources, D.-U.L.; data curation, S.-H.J. and D.-U.L.; writing—original draft preparation, G.-S.P.; writing—review and editing, G.-S.P., S.-H.J., and H.M.S.; visualization, H.S.; supervision, S.-H.J.; project administration, D.-U.L.; funding acquisition, D.-U.L. All authors have read and agreed to the published version of the manuscript.

Funding: This work was supported by the Korea Institute of Planning and Evaluation for Technology in Food, Agriculture, Forestry (IPET) through the High Value-added Food Technology Development Program funded by the Ministry of Agriculture, Food, and Rural Affairs (MAFRA) (332018-04).

Institutional Review Board Statement: Not applicable.

Informed Consent Statement: Not applicable.

Data Availability Statement: The original contributions presented in this study are included in the article/Supplementary Materials. Further inquiries can be directed to the corresponding author.

Acknowledgments: This research was supported by the Chung-Ang University Graduate Research Scholarship in 2023.

Conflicts of Interest: The authors declare no conflicts of interest.

References

1. Hannan, M.A.; Habib, K.A.; Shahabuddin, A.; Haque, M.A.; Munir, M.B. *Post-Harvest Processing, Packaging and Inspection of Frozen Shrimp: A Practical Guide*; Springer Nature: Berlin, Germany, 2022.
2. Jory, D. *Annual Farmed Shrimp Production Survey: A Slight Decrease in Production Reduction in 2023 with Hopes for Renewed Growth in 2024*; Global Seafood Alliance: Portsmouth, NH, USA, 2023.
3. Yang, X.; Hao, S.; Pan, C.; Li, L.; Huang, H.; Yang, X.; Wang, Y. A quantitative method to analysis shrimp peelability and its application in the shrimp peeling process. *J. Food Process. Preserv.* **2020**, *44*, e14882. [CrossRef]
4. Dang, T.T.; Feyissa, A.H.; Gringer, N.; Jessen, F.; Olsen, K.; Bøknæs, N.; Orlien, V. Effects of high pressure and ohmic heating on shell loosening, thermal and structural properties of shrimp (*Pandalus borealis*). *Innov. Food Sci. Emerg. Technol.* **2020**, *59*, 102246. [CrossRef]
5. Shuxian, H.; Xiaojie, Y.; Hui, H.; Laihao, L.; Chuang, P.; Xianqing, Y.; Jianwei, C. Status of shrimp peeling and pretreatment technology for facilitating peeling. *South China Fish. Sci.* **2020**, *16*, 121–128.
6. Shen, L.; Qiu, W.; Du, L.; Zhou, M.; Qiao, Y.; Wang, C.; Wang, L. Effects of high hydrostatic pressure on peelability and quality of crayfish (*Procambarus clarkii*). *J. Sci. Food Agric.* **2024**, *104*, 611–619. [CrossRef]
7. Dang, T.T.; Gringer, N.; Jessen, F.; Olsen, K.; Bøknæs, N.; Nielsen, P.L.; Orlien, V. Facilitating shrimp (*Pandalus borealis*) peeling by power ultrasound and proteolytic enzyme. *Innov. Food Sci. Emerg. Technol.* **2018**, *47*, 525–534. [CrossRef]
8. Shahbaz, H.M.; Javed, F.; Park, J. *Advances in Food Applications for High Pressure Processing Technology*; Springer: Berlin/Heidelberg, Germany, 2023.
9. Basiri, S.; Shekarforoush, S.S.; Aminlari, M.; Akbari, S. The effect of pomegranate peel extract (PPE) on the polyphenol oxidase (PPO) and quality of Pacific white shrimp (*Litopenaeus vannamei*) during refrigerated storage. *LWT-Food Sci. Technol.* **2015**, *60*, 1025–1033. [CrossRef]
10. Giancaterino, M.; Jaeger, H. Impact of pulsed electric fields (PEF) treatment on the peeling ability of tomatoes and kiwi fruits. *Front. Food Sci. Technol.* **2023**, *3*, 1152111. [CrossRef]
11. Koch, Y.; Witt, J.; Lammerskitten, A.; Siemer, C.; Toepfl, S. The influence of Pulsed Electric Fields (PEF) on the peeling ability of different fruits and vegetables. *J. Food Eng.* **2022**, *322*, 110938. [CrossRef]
12. Niu, D.; Zeng, X.-A.; Ren, E.-F.; Xu, F.-Y.; Li, J.; Wang, M.-S.; Wang, R. Review of the application of pulsed electric fields (PEF) technology for food processing in China. *Food Res. Int.* **2020**, *137*, 109715. [CrossRef] [PubMed]
13. Caminiti, I.M.; Palgan, I.; Noci, F.; Muñoz, A.; Whyte, P.; Cronin, D.A.; Morgan, D.J.; Lyng, J.G. The effect of pulsed electric fields (PEF) in combination with high intensity light pulses (HILP) on Escherichia coli inactivation and quality attributes in apple juice. *Innov. Food Sci. Emerg. Technol.* **2011**, *12*, 118–123. [CrossRef]
14. Alam, M.R.; Lyng, J.G.; Frontuto, D.; Marra, F.; Cinquanta, L. Effect of pulsed electric field pretreatment on drying kinetics, color, and texture of parsnip and carrot. *J. Food Sci.* **2018**, *83*, 2159–2166. [CrossRef] [PubMed]
15. Lin, T.; Wang, J.J.; Li, J.B.; Liao, C.; Pan, Y.J.; Zhao, Y. Use of acidic electrolyzed water ice for preserving the quality of shrimp. *J. Agric. Food Chem.* **2013**, *61*, 8695–8702. [CrossRef] [PubMed]

16. Boonsumrej, S.; Chaiwanichsiri, S.; Tantratian, S.; Suzuki, T.; Takai, R. Effects of freezing and thawing on the quality changes of tiger shrimp (Penaeus monodon) frozen by air-blast and cryogenic freezing. *J. Food Eng.* **2007**, *80*, 292–299. [CrossRef]
17. Xu, N.; Shi, W.; Wang, X.; Wang, Z. Effect of ice water pretreatment on the quality of Pacific White Shrimps (*Litopenaeus vannamei*). *Food Sci. Nutr.* **2019**, *7*, 645–655. [CrossRef] [PubMed]
18. Gómez, B.; Munekata, P.E.; Gavahian, M.; Barba, F.J.; Martí-Quijal, F.J.; Bolumar, T.; Campagnol, P.C.B.; Tomasevic, I.; Lorenzo, J.M. Application of pulsed electric fields in meat and fish processing industries: An overview. *Food Res. Int.* **2019**, *123*, 95–105. [CrossRef] [PubMed]
19. Arroyo, C.; Eslami, S.; Brunton, N.P.; Arimi, J.M.; Noci, F.; Lyng, J.G. An assessment of the impact of pulsed electric fields processing factors on oxidation, color, texture, and sensory attributes of turkey breast meat. *Poult. Sci.* **2015**, *94*, 1088–1095. [CrossRef] [PubMed]
20. Luo, Q.; Hamid, N.; Oey, I.; Leong, S.Y.; Kantono, K.; Alfaro, A.; Lu, J. Physicochemical changes in New Zealand abalone (*Haliotis iris*) with pulsed electric field (PEF) processing and heat treatments. *LWT* **2019**, *115*, 108438. [CrossRef]
21. Heu, M.-S.; Kim, J.-S.; Shahidi, F. Components and nutritional quality of shrimp processing by-products. *Food Chem.* **2003**, *82*, 235–242. [CrossRef]
22. Rødde, R.H.; Einbu, A.; Vårum, K.M. A seasonal study of the chemical composition and chitin quality of shrimp shells obtained from northern shrimp (*Pandalus borealis*). *Carbohydr. Polym.* **2008**, *71*, 388–393. [CrossRef]

foods

MDPI

Article

Changes to Pork Bacterial Counts and Composition After Dielectric Barrier Discharge Plasma Treatment and Storage in Modified-Atmosphere Packaging

Yi Zhou [1], Huixin Zuo [1], Zhaoqi Dai [2], Zonglin Guo [3], Benjamin W. B. Holman [4], Yanqin Ding [5], Jingying Shi [1], Xiaoxiao Ding [1], Mingming Huang [1,*] and Yanwei Mao [1]

[1] Key Laboratory of Food Processing Technology and Quality Control in Shandong Province, College of Food Science and Engineering, Shandong Agricultural University, Tai'an 271018, China; ytuzy2023@163.com (Y.Z.); hxzuo@sdau.edu.cn (H.Z.); jyshi80@163.com (J.S.); 19806101565@163.com (X.D.); maoyanwei@163.com (Y.M.)

[2] College of Biotechnology, Jiangsu University of Science and Technology, Zhenjiang 212100, China; dai0003199@jsafc.edu.cn

[3] College of Food Science, South China Agricultural University, Guangzhou 510642, China; gzl072810@scau.edu.cn

[4] Wagga Wagga Agricultural Institute, NSW Department of Primary Industries, Wagga Wagga, NSW 2650, Australia; benjamin.holman@dpi.nsw.gov.au

[5] College of Biotechnology, Shandong Agricultural University, Tai'an 271018, China; dyq@sdau.edu.cn

[*] Correspondence: hades3709@126.com

Abstract: The aim of this study was to compare the succession of natural microbiota in pork held under refrigerated storage for up to 10 days after dielectric barrier discharge (DBD) plasma treatment. Two methods were used to assess the impact of DBD on microorganisms. Firstly, traditional selective media (SM) were employed to detect the bactericidal effects of DBD on *Pseudomonas* spp., *Enterobacteriaceae*, *Lactic acid bacteria* (LAB), and *Brochothrix thermosphacta*. Secondly, the thin agar layer (TAL) method was used to further evaluate the bactericidal effects of DBD. In addition, the Baranyi and Roberts model was applied to explore the kinetic parameters of *Pseudomonas* spp., *Enterobacteriaceae*, LAB, and *B. thermosphacta* during storage. Finally, the modified Lotka–Volterra model was used to describe the interactions between each microorganism. The study found that when using traditional selective media (SM), 85 kV DBD had a significant bactericidal effect on *Pseudomonas* spp., *Enterobacteriaceae*, LAB, and *Brochothrix thermosphacta*. However, when using the thin agar layer (TAL) method, the results suggested that DBD had no significant bactericidal effect, suggesting that DBD caused sublethal damage to the natural microorganisms on pork. Analysis with the Baranyi and Roberts model showed that DBD treatment significantly extended the lag phase of these four types of microorganisms and significantly reduced the μ_{max} of all microorganisms except LAB. The analysis results of the modified Lotka–Volterra model showed that LAB had a greater impact on *Pseudomonas* spp., *Enterobacteriaceae*, and *B. thermosphacta* ($a_{21} > a_{12}$). In conclusion, DBD treatment was shown to have a significant sublethal bactericidal effect that impacted both the count and composition of natural microorganisms found on pork.

Keywords: cold plasma; pork; sublethal injury; refrigerated storage; indigenous microbiota populations; spoilage bacteria interaction; modified culture methods

1. Introduction

Demand for high-quality and safe meat has grown over the past 50 years, and in response, the production of meat has more than tripled to ~340 million tons per year. The same trend is apparent for pork, with consumers preferencing its nutritional and organoleptic properties—although these same properties, its high level of water activity, and moderate pH make pork highly susceptible to microbial contamination and proliferation [1]. Microorganisms can cause the degradation of proteins, carbohydrates, and other physiochemical components of pork and thereby result in the development of malodors, structure/texture degradation, and discoloration [2]. Microorganisms on pork also represent a potential risk to the consumer, as many microorganisms are associated with foodborne disease and health complications, and infer a socioeconomic cost when ingested [3]. For these reasons, consumers are increasingly conscious about food safety and its processing methods being free from chemical sanitizers and artificial additives. Non-thermal sterilization technologies should therefore be considered novel methods for controlling microorganisms on pork and pork products [4].

Cold plasma has emerged as a sanitizing technology with considerable potential in its application to the preservation of foods [5,6]. Cold plasma is an ionized gas that is highly energetic and composed of ions, electrons, neutral atoms, and free radicals. Among the various constituents in cold plasma, the reactive species (reactive oxygen species and reactive nitrogen species) are regarded as the major agents for plasma-induced bactericidal effects [7]. It is generally believed that these reactive species, independently or synergistically, cause the destruction of the cell membrane, the dysfunction of proteins, damage to DNA, the peroxidation of lipid barriers, and disturbances to cellular homeostasis—actions that can effectively sterilize a food product [8–10]. Cold plasma devices have low energy inputs, high sterilization efficiency, and non-toxic residue characteristics, allow for the good retention of produce quality attributes, and are operated at room temperature [10–13]. These characteristics promote its application in the processing of meat products. Jayasena et al. [14] reported that exposure to flexible thin-layer cold plasma allowed for the effective inactivation of pathogens inoculated on beef jerky. Likewise, Moutiq, Misra, Mendonca and Keener [15] found that exposure to atmospheric cold plasma could decontaminate chicken breasts and deliver a 2 log CFU/g reduction in microorganism levels. Although cold plasma has been verified to efficiently inactivate microorganisms, its sublethal effect on microorganisms should also be considered.

Sublethal injury refers to microorganisms that have been exposed to chemical or physical stress and are consequentially damaged but not killed, the damage being structural (i.e., cell morphology, membrane integrity), metabolic (i.e., cellular homeostasis, membrane potential), or a combination of both [16–18]. Conventional analytical methods used in food-microbial diagnostics contain a variety of selective compounds that may be harmful to injured cells and contribute to type II error. On the other hand, under optimal environmental conditions, the rejuvenation of injured bacteria can increase the risk of underestimating the microbiological quality and safety of products during storage and distribution [19]. Initially, the inactivation of microorganisms triggered by membrane damage, due to non-thermal treatment, were deemed 'all-or-nothing', which means that cells without sublethal injury are generated [8]. However, research has demonstrated that cold plasma treatment could induce reversible cellular damage to microorganisms. Laroussi, Richardson and Dobbs [20], for example, reported that normal cell functioning is altered upon exposure to plasma and may return after a period of recovery, demonstrating reversible damage in cells exposed to plasma. Critzer, Kelly-Wintenberg, South and Golden [21] also found that the recovery of plasma-treated *Salmonella* on a selective medium (SM, xylose lysine tergitol-4 agar) was poorer than its recovery on the non-selective medium (TSAN). Microorganisms with sublethal injury often lose the ability to form visible colonies on a selective medium (SM), but they may recover and subsequently reproduce on a non-selective medium. This is important as previous research shows that cold plasma treatment can induce the sublethal injury of microorganisms [16,22–24]. Most of these studies have investigated microorganisms in vitro or via an inoculated food matrix. Furthermore, the sublethal effect of cold plasma on indigenous microorganisms of food remains to be understood—specially, there is limited information on the dynamic changes in the microbial communities of cold-plasma-treated pork held under refrigerated storage conditions and in modified-atmosphere packaging [25].

This study aimed to compare the effect of dielectric barrier discharge (DBD) plasma treatment on the indigenous microorganisms of pork during its refrigerated storage. The sublethal effects of plasma treatment on microorganisms were examined using the SM and thin agar layer (TAL) culture methods. The kinetics parameters of the indigenous microorganisms were determined using a modified Baranyi and Roberts model, and the antagonistic activities among different microorganisms were explored using a modified Lotka–Volterra model.

2. Materials and Methods

2.1. Meat Sample Preparation

Figure 1 provides a schematic diagram of the experimental factors applied to each of the three replicates (chronological repeats). Within each replicate, three *M. longissimus thoracis et lumborum* (LTL, average weight: 6.5 kg) were purchased from a commercial retail outlet (Suguo Supermarket Co., Ltd., Nanjing, China). These were refrigerated overnight at 4 °C before trimming under aseptic conditions to remove all the external fat. From each LTL (216 per replicate), a total of 72 sample cores (weight: ~50 g, thickness: 2.0 cm, diameter: 5.5 cm) were prepared, under aseptic conditions and using a hollow stainless-steel cylinder. These 72 sample cores were randomly allocated into groups of 4, which were packaged together (18 trays per LTL) in a polypropylene tray (HS-6; Shanghai Chuo Kagaku Co., Ltd., Shanghai, China). The trays were flushed with an 80% O_2 and 20% N_2 gas mixture and heat sealed with polyamide/polyethylene barrier film (oxygen transmission rate of 3 $cm^3/m^2/24$ h) using a modified-atmosphere packaging machine (Senrui H360, Suzhou Senrui Fresh Keeping Equipment Co., Ltd., Suzhou, China). The in-pack gas compositions were verified with a gas analyzer (Check Point-Handheld Gas Analyzer, Ringsted, Denmark); the headspace compositions were found to be 80.1 ± 2.3% O_2, and 19.9 ± 2.4% N_2. All the sealed trays were placed in a refrigerated room for 2 h, utilizing the RH/Temp data logger R-4HC to ensure that in-pack relative humidity could stabilize at >80% (Elitech Inc., Fujian, China).

Within the replicates (54 trays per replicate) and balanced by LTL (18 trays per LTL), 3 trays were randomly allocated to each combination of 3 DBD treatments (0 kV (control) 60 kV, or 85 kV DBD) and 6 storage periods (0, 2, 4, 6, 8, or 10 days). The DBD treatments were applied using a DBD plasma system comprising a BK-130, step-up transformer (Phenix Technologies, Accident, MD, USA), two aluminum cyclic annular electrodes (150 mm diameter), and two dielectric barriers layers (polypropylene sheets). The trays were placed in the discharge area between two dielectric barriers and, when appropriate, treated for 60 s. The trays were stored at 4 °C for the duration of their storage period.

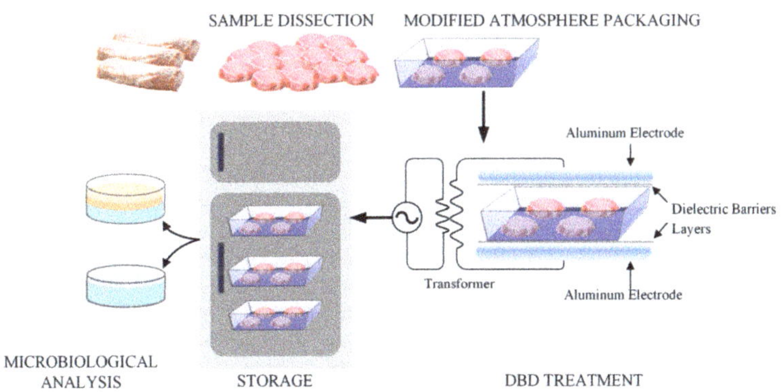

Figure 1. A schematic diagram of the experimental design used for the cold plasma treatment of pork samples held under refrigerated storage for up to 10 days, post-treatment. After the meat is trimmed, it undergoes modified-atmosphere packaging (MAP) with a composition of (80% O_2 and 20% N_2). Following the MAP, the products are treated with plasma at different voltages. The treated samples are then stored and sampled at various points for microbial analysis.

2.2. Microbiological Analysis

As separate batches, 4 sample cores per tray were minced together under aseptic conditions. From this preparation, 25 g of minced meat was weighed and aseptically transferred into a sterile stomacher bag with 225 mL of sterile 0.1% peptone PBS solution, and then blended with a BagMixer® 400 stomacher (Interscience Ind., Puycapel, France) for 2 min. The resulting suspension was serially diluted with 0.1% peptone PBS (Phosphate-Buffered Saline) solution. To enumerate the non-injured cells, 100 μL aliquots were spread on the following agars: (1) Tryptic Soy Agar (TSA, Hope Bio-Technology Co., Ltd., Qingdao, China) incubated at 37 °C for 48 h, for the enumeration of viable aerobic flora; (2) *Pseudomonas* Cetrimide-Fusidin-Cephaloridine Selective Agar (CFC, Hope Bio-Technology Co. Ltd., Qingdao, China) incubated at 25 °C for 48 h, for the enumeration of *Pseudomonas* spp.; (3) Violet Red Bile Glucose Agar (VRBGA, Hope Bio-Technology Co., Ltd., Qingdao, China) incubated at 37 °C for 48 h, for the enumeration of *Enterobacteriaceae*; (4) De Man-Rogosa-Sharpe Agar (MRS, Hope Bio-Technology Co., Ltd., Qingdao, China) incubated at 30 °C for 48 h, for the enumeration of LAB; and (5) Streptomycin Sulphate Thallous Acetate cycloheximide (actidione) Agar (STAA, Hope Bio-Technology Co., Ltd., Qingdao, China) incubated at 25 °C for 48 h, for the enumeration of *B. thermosphacta*.

To enumerate the total cells, i.e., both injured and non-injured cells, on pork, the TAL method was employed [26]. The TAL method involves usinh a selective medium (SM) overlaid by a non-selective medium. Here, 14 mL (7 mL/7 mL; two times overlay) of melted tryptic Soy Agar was overlaid onto 25 mL of a pre-poured and solidified specific selective medium. The injured cells can resuscitate and grow on top layer (TSA) within the first few hours of incubation; then, the selective agents from the bottom layer (specific selective medium) can diffuse to the TSA layer and the resuscitated target cells can react with the selective agents to develop typical reactions, while other cells are inhibited by the selective agents [27,28]. Aliquots of 1000 or 100 μL were directly spread onto the TAL medium and incubated at corresponding conditions before counting the colonies. The sublethal ratios of bacteria after being exposed to DBD plasma were calculated as per Equation (1):

$$\text{Sublethal ratio } (\%) = \frac{\text{counts on TAL medium} - \text{counts on selective medium}}{\text{counts on TAL medium}} \quad (1)$$

2.3. Modeling Growth Kinetics

To model growth kinetics of bacteria growth on pork during storage, the Baranyi and Roberts model [29] was applied (Equation (2)):

$$y_t = y_0 + \mu_{max} \times A(t) - \ln\left(1 + \frac{e^{\mu_{max} \times A(t)} - 1}{e^{y_{max} - y_0)}}\right) \quad (2)$$

where y_t is the cell concentration in $\log_{10}$ CFU/g at time t, y_0 represents the initial cell concentration; y_{max} is the maximum cell concentration; μ_{max} is the maximum specific growth rate in $\log_{10}$ CFU/d; A(t) is the lag phase described by Equation (3):

$$A(t) = t + \frac{1}{v} \times \ln\left(e^{-v \times t} + e^{-h_0} - e^{(-v \times t - h_0)}\right) \quad (3)$$

where vis the rate of increase in the limiting substrate, assumed to be equal to μ_{max}; h_0 is the product of μ_{max} and λ; λ is the lag phase duration and represents the cells' adjustment to the new environment [30]. The goodness of fit for the data was evaluated via the coefficient of determination (R^2) and standard error of fit (*SE of Fit*), which was provided by DMFit.

The generation time was calculated in accordance with Reid et al. [31], and the calculation is described in Equation (4):

$$G = \frac{t}{3.3 \times \log_{10}^{\frac{b}{B}}} \quad (4)$$

where G = generation time, t = time interval in days, b = number of bacteria at the end of t, and B = number of bacteria at the beginning of t.

2.4. Modified Lotka–Volterra Model

The modified Lotka–Volterra competition model is valuable in predictive microbiology for analyzing mixed microbial populations in homogeneous food products [32]. This is a straightforward model which can be represented using the following equations, Equations (5) and (6):

$$\frac{dN_A}{dt} = \mu_{maxA}\frac{Q_A}{1+Q_A} \times \frac{N_A}{N_{maxA}} \times (N_{maxA} - N_A - \alpha_{AB}N_B) \tag{5}$$

$$\frac{dN_B}{dt} = \mu_{maxB}\frac{Q_B}{1+Q_B} \times \frac{N_B}{N_{maxB}} \times (N_{maxB} - N_B - \alpha_{BA}N_A) \tag{6}$$

where N represent the quantities of colonies, μ_{max} represents the maximum specific growth rate, Q represent the physiological state of the cell, N_{max} represents the maximum population density and α_{AB} is a coefficient of interaction measuring the effects of cell B on cell A. At an interval $[t_{i-1}, t_i]$, the model could be described using the following equations, Equations (7) and (8) [33,34]:

$$\ln N_{Ai} - \ln N_{Ai-1} = \mu_{maxA} \times \frac{Q_A}{1+Q_A} \times (t_i - t_{i-1}) - \frac{\mu_{maxA}}{N_{maxA}} \times \int_{i-1}^{i} N_A(t)dt - \frac{\mu_{maxA}\alpha_{AB}}{N_{maxA}}\int_{i-1}^{i} N_B(t)dt \tag{7}$$

$$\ln N_{Bi} - \ln N_{Bi-1} = \mu_{maxB} \times \frac{Q_B}{1+Q_B} \times (t_i - t_{i-1}) - \frac{\mu_{maxB}}{N_{maxB}} \times \int_{i-1}^{i} N_B(t)dt - \frac{\mu_{maxB}\alpha_{BA}}{N_{maxB}}\int_{i-1}^{i} N_A(t)dt \tag{8}$$

where N represents the quantities of colonies, μ_{max} represents the maximum specific growth rate, Q represents the physiological state of the cell, N_{max} represents the maximum population density and α_{AB} is a coefficient of interaction measuring the effects of cell B on cell A.

2.5. Statistical Analysis

Data were analyzed in SPSS (Version 20.0, SPSS, New York, NY, USA) using multiple one-way analysis of variance (ANOVA) and general linear regression (GLM) models. DBD treatment, storage period, culture methods, and their interactions were fitted as fixed effects, and a replicate was fitted as a random effect. Duncan's multiple-range tests were applied to identify means of significant difference. The level of significance was set at 5% ($p < 0.05$).

The performance of models was evaluated by using the coefficient of determination (R^2) and standard error of fit (*SE of Fit*). Generally, the closer R^2 (Equation (9)) gets to 1, the higher the fitting degree of the equation. The *SE of Fit* (Equation (10)) quantifies the accuracy or precision of how well a mathematical model fits the observed data. It provides information about how closely the predicted values from the model align with the actual measurements. Essentially, a smaller *SE of Fit* indicates a better fit of the model to the data.

$$R^2 = 1 - \frac{\sum_{i=1}^{n}\left(y_o - y_p\right)^2}{\sum_{i=1}^{n}\left(y_o - y_m\right)^2} \tag{9}$$

$$SE\ of\ Fit = \sqrt{\frac{\sum_{i=1}^{n}\left(y_o - y_p\right)^2}{n-f}} \tag{10}$$

where n is the number of observed points, f is the number of parameters estimated in the model, y_o and y_p represent the observed value and predicted value, and y_m represents the mean values of all samples at each detection point.

3. Results and Discussion

3.1. Variations in the Total Viable Counts of DBD-Treated Pork During Refrigerated Storage

The objective of the microbiological analysis was to quantify the sublethal and lethal effect of DBD treatment on the indigenous microbiota population in the pork samples. The corresponding growth curves for TVC on pork during storage are presented in Figure 2. These show that, between Day 2 and Day 10 of the storage period, TVC was significant lower for the DBD-treated pork, and that an increase in DBD voltage led to more prominent inhibitory effects on the growth and reproduction of microorganisms. The TVC for the control, 60 kV, and 85 kV DBD-treated samples were 7.8, 7.0 and 5.4 log CFU/g, respectively, on Day 8 of the storage period. It is widely acknowledged that ROS and RNS, generated from cold plasma discharge, play a key role in the bactericidal effect of this non-thermal sterilization technology [35–37]. Under a higher voltage, the concentrations of ROS and RNS are also higher [38], meaning their bactericidal efficacy is increased. This outcome is the result of reactive species altering the permeability of the microorganism's cell membranes and causing the oxidation of intracellular biomacromolecules [10,39]. However, differences between the TVC of control and DBD-treated samples were not observed at Day 0, confirming the results of Huang et al. [22] and Wang, Zhuang, Lawrence and Zhang [40].

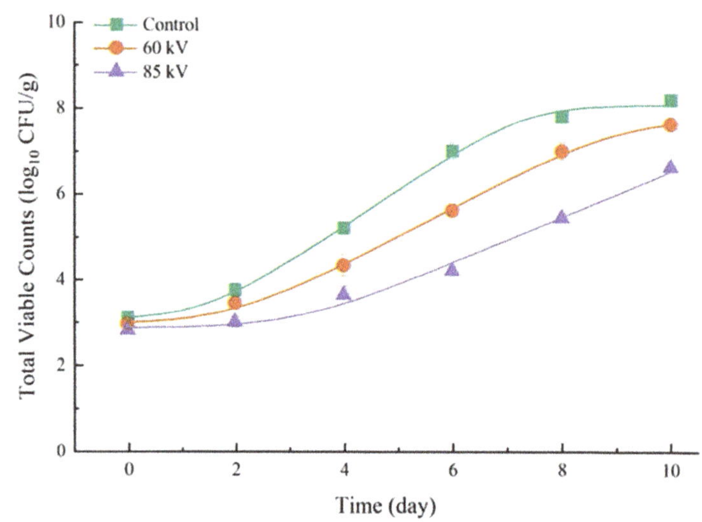

Figure 2. Total viable counts obtained for the pork loins exposed to different DBD treatments and held under refrigerated storage for up to 10 days (post-treatment). Plotted means (standard error shown as error bars) were calculated using the Baranyi and Roberts Model [41].

The rough surface of meat can provide protective sites for bacteria to evade bactericidal treatments [42]. Microorganisms could migrate from the surface to depths up to 140 μm, through feather follicles, capillaries, and the routes formatted by the radial shrinkage of muscle fiber, where they are largely unaffected by bactericidal treatments [43,44]. From the Day 0 TVC results, it seems that DBD treatment does not offer any immediate bactericidal effect ($p > 0.05$). The growth rate of native bacteria on pork was, however, significantly inhibited post-treatment. This is evidenced in Table 1, which includes the pork TVC growth kinetic parameters (initial values, lag time, μ_{max}). The R^2 and *SE of Fit* were shown to represent the degree of kinetics compared to the reality. Samples exposed to 85 kV DBD treatments had lag times almost double those of control samples; the former's μ_{max} was also significantly decreased when compared with that of the control sample ($p < 0.05$). In view of these results, sublethal damage to microorganisms, caused by DBD treatments, and the subsequent repair mechanism are likely to be responsible for the change in the dynamic of growth of TVC [37].

Table 1. Lag time (λ) and maximum specific growth rate (μ_{max}) for total viable counts according to the Baranyi and Roberts model [41].

DBD [A] (kV)	Initial Value (log_{10}CFU/g)	λ [B] (d)	μ_{max} [C] (d^{-1})	R^2 [D]	*SE of Fit* [E]
0	3.12 ± 0.09 [a]	1.52 ± 0.08 [b]	0.86 ± 0.02 [a]	0.995	0.151
60	2.95 ± 0.12 [a]	1.82 ± 0.14 [b]	0.64 ± 0.03 [b]	0.996	0.116
85	2.81 ± 0.11 [a]	3.12 ± 0.13 [a]	0.53 ± 0.01 [c]	0.986	0.177

[A] DBD: dielectric barrier discharge plasma treatment time; [B] λ: the lag phase duration; [C] μ_{max}: the maximum specific growth rate. [D] R^2: the coefficient of determination; [E] *SE of Fit:* standard error of fit. Different lower case letters, [a,b,c], denote significant differences ($p < 0.05$) between different treatment groups. Results are expressed as the means ± standard error.

3.2. *Pseudomonas spp., Enterobacteriaceae, LAB, and B. thermosphacta Counts in DBD-Treated Pork During Refrigerated Storage*

Pseudomonas spp. and *B. thermosphacta* grow readily under packaging conditions with relatively high oxygen concentrations and are the main spoilage organisms in meat products [45]. In addition, bacteria of the *Enterobacteriaceae* family, especially some cryophiles, may become dominant spoilage organisms during the cooling and storage of raw meat products. It has been shown that spoilage bacteria of the genus LAB are mainly found in fresh meat products packaged in vacuum or air-conditioned packaging [46]. *Lactobacillus sakeus, Lactobacillus flexneri, Lactobacillus marinus, Lactobacillus reuteri,* and *Lactobacillus oligomerus* have been shown to be associated with the spoilage of certain fresh meats, and that these *Lactic acid bacteria* can cause severe acidification of meats, which can result in a rancid, sour taste and spoilage of the meat products [47]. Therefore, these four genera were chosen as the primary targets for microbial enumeration in this study during the refrigerated storage of pork.

The precise composition of indigenous microorganisms on meat and meat products is influenced by the type of meat, processing method, and storage conditions. This point withstanding, *Pseudomonas* spp., *Enterobacteriaceae, B. thermosphacta*, LAB were found to be the dominant microorganisms in the pork samples—aligning with previous research of meat held under low temperatures and in modified-atmosphere packaging [48–51].

For a more comprehensive insight into microorganisms' response to the DBD treatment of pork, TAL and SM methods were applied to quantify the extent of sublethal injury of *Pseudomonas* spp., *Enterobacteriaceae*, LAB and *B. thermosphacta*. Figure 3A shows that DBD treatment had no significant bactericidal effect on *Pseudomonas* spp., *Enterobacteriaceae*, LAB, and *B. thermosphacta* when measured using the TAL method. Interestingly, when the SM method was employed, the 85 kV DBD treatment significantly decreased the population of *Pseudomonas* spp., *Enterobacteriaceae*, LAB and *B. thermosphacta* on pork by 0.4, 0.8, 0.5, and 0.5 log CFU/g, respectively. These results indicate a significant bactericidal effect of 85 kV DBD on pork for these four microorganisms (Figure 3B). Meanwhile, counts on DBD treated pork that were obtained using the SM method, irrespective of voltage and microorganism, were significantly lower than was observed using the TAL method. This finding may partially elucidate the seemingly contradictory research outcomes observed in previous studies, where DBD treatment did not demonstrate a significant bactericidal effect on indigenous microorganisms in meat but did demonstrate a sterilization effect on inoculated microorganism strains [44,52]. Within this context, the choice of cultivation method might play a substantive role.

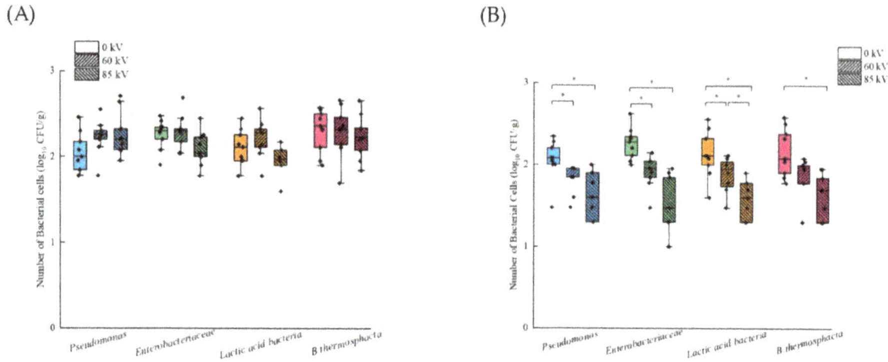

Figure 3. The *Pseudomonas* spp., *Enterobacteriaceae*, *Lactic acid bacteria* and *B. thermosphacta* counts on pork exposed to different levels of DBD. (**A**) The bacteria counts obtained by TAL method, and (**B**) the bacteria counts obtained on SM. * denotes significant differences ($p < 0.05$) between different treatment groups of the same strain.

Perni, Liu, Shama and Kong [53] found that cold plasma did not have a bactericidal effect when *E. coli* were recovered on TSA medium. However, when recovered on eosin methylene blue (EMB) medium, it was found that cold plasma could reduce *E. coli* by 1.0 log CFU/g, whereas on a more stringent medium, a reduction of 1.5 log CFU/g was achieved. Thus, it is inferred that cold plasma treatment can induce sublethal injury to microorganisms [53]. Smet et al. [54] systematically studied the effect of cold plasma on sublethal damage to *Salmonella* under different intrinsic factors (pH and salt concentration) and support system (liquid carrier and solid surface). Their results demonstrate the minor influence of the support system on *Salmonella* sublethal injury caused by cold plasma and the significant effects of the intrinsic factors to sublethal injury. Specifically, *Salmonella* cultured under incubation conditions of pH 5.5 and 6% NaCl suffered sensitive sublethal injury from cold plasma treatment than *Salmonella* cultured under incubation conditions of pH 7.0 and 0% NaCl [54]. It is apparent, however, that much focus has been on the sublethal injury of inoculated microorganisms, with few studies available on the sublethal injury of indigenous microorganism on food products caused by cold plasma treatment.

The recovery and growth of *Pseudomonas*, *Enterobacteriaceae*, LAB, *B. thermosphacta* in pork exposed to different DBD treatments and stored for up to 10 days, post-treatment, determined by the SM and TAL method, are shown in Figure 4. The growth curves for these four microorganisms show a similar trend to TVC, whereby their growth and reproduction are inhibited by DBD, and this effect is more prominent under the higher-voltage treatment. It was also found that the difference in microorganism population, determined by the SM and TAL methods, decreased as the storage period increased, before somewhat equalizing in count by Day 10 of the storage period. This result suggests that DBD treatment induced the broad-spectrum sublethal injury of microorganisms on pork and that the injured microorganisms regained viability during the storage period. This shows a real risk of underestimating the presence of foodborne microorganisms, following DBD treatment, when using the SM method.

To confirm the effects of DBD treatment on the sublethal injury of *Pseudomonas* spp., *Enterobacteriaceae*, LAB and *B. thermosphacta* on pork, the variations in the sublethal injury rates of these four microorganisms were monitored across the 10-day storage period (Figure 5). For the 60 kV DBD treatment, the sublethal injury rate of *Pseudomonas* spp. and *B. thermosphacta* decreased with an increase in storage period. The injured *B. thermosphacta* persisted for at least 2 days into the total storage period, the injured *Pseudomonas* spp. persisted for at least 4 days, while the injured *Enterobacteriaceae* and LAB persisted for at least 6 days. At the 85 kV DBD treatment, the sublethal injury durations of these four microorganisms were longer than observed for the 60 kV DBD treatment. By Day 6, the sublethal injury rate decreased for *Pseudomonas* spp. (78.4 to 32.6%), *Enterobacteriaceae* (75.7 to 42.9%), LAB (55.9 to 27.7%), and *B. thermosphacta* (75.9 to 39.5%). Regardless of whether the DBD treatment was conducted at 60 kV or 85 kV, the sublethal rates of the four microorganisms decreased with increases in the storage period. This finding indicates that microorganism with sublethal injury have the capacity for self-renewal.

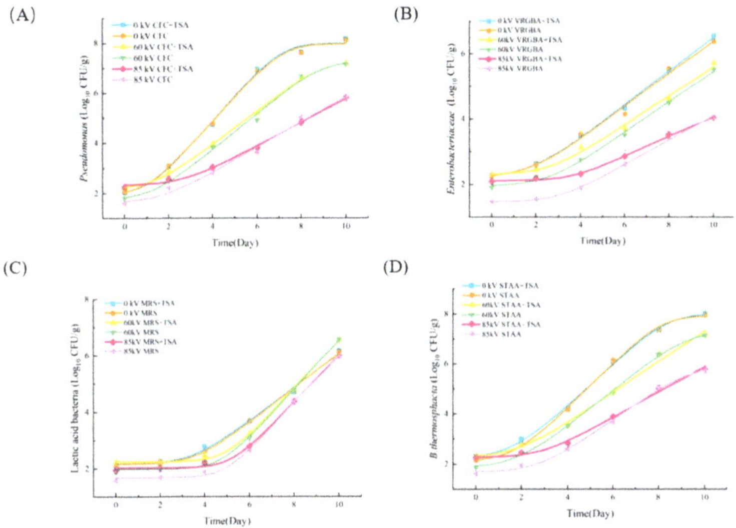

Figure 4. The recovery and growth of (**A**) *Pseudomonas*, (**B**) *Enterobacteriaceae*, (**C**) *Lactic acid bacteria*, and (**D**) *B. thermosphacta* on pork with and without DBD treatment during refrigerated storage for up to 10 days (post-treatment). All points are actual values, and all lines are fitted values of the model. Error bars represent the standard error.

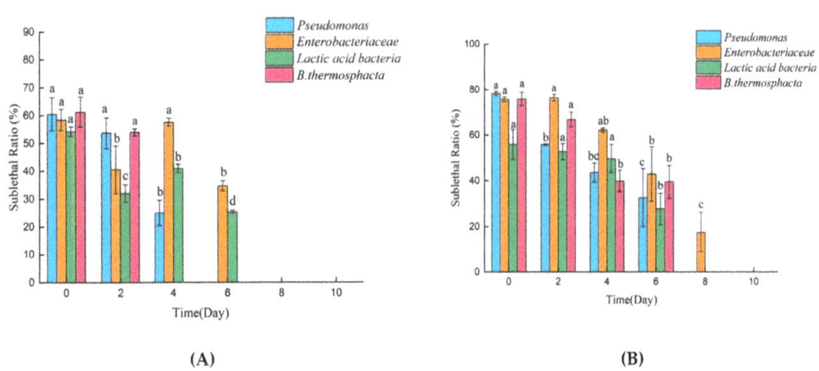

Figure 5. The changes in sublethal rates of *Pseudomonas* spp., *Enterobacteriaceae*, *Lactic acid bacteria* and *B. thermosphacta* on (**A**) 60 kV-DBD- and (**B**) 85 kV-DBD-treated pork held under refrigerated storage for up to 10 days (post-treatment). Error bars represent the standard error. Within the same bacterial genus, different lowercase letters at different times indicate significant differences ($p < 0.05$).

Following DBD treatment, microorganisms will be either alive, dead, or in a sublethal injured state. Plasma treatment causes injury to the structure or function of bacterial cell membranes, or a combination of both, including the destruction of membrane integrity and increased membrane permeability [55,56]. When a membrane is perforated but its metabolic activity is still active, a sublethal condition is present. In this case, the injured subpopulation can subsequently develop into a dead or alive microorganism, and the key to how they develop depends on the external environment. When external stressors are removed and microorganisms are in a nutrient-rich and favorable environment, some of the genes related to SOS response, nitrosative stress, the cell envelop-related response, and metabolism are found to be up-regulated, which contributes to the repair of damaged membranes [57,58]. Meanwhile, OxyR is an important regulatory protein. This protein is activated by ROS and upregulates the expression of antioxidant genes to prevent further damage by ROS to cells [55]. Further research has found that the during the SOS response, YneA accumulates within the cell, which inhibits septum formation and prevents cellular division [59–61]. This may be one of the reasons why sublethally injured microorganisms have an elongated lag phase. Sublethally injured microorganisms require energy to repair the damaged membrane, maintain osmotic pressure, and express selective genes [62]. Therefore, sufficient and complex nutrients are required for their recovery [63]. In addition to nutrients, certain minerals have been proposed to be beneficial for promoting the recovery of injured microorganisms. For example, Mn cations participate in the adjustment of the proteome, and the adjusted proteins participate in DNA repair, oxidoreductase activity, and the remodeling of gene expression [64]. Zn cations can enhance the activity of Cu/Zn superoxide dismutase, alleviate oxidative stress, and likewise promote microorganism recovery [65].

The lag phase is an adjustment period during which the microorganisms undergo some regulation to survive and thrive better in the new environment [66]. This parameter indicates the physiological state of the microorganisms, and thus can provide a better insight into their basic state [67]. DBD treatment caused an extension to the lag phase of *Pseudomonas* spp., *Enterobacteriaceae*, LAB and *B. thermosphacta*. Furthermore, the higher the treatment voltage, the longer the lag phase observed. These results suggest that DBD treatment prolonged the time required for these four microorganisms to adapt to the new environment. For the DBD-treated samples, the lag phase of the microorganisms cultured by the TAL method was longer than that of the microorganisms cultured by the SM method, and this phenomenon was more pronounced under higher-voltage treatment. During the 85 kV DBD treatment, the lag phases of *Pseudomonas* spp., *Enterobacteriaceae*, LAB and *B. thermosphacta* cultured by the TAL method were 2.5, 3.7, 4.3, and 3.0 days, respectively, while the lag phases of *Pseudomonas* spp., *Enterobacteriaceae*, LAB and *B. thermosphacta* cultured by the SM method were 1.6, 3.0, 3.6, and 2.4 days, respectively.

The maximum specific growth rate is a critical growth kinetic parameter and used to describe the growth behavior of microorganisms on food [68]. The maximum specific growth rate can vary depending on the type of microorganism, the environmental conditions, and the limiting substrate [49,69]. Different microorganisms have different maximum specific growth rates, which reflect their physiological and metabolic capabilities [70]. Table 2 shows that DBD treatment significantly affects the maximum specific growth rate of *Pseudomonas* spp., *Enterobacteriaceae* and *B. thermosphacta*; and this effect becomes more obvious with increasing treatment voltage. Using the TAL method, the maximum specific growth rates of *Pseudomonas* spp., *Enterobacteriaceae* and *B. thermosphacta* in the control group were found to be 1.0, 0.5, and 0.9, respectively. After 60 kV DBD treatment, the maximum specific growth rates of *Pseudomonas* spp., *Enterobacteriaceae* and *B. thermosphacta* in pork were shown to significantly decrease to 0.7, 0.5, and 0.7, respectively. As the DBD voltage increased to 85 kV, the maximum specific growth rates of *Pseudomonas* spp., *Enterobacteriaceae* and *B. thermosphacta* decreased significantly to 0.4, 0.38, and 0.6, respectively. The results obtained using the TAL method and the SM method did not significantly differ, except for the *Enterobacteriaceae* exposed to 85 kV DBD. However, DBD treatment had no significant effect on the maximum specific growth rate of LAB ($p > 0.05$).

Table 2. Lag time (λ), maximum specific growth rate (μ_{max}), and generation time for *Pseudomonas*, *Enterobacteriaceae*, LAB, and *B. thermosphacta* according to the Baranyi and Roberts model [41].

DBD [A] (kV)		Medium	λ [B] (d)	μ_{max} [C] (d^{-1})	Generation Time (d)	R^2 [D]	*SE of Fit* [E]
Pseudomonas	0	TAL	1.14 ± 0.13 [b]	1 ± 0.02 [a]	0.48 ± 0.01 [c]	0.988	0.274
		SM	1.19 ± 0.05 [b]	0.95 ± 0.01 [a]	0.51 [bc]	0.99	0.229
	60	TAL	1.35 ± 0.12 [b]	0.67 ± 0.01 [b]	0.57 ± 0.01 [b]	0.997	0.111
		SM	1.28 ± 0.03 [b]	0.72 ± 0.03 [b]	0.55 ± 0.01 [b]	0.983	0.275
	85	TAL	2.47 ± 0.18 [a]	0.44 ± 0 [c]	0.77 ± 0.02 [a]	0.994	0.099
		SM	1.61 ± 0.32 [b]	0.5 ± 0.03 [c]	0.73 ± 0.03 [a]	0.971	0.263
Enterobacteriaceae	0	TAL	2.31 ± 0.06 [cd]	0.51 ± 0.01 [a]	0.63 ± 0.01 [c]	0.992	0.142
		SM	1.96 ± 0.13 [d]	0.5 ± 0.01 [a]	0.66 ± 0.01 [c]	0.985	0.192
	60	TAL	3.54 ± 0.06 [a]	0.46 ± 0.01 [b]	0.81 ± 0.13 [bc]	0.987	0.14
		SM	2.56 ± 0.18 [bc]	0.47 ± 0.01 [b]	0.68 ± 0.01 [c]	0.996	0.093
	85	TAL	3.72 ± 0.05 [a]	0.28 ± 0.01 [d]	1.21 ± 0.03 [a]	0.978	0.104
		SM	2.96 ± 0.14 [b]	0.36 ± 0.01 [c]	0.94 ± 0.01 [b]	0.99	0.102
LAB [F]	0	TAL	2.34 ± 0.25 [d]	0.77 ± 0.07 [a]	0.53 ± 0.01 [a]	0.991	0.213
		SM	2.54 ± 0.08 [d]	0.74 ± 0.03 [a]	0.53 ± 0.01 [a]	0.987	0.244
	60	TAL	3.56 ± 0.03 [b]	0.83 ± 0.01 [a]	0.49 ± 0.01 [b]	0.994	0.164
		SM	3.03 ± 0.05 [c]	0.81 ± 0.02 [a]	0.54 ± 0.01 [a]	0.985	0.28
	85	TAL	4.31 ± 0.14 [a]	0.79 ± 0.04 [a]	0.46 ± 0.01 [c]	0.99	0.178
		SM	3.55 ± 0.18 [b]	0.74 ± 0.01 [a]	0.47 ± 0.01 [bc]	0.991	0.184
B. thermosphacta	0	TAL	1.75 ± 0.06 [c]	0.87 ± 0.01 [a]	0.47 ± 0.01 [b]	0.997	0.126
		SM	1.79 ± 0.12 [c]	0.93 ± 0.02 [a]	0.46 [b]	0.998	0.101
	60	TAL	2.46 ± 0.16 [b]	0.73 ± 0.03 [b]	0.49 ± 0.01 [b]	0.988	0.225
		SM	1.8 ± 0.05 [c]	0.72 ± 0.01 [b]	0.47 ± 0.01 [b]	0.999	0.083
	85	TAL	3.01 ± 0.17 [a]	0.59 ± 0.04 [c]	0.64 ± 0.01 [a]	0.997	0.081
		SM	2.36 ± 0.33 [b]	0.58 ± 0.08 [c]	0.61 ± 0.02 [a]	0.976	0.272

[A] DBD: dielectric barrier discharge plasma treatment time; [B] λ: the lag phase duration; [C] μ_{max}: the maximum specific growth rate; [D] R^2: the coefficient of determination; [E] *SE of Fit:* standard error of fit; [F] LAB: lactic acid bacteria. Different lower case letters, [a,b,c,d], denote significant differences ($p < 0.05$) between different treatment groups of the same strain. Results are expressed as the means ± standard error.

The lag phase and the maximum specific growth rate contribute to our understanding of the physiological state of microorganisms. Pina-Perez, Martinet, Palacios-Gorba, Ellert and Beyrer [37] evaluated the influence of plasma treatment on the dynamics of growth of *Bacillus subtilis* using the Gompertz model. Their results indicated significant differences in the lag phase and maximum specific growth rate between the treatment and control groups, with the lag phase differences becoming more evident as the plasma power density increased [37]. Han, Boehm, Patil, Cullen and Bourke [57] revealed that the maximum specific growth rate of plasma-treated *E. coli*, *L. monocytogenes*, and *S. aureus* decreased as plasma treatment times increased [71]. In general, plasma treatment results in lipid and protein oxidation, with implications on the changes in cell membrane fluidity and enzyme activity [72]. These alterations are considered to impact the lag phase and the maximum specific growth rate of microorganisms [73,74]. Table 2 shows that the maximum specific growth rate of LAB was not affected by DBD treatment, unlike that of *Pseudomonas* spp., *Enterobacteriaceae* and *B. thermosphacta*. DBD treatment did, however, significantly reduced the generation time of LAB on pork. This raised the following question: does DBD treatment improve the ability of LAB to adapt to colonies and enhance its competitiveness?

3.3. *Interactions Between Pseudomonas spp., Enterobacteriaceae, LAB and B. thermosphacta in DBD-Treated Pork*

Modified Lotka–Volterra Models were used to evaluate the dynamic interactions between the *Pseudomonas* spp., *Enterobacteriaceae*, LAB, and *B. thermosphacta* (Figure 6). For control samples, the coefficients of interaction of LAB on *Pseudomonas* spp., *Enter-*

obacteriaceae, and *B. thermosphacta* were $\alpha_{12} = 0.7$, 1.0, and 0.7, respectively, and $\alpha_{21} = 2.3$, 1.1, and 1.4, respectively. As $\alpha_{12} < \alpha_{21}$, it was concluded that the influence of LAB on *Pseudomonas* spp., *Enterobacteriaceae*, and *B. thermosphacta* was much lower than that of *Pseudomonas* spp., *Enterobacteriaceae*, and *B. thermosphacta* on LAB. For pork exposed to 60 kV DBD, the coefficient of interaction of LAB on *Pseudomonas* spp., *Enterobacteriaceae*, and *B. thermosphacta*, α_{12} decreased to 1.9, 2.6 and 1.2, respectively, and α_{21} increased to 1.7, 0.9 and 0.9, respectively. These indicate the influence of LAB on *Pseudomonas* spp.; *Enterobacteriaceae*, and *B. thermosphacta* was enhanced, while the influence of *Pseudomonas* spp., *Enterobacteriaceae*, and *B. thermosphacta* on LAB was lessened. These results present that DBD enhances the antagonistic activity of LAB during pork storage—an outcome confirmed by Zhao et al. [75], who found the proportion of *Pseudomonas* in plasma-treated chicken meat increased during storage, while the proportion of *Carnobacterium* decreased. ROS generated by plasma can oxidize the lipids in the outer membrane, leading to the inactivation of Gram-negative bacteria and an increase in the proportion of Gram-positive bacteria, and the higher the initial load of microorganisms, the higher the proportion in the latter storage periods [75]. Previous studies have shown that plasma treatment can cause a decrease in surface pH [15,76,77], which may create an environment that is more suited for LAB compared to other microorganisms. In addition, when grown in associated cultures, LAB exerted an antagonistic action on the growth of *Staphylococcus aureus*, *Salmonella typhimurium* and *E. coli* [67]. The metabolic product of LAB, lactic acid, also has a bacteriostatic effect, and the bacteriostatic effect of lactic acid on plasma-treated microorganisms is enhanced [20,78]. However, it should be noted that CFC, STAA, and VRBGA media all contain one or more bactericidal agents (i.e., cetrimide, streptomycin sulfate, bile salts), while MRS media primarily select LAB via pH adjustments, which may lead to a higher detection rate of LAB.

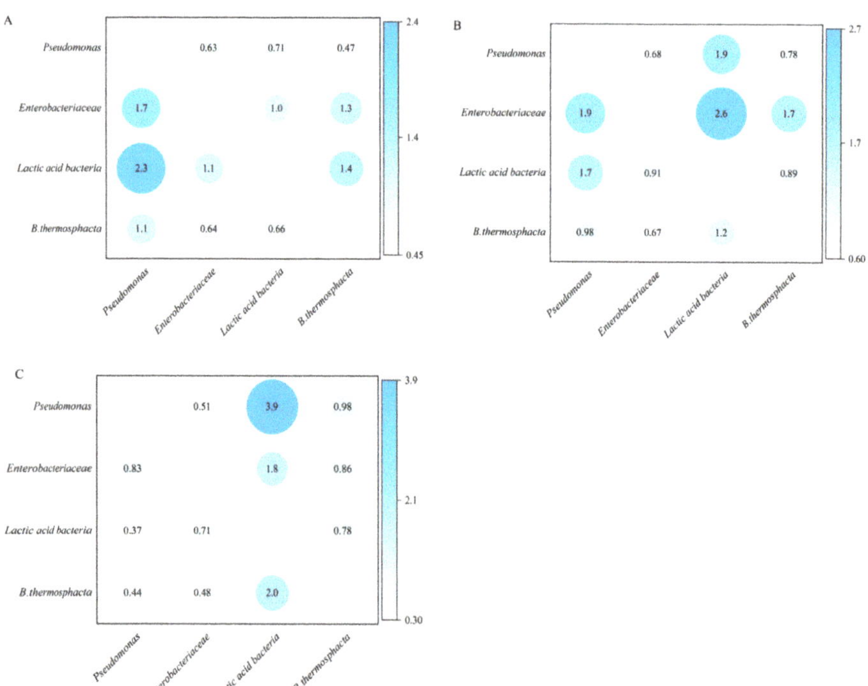

Figure 6. Interaction of *Pseudomonas* spp., *Enterobacteriaceae*, *Lactic acid bacteria* and *B. thermosphacta* on (**A**) control, (**B**) 60 kV DBD-treated pork, and (**C**) 85 kV DBD pork based on the Lotka–Volterra model.

4. Conclusions

This study proves that DBD treatment will cause damage to the indigenous microorganisms on pork and will extend its shelf-life when refrigerated and in modified-atmosphere packaging. However, when using conventional detection methods, this sublethal damage can lead to type II error. Adopting modified culture methods can help avoid this problem, to some extent, and make the detection results more accurate. DBD treatment also causes changes in the succession of the microorganisms in the pork during storage, decreasing the proportion of *Pseudomonas* spp. and *B. thermosphacta*, and increasing the proportion of LAB. This bactericidal characteristic as well as its effects on the preservation and increase in the proportion of LAB suggests that DBD treatment can likewise prolong the preservation of fermented meat products. Additional research is, nonetheless, advisable given the current experimental conditions and the extrinsic and intrinsic factors that impact microorganism growth and population.

Author Contributions: Y.Z.: writing—original draft. H.Z.: methodology, data curation. Z.D.: methodology, data curation. Z.G.: writing—review and editing. B.W.B.H.: supervision, writing—review and editing. Y.D.: methodology. J.S.: methodology. X.D.: methodology. M.H.: project administration, funding acquisition. Y.M.: funding acquisition, supervision, writing—review and editing. All authors have read and agreed to the published version of the manuscript.

Funding: This research was funded by National Natural Science Foundation of China (32202142), Shandong Province Pig Industry Technology System (SDAIT-08-10), Shandong College and Universities Plan of Youth Innovation Science and Technology Program (2020KJE007).

Data Availability Statement: Data will be made available on request.

Conflicts of Interest: The authors declare no conflicts of interest.

References

1. Zhu, Y.L.; Li, C.Z.; Cui, H.Y.; Lin, L. Plasma enhanced-nutmeg essential oil solid liposome treatment on the gelling and storage properties of pork meat batters. *J. Food Eng.* **2020**, *266*, 109696. [CrossRef]
2. Subrahmanyam, K.; Gul, K.; Sehrawat, R.; Tiwari, B.K.; Sahoo, S. Cold plasma-mediated inactivation of microorganisms for the shelf-life extension of animal-based foods: Efficiency, mechanism of inactivation, and impact on quality attributes. *Food Control.* **2024**, *162*, 110464. [CrossRef]
3. Xiang, Q.S.; Liu, X.F.; Li, J.G.; Ding, T.; Zhang, H.; Zhang, X.S.; Bai, Y.H. Influences of cold atmospheric plasma on microbial safety, physicochemical and sensorial qualities of meat products. *J. Food Sci. Technol.* **2018**, *55*, 846–857. [CrossRef] [PubMed]
4. Allai, F.M.; Azad, Z.R.A.A.; Mir, N.A.; Gul, K. Recent advances in non-thermal processing technologies for enhancing shelf life and improving food safety. *Appl. Food Res.* **2023**, *3*, 100258. [CrossRef]
5. Li, X.; Farid, M. A review on recent development in non-conventional food sterilization technologies. *J. Food Eng.* **2016**, *182*, 33–45. [CrossRef]
6. Mendes-Oliveira, G.; Jensen, J.L.; Keener, K.M.; Campanella, O.H. Modeling the inactivation of *Bacillus subtilis* spores during cold plasma sterilization. *Innov. Food Sci. Emerg. Technol.* **2019**, *52*, 334–342. [CrossRef]
7. Xu, H.B.; Ma, R.N.; Zhu, Y.P.; Du, M.R.; Zhang, H.; Jiao, Z. A systematic study of the antimicrobial mechanisms of cold atmospheric-pressure plasma for water disinfection. *Sci. Total Environ.* **2020**, *703*, 134965. [CrossRef]
8. Pan, Y.; Cheng, J.H.; Lv, X.; Sun, D.W. Assessing the inactivation efficiency of Ar/O_2 plasma treatment against Listeria monocytogenes cells: Sublethal injury and inactivation kinetics. *LWT Food Sci Technol.* **2019**, *111*, 318–327. [CrossRef]
9. Winter, S.; Meyer-Lindenberg, A.; Wolf, G.; Reese, S.; Nolff, M.C. In vitro evaluation of the decontamination effect of cold atmospheric argon plasma on selected bacteria frequently encountered in small animal bite injuries. *J. Microbiol. Methods.* **2020**, *169*, 105728. [CrossRef]
10. Ziuzina, D.; Misra, N.N.; Han, L.; Cullen, P.J.; Moiseev, T.; Mosnier, J.P.; Keener, K.; Gaston, E.; Vilaró, I.; Bourke, P. Investigation of a large gap cold plasma reactor for continuous in-package decontamination of fresh strawberries and spinach. *Innov. Food Sci. Emerg. Technol.* **2020**, *59*, 102229. [CrossRef]
11. Batt, H.; Fahey, J.W. Cold plasma technology: Does it have a place in food processing? *Crit. Rev. Food Sci. Nutr.* **2024**. [CrossRef] [PubMed]
12. Lu, X.; Naidis, G.V.; Laroussi, M.; Reuter, S.; Graves, D.B.; Ostrikov, K. Reactive species in non-equilibrium atmospheric-pressure plasmas: Generation, transport, and biological effects. *Phys. Rep.* **2016**, *630*, 1–84. [CrossRef]
13. Wang, Y.; Wang, Z.R.; Zhu, X.; Yuan, Y.H.; Gao, Z.P.; Yue, T.L. Application of electrical discharge plasma on the inactivation of Zygosaccharomyces rouxii in apple juice. *LWT Food Sci. Technol.* **2020**, *121*, 108974. [CrossRef]

14. Jayasena, D.D.; Kim, H.J.; Yong, H.I.; Park, S.; Kim, K.; Choe, W.; Jo, C. Flexible thin-layer dielectric barrier discharge plasma treatment of pork butt and beef loin: Effects on pathogen inactivation and meat-quality attributes. *Food Microbiol.* **2015**, *46*, 51–57. [CrossRef]
15. Moutiq, R.; Misra, N.N.; Mendonca, A.; Keener, K. In-package decontamination of chicken breast using cold plasma technology: Microbial, quality and storage studies. *Meat Sci.* **2020**, *159*, 107942. [CrossRef]
16. Liao, X.; Xiang, Q.S.; Liu, D.H.; Chen, S.G.; Ye, X.Q.; Ding, T. Lethal and sublethal effect of a dielectric barrier discharge atmospheric cold plasma on *Staphylococcus aureus*. *J. Food Prot.* **2017**, *80*, 928–932. [CrossRef]
17. Verheyen, D.; Baka, M.; Akkermans, S.; Skara, T.; Van Impe, J.F. Effect of microstructure and initial cell conditions on thermal inactivation kinetics and sublethal injury of *Listeria monocytogenes* in fish-based food model systems. *Food Microbiol.* **2019**, *84*, 103267. [CrossRef]
18. Wang, X.; Devlieghere, F.; Geeraerd, A.; Uyttendaele, M. Thermal inactivation and sublethal injury kinetics of *Salmonella enterica* and *Listeria monocytogenes* in broth versus agar surface. *Int. J. Food Microbiol.* **2017**, *243*, 70–77. [CrossRef]
19. Izumi, H.; Inoue, A. Viability of sublethally injured coliform bacteria on fresh-cut cabbage stored in high CO_2 atmospheres following rinsing with electrolyzed water. *Int. J. Food Microbiol.* **2018**, *266*, 207–212. [CrossRef]
20. Laroussi, M.; Richardson, J.P.; Dobbs, F.C. Effects of nonequilibrium atmospheric pressure plasmas on the heterotrophic pathways of bacteria and on their cell morphology. *Appl. Phys. Lett.* **2002**, *81*, 772–774. [CrossRef]
21. Li, P.R.; Zhang, H.A.; Tian, C.Q.; Zou, H.M. Experimental Investigation of Bacterial Inactivation of Beef Using Indirect Cold Plasma in Cold Chain and at Room Temperature. *Foods* **2024**, *13*, 2846. [CrossRef] [PubMed]
22. Huang, M.M.; Zhuang, H.; Wang, J.M.; Yan, W.J.; Zhao, J.Y.; Zhang, J.H. Inactivation kinetics of *Salmonella typhimurium* and *Staphylococcus aureus* in different media by dielectric barrier discharge non-thermal plasma. *Appl. Sci.* **2018**, *8*, 2087. [CrossRef]
23. Smet, C.; Noriega, E.; Rosier, F.; Walsh, J.L.; Valdramidis, V.P.; Van Impe, J.F. Influence of food intrinsic factors on the inactivation efficacy of cold atmospheric plasma: Impact of osmotic stress, suboptimal pH and food structure. *Innov. Food Sci. Emerg. Technol.* **2016**, *38*, 393–406. [CrossRef]
24. Thomas-Popo, E.; Mendonça, A.; Misra, N.N.; Little, A.; Wan, Z.; Moutiq, R.; Coleman, S.; Keener, K. Inactivation of Shiga-toxin-producing *Escherichia coli*, *Salmonella enterica* and natural microflora on tempered wheat grains by atmospheric cold plasma. *Food Control.* **2019**, *104*, 231–239. [CrossRef]
25. Bezerra, J.D.; Lamarao, C.V.; Sanches, E.A.; Rodrigues, S.; Fernandes, F.A.N.; Ramos Glpa Esmerino, E.A.; Cruz, A.G.; Campelo, P.H. Cold plasma as a pre-treatment for processing improvement in food: A review. *Food Res. Int.* **2023**, *167*, 1873–7145. [CrossRef]
26. Kang, D.H.; Fung DY, C. Application of thin agar layer method for recovery of injured *Salmonella typhimurium*. *Int. J. Food Microbiol.* **2000**, *54*, 127–132. [CrossRef]
27. Tian, X.J.; Shao, L.L.; Yu, Q.Q.; Yang, H.; Li, X.M.; Dai, R.T. Comparative study of survival of *Escherichia coli* O_{157}:H_7 inoculated in pork batter after ohmic cooking and water bath cooking. *Int. J. Food Microbiol.* **2019**, *304*, 11–18. [CrossRef]
28. Kitsiou, M.; Wantock, T.; Sandison, G.; Harle, T.; Gutierrez-Merino, J.; Klymenko, O.V.; Karatzas, K.A.; Velliou, E. Determination of the combined effect of grape seed extract and cold atmospheric plasma on foodborne pathogens and their environmental stress knockout mutants. *Appl. Environ.* **2024**, *90*, e00177-24. [CrossRef]
29. Benavides-López, S.; Ortega-Quintanab, F.A.; Alvarez, H. Phenomenological-based semi-physical model characterizing the central metabolic response of *Escherichia coli* to changes in batch culture temperature. *Biochem. Eng. J.* **2024**, *203*, 109183. [CrossRef]
30. Koutsoumanis, K.P.; Stamatiou, A.P.; Drosinos, E.H.; Nychas, G.J. Control of spoilage microorganisms in minced pork by a self-developed modified atmosphere induced by the respiratory activity of meat microflora. *Food Microbiol.* **2008**, *25*, 915–921. [CrossRef]
31. Reid, R.; Fanning, S.; Whyte, P.; Kerry, J.; Lindqvist, R.; Yu, Z.; Bolton, D. The microbiology of beef carcasses and primals during chilling and commercial storage. *Food Microbiol.* **2017**, *61*, 50–57. [CrossRef] [PubMed]
32. Dens, E.J.; Vereecken, K.M.; Van Impe, J.F. A prototype model structure for mixed microbial populations in homogeneous food products. *J. Theor. Biol.* **1999**, *201*, 159–170. [CrossRef] [PubMed]
33. Liu, F.; Guo, Y.Z.; Li, Y.F. Interactions of microorganisms during natural spoilage of pork at 5 °C. *J. Food Eng.* **2006**, *72*, 24–29. [CrossRef]
34. Ye, K.P.; Wang, H.H.; Jiang, Y.; Xu, X.L.; Cao, J.X.; Zhou, G.H. Development of interspecific competition models for the growth of *Listeria monocytogenes* and *Lactobacillus* on vacuum-packaged chilled pork by quantitative real-time PCR. *Food Res. Int.* **2014**, *64*, 626–633. [CrossRef]
35. Choi, M.S.; Jeon, E.B.; Kim, J.Y.; Choi, E.H.; Lim, J.S.; Choi, J.; Park, S.Y. Impact of non-thermal dielectric barrier discharge plasma on *Staphylococcus aureus* and *Bacillus cereus* and quality of dried blackmouth angler (*Lophiomus setigerus*). *J. Food Eng.* **2020**, *278*, 109952. [CrossRef]
36. Frias, E.; Iglesias, Y.; Alvarez-Ordonez, A.; Prieto, M.; Gonzalez-Raurich, M.; Lopez, M. Evaluation of cold atmospheric pressure plasma (CAPP) and plasma-activated water (PAW) as alternative non-thermal decontamination technologies for tofu: Impact on microbiological, sensorial and functional quality attributes. *Food Res. Int.* **2020**, *129*, 108859. [CrossRef]
37. Pina-Perez, M.C.; Martinet, D.; Palacios-Gorba, C.; Ellert, C.; Beyrer, M. Low-energy short-term cold atmospheric plasma: Controlling the inactivation efficacy of bacterial spores in powders. *Food Res. Int.* **2020**, *130*, 108921. [CrossRef]
38. Wan, Z.; Misra, N.N.; Li, G.; Keener, K.M. High voltage atmospheric cold plasma treatment of *Listeria innocua* and *Escherichia coli* K-12 on Queso Fresco (fresh cheese). *LWT Food Sci. Technol.* **2021**, *146*, 111406. [CrossRef]

39. Dong, X.Y.; Xiu, Z.L.; Li, S.; Hou, Y.M.; Zhang, D.J.; Ren, C.S. Dielectric barrier discharge plasma as a novel approach for improving 1,3-propanediol production in *Klebsiella pneumoniae*. *Biotechnol. Lett.* **2010**, *32*, 1245–1250. [CrossRef]
40. Wang, J.M.; Zhuang, H.; Lawrence, K.; Zhang, J.H. Disinfection of chicken fillets in packages with atmospheric cold plasma: Effects of treatment voltage and time. *J. Appl. Microbiol.* **2018**, *124*, 1212–1219. [CrossRef]
41. Baranyi, J.; Roberts, T.A. A dynamic approach to predicting bacterial growth in food. *Int. J. Food Microbiol.* **1994**, *23*, 277–294. [CrossRef] [PubMed]
42. Song, H.P.; Kim, B.; Choe, J.H.; Jung, S.; Moon, S.Y.; Choe, W.; Jo, C. Evaluation of atmospheric pressure plasma to improve the safety of sliced cheese and ham inoculated by 3-strain cocktail *Listeria monocytogenes*. *Food Microbiol.* **2009**, *26*, 432–436. [CrossRef] [PubMed]
43. Jayasena, D.D.; Kang, T.; Wijayasekara, K.N.; Jo, C. Innovative Application of Cold Plasma Technology in Meat and Its Products. *Food Sci. Anim. Resour.* **2023**, *43*, 1087–1110. [CrossRef] [PubMed]
44. Noriega, E.; Shama, G.; Laca, A.; Diaz, M.; Kong, M.G. Cold atmospheric gas plasma disinfection of chicken meat and chicken skin contaminated with *Listeria innocua*. *Food Microbiol.* **2011**, *28*, 1293–1300. [CrossRef]
45. Lahiri, D.; Nag, M.; Sarkar, T.; Ray, R.R.; Shariati, M.A.; Rebezov, M.; Bangar, S.P.; Lorenzo, J.M.; Domínguez, R. *Lactic Acid Bacteria* (LAB): Autochthonous and Probiotic Microbes for Meat Preservation and Fortification. *Foods* **2022**, *11*, 2792. [CrossRef]
46. Marcelli, V.; Osimani, A.; Aquilanti, L. Research progress in the use of *lactic acid bacteria* as natural biopreservatives against *Pseudomonas* spp. in meat and meat products: A review. *Food Res. Int.* **2024**, *196*, 115129. [CrossRef]
47. Baillo, A.A.; Cisneros, L.; Villena, J.; Vignolo, G.; Fadda, S. Bioprotective *Lactic Acid Bacteria* and Lactic Acid as a Sustainable Strategy to Combat *Escherichia coli* O_{157}:H_7 in Meat. *Foods* **2023**, *12*, 231. [CrossRef]
48. Ajaykumar, V.J.; Mandal, P.K. *Modern Concept and Detection of Spoilage in Meat and Meat Products*; Academic Press Inc.: New York, NY, USA, 2020; pp. 335–349.
49. Nychas GJ, E.; Skandamis, P.N.; Tassou, C.C.; Koutsoumanis, K.P. Meat spoilage during distribution. *Meat Sci.* **2008**, *78*, 77–89. [CrossRef]
50. Pellissery, A.J.; Vinayamohan, P.G.; Amalaradjou MA, R.; Venkitanarayanan, K. *Spoilage Bacteria and Meat Quality*; Academic Press Inc.: New York, NY, USA, 2020; pp. 307–334.
51. Zagorec, M.; Champomier-Vergès, M.C. *Meat Microbiology and Spoilage*; Woodhead Publishing Inc.: Cambridge, UK, 2023; pp. 195–218.
52. Hertwig, C.; Reineke, K.; Ehlbeck, J.; Knorr, D.; Schlüter, O. Decontamination of whole black pepper using different cold atmospheric pressure plasma applications. *Food Control.* **2015**, *55*, 221–229. [CrossRef]
53. Perni, S.; Liu, D.W.; Shama, G.; Kong, M.G. Cold atmospheric plasma decontamination of the pericarps of fruit. *J. Food Prot.* **2008**, *71*, 302–308. [CrossRef]
54. Smet, C.; Baka, M.; Steen, L.; Fraeye, I.; Walsh, J.L.; Valdramidis, V.P.; Van Impe, J.F. Combined effect of cold atmospheric plasma, intrinsic and extrinsic factors on the microbial behavior in/on (food) model systems during storage. *Innov. Food Sci. Emerg. Technol.* **2019**, *53*, 3–17. [CrossRef]
55. Huang, M.M.; Zhuang, H.; Zhao, J.Y.; Wang, J.M.; Yan, W.J.; Zhang, J.H. Differences in cellular damage induced by dielectric barrier discharge plasma between *Salmonella Typhimurium* and *Staphylococcus aureus*. *Bioelectrochemistry* **2020**, *132*, 107445. [CrossRef] [PubMed]
56. Qian, J.; Ma, L.J.; Yan, W.J.; Zhuang, H.; Huang, M.M.; Zhang, J.H.; Wang, J.M. Inactivation kinetics and cell envelope damages of foodborne pathogens *Listeria monocytogenes* and *Salmonella enteritidis* treated with cold plasma. *Food Microbiol.* **2022**, *101*, 103891. [CrossRef] [PubMed]
57. Han, L.; Boehm, D.; Patil, S.; Cullen, P.J.; Bourke, P. Assessing stress responses to atmospheric cold plasma exposure using *Escherichia coli* knock-out mutants. *J. Appl. Microbiol.* **2016**, *121*, 352–363. [CrossRef]
58. Joshi, S.G.; Yost, A.; Joshi, S.S.; Addya, S.; Ehrlich, G.; Brooks, A. Microarray analysis of transcriptomic response of *Escherichia coli* to nonthermal plasma-treated PBS solution. *ABB.* **2015**, *6*, 49–62. [CrossRef]
59. Jones, T.H.; Vail, K.M.; McMullen, L.M. Filament formation by foodborne bacteria under sublethal stress. *Int. J. Food Microbiol.* **2013**, *165*, 97–110. [CrossRef]
60. Kawai, Y.; Moriya, S.; Ogasawara, N. Identification of a protein, YneA, responsible for cell division suppression during the SOS response in *Bacillus subtilis*. *Mol Microbiol.* **2003**, *47*, 1113–1122. [CrossRef]
61. Winterhalter, C.; Pelliciari, S.; Stevens, D.; Fenyk, S.; Marchand, E.; Cronin, N.B.; Soultanas, P.; Costa TR, D.; Ilangovan, A.; Murray, H. The DNA replication initiation protein DnaD recognises a specific strand of the *Bacillus subtilis* chromosome origin. *Nucleic Acids Res.* **2023**, *51*, 4322–4340. [CrossRef]
62. Shao, L.L.; Liu, Y.; Tian, X.J.; Zou, B.; Zhao, Y.J.; Li, X.M.; Dai, R.T. Global proteomic responses of sublethally injured *Staphylococcus aureus* induced by ohmic heating. *Food Control.* **2021**, *127*, 108106. [CrossRef]
63. Lan, L.; Zhang, R.; Zhang, X.; Shi, H. Sublethal injury and recovery of *Listeria monocytogenes* and *Escherichia coli* O_{157}:H_7 after exposure to slightly acidic electrolyzed water. *Food Control.* **2019**, *106*, 106746. [CrossRef]
64. Ruas FA, D.; Barboza, N.R.; Castro-Borges, W.; Guerra-Sa, R. Manganese alters expression of proteins involved in the oxidative stress of *Meyerozyma guilliermondii*. *J. Proteom.* **2019**, *196*, 173–188. [CrossRef] [PubMed]
65. Zhang, R.; Lan, L.S.; Shi, H. Sublethal injury and recovery of Escherichia coli O_{157}:H_7 after freezing and thawing. *Food Control.* **2021**, *120*, 107488. [CrossRef]

66. Swinnen, I.A.M.; Bernaerts, K.; Dens, E.J.J.; Geeraerd, A.H.; Van Impe, J.F. Predictive modelling of the microbial lag phase: A review. *Int. J. Food Microbiol.* **2004**, *94*, 137–159. [CrossRef] [PubMed]
67. Schlei, K.P.; Angioletti, B.L.; Fernandes de Carvalho, L.; Bertoli, S.L.; Ratto Reiter, M.G.; Krebs de Souza, C. Influence of temperature on microbial growth during processing of kochkäse cheese made from raw and pasteurised milk. *Int. Dairy J.* **2020**, *109*, 104786. [CrossRef]
68. Tarlak, F.; Ozdemir, M.; Melikoglu, M. Predictive modelling for the growth kinetics of *Pseudomonas* spp. on button mushroom (*Agaricus bisporus*) under isothermal and non-isothermal conditions. *Food Res. Int.* **2020**, *130*, 108912. [CrossRef]
69. Dey, A.; Bokka, V.; Sen, S. Dependence of bacterial growth rate on dynamic temperature changes. *IET Syst. Biol.* **2020**, *14*, 68–74. [CrossRef]
70. Veshareh, M.J.; Nick, H.M. A novel relationship for the maximum specific growth rate of a microbial guild. *FEMS Microbiol. Lett.* **2021**, *368*, fnab064. [CrossRef]
71. Oliulla, H.; Mizan MF, R.; Ashrafudoulla, M.; Meghla, N.S.; Ha AJ, W.; Park, S.H.; Ha, S.D. The challenges and prospects of using cold plasma to prevent bacterial contamination and biofilm formation in the meat industry. *Meat Sci.* **2024**, *217*, 1873–4138. [CrossRef]
72. Hensel, K.; Kucerova, K.; Tarabova, B.; Janda, M.; Machala, Z.; Sano, K.; Mihai, C.T.; Ciorpac, M.; Gorgan, L.D.; Jijie, R.; et al. Effects of air transient spark discharge and helium plasma jet on water, bacteria, cells, and biomolecules. *Biointerphases* **2015**, *10*, 029515. [CrossRef]
73. Madigan, M.T.; Martinko, J.M.; Bender, K.S.; Buckley, D.H.; Stahl, D.A.; Brock, T. *Brock's Biology of Microorganisms*, 14th ed.; Pearson: Madrid, Spain, 2014; pp. 184–201.
74. Wong, W.W.; Liao, J.C. Microbial maximal specific growth rate as a square-root function of biomass yield and two kinetic parameters. *Metab Eng.* **2009**, *11*, 409–414. [CrossRef]
75. Zhao, Y.J.; Meng, Z.Y.; Shao, L.L.; Dai, R.T.; Li, X.M.; Jia, F. Employment of cold atmospheric plasma in chilled chicken breasts and the analysis of microbial diversity after the shelf-life storage. *Food Res. Int.* **2022**, *162 Pt A*, 111934. [CrossRef] [PubMed]
76. Fröhling, A.; Durek, J.; Schnabel, U.; Ehlbeck, J.; Bolling, J.; Schlüter, O. Indirect plasma treatment of fresh pork: Decontamination efficiency and effects on quality attributes. *Innov. Food Sci. Emerg. Technol.* **2012**, *16*, 381–390. [CrossRef]
77. Rothrock, M.J.; Jr Zhuang, H.; Lawrence, K.C.; Bowker, B.C.; Gamble, G.R.; Hiett, K.L. In-package inactivation of pathogenic and spoilage bacteria associated with poultry using dielectric barrier discharge-cold plasma treatments. *Curr. Microbiol.* **2017**, *74*, 149–158. [CrossRef] [PubMed]
78. Shi, H.; Chen, Z.Z.; Chen, D.; Kan, J.Q. Sublethal injury and recovery of *Escherichia coli* O_{157}:H_7 and K-12 after exposure to lactic acid. *Food Control.* **2017**, *82*, 190–195. [CrossRef]

foods

MDPI

Article

Effect of High-Pressure Processing Pretreatment on the Textural Properties of Cooked *Nuovo Maratelli* Rice

Cristina Arroqui, Sandra Horvitz, María José Noriega, Idoya Fernández-Pan, Francisco C. Ibañez * and Paloma Vírseda

Institute for Sustainability & Food Chain Innovation—ISFOOD, Universidad Pública de Navarra, Campus Arrosadia s/n, 31006 Pamplona, Spain; cristina.arroqui@unavarra.es (C.A.); sandra.horvitz@unavarra.es (S.H.); nodo@unavarra.es (M.J.N.); idoya.fernandez@unavarra.es (I.F.-P.); virseda@unavarra.es (P.V.)

* Correspondence: pi@unavarra.es

Abstract: *Nuovo Maratelli*, a *japonica* rice with an intermediate amylose content, is suitable for *paella* (a traditional Spanish dish) due to its ability to withstand cooking and absorb flavors. In this study, high-pressure processing (HPP) at 400 and 600 MPa (10 min) was used as a pretreatment to improve the properties of rice cooked by either boiling or microwaving. The microstructure and pasting properties of unpressurized and pressurized rice were examined. Also, the cooking time and cooking kinetics were determined for each cooking method. Overall, the pasting properties of the rice were not impacted by the HPP treatments, but the typical polyhedral form of the rice starch granules was lost, especially at 600 MPa. Cooking times were reduced from 14 and 10 min for unpressurized samples to 12 and 8 min (400 MPa) and 8 and 6 min (600 MPa) for boiling- and microwave-cooked rice, respectively. The rice pretreated at 400 MPa (10 min) and microwaved (8 min) had lower hardness and adhesiveness, which was linked to the release of amylose during cooking. In summary, HPP could be an effective pretreatment for the improvement of the cooking and textural properties of *Nuovo Maratelli* rice, particularly when cooked by microwaving.

Keywords: amylose; pressurization; cooking method; microstructure; instrumental texture

1. Introduction

The production of rice in Europe may be considered low in comparison with world production. However, its use has a high socio-cultural value, and several Mediterranean countries have developed world-famous dishes based on rice. This is the case of *risotto* in Italy or *paella* in Spain. These dishes are traditionally prepared using *japonica* rice cultivars with a lower amylose content than *indica* rice cultivars, which in turn, contributes to their unique cooking and sensory properties [1]. In the preparation of *paella*, pearled (white-core) *japonica* rice cultivars are employed due to their capacity to withstand cooking and absorb flavors [2]. *Bomba* rice is the preferred cultivar for this dish due to its medium amylose content (18%); however, it presents a low-yielding variety, resulting in a higher cost. Consequently, other cultivars with a lower amylose content (14%) are utilized. Thus, in addition to selecting the most suitable cultivars, different processing technologies, like soaking or high-pressure processing (HPP), can be applied to enhance the culinary and cooking quality of rice [3]. Moreover, in comparison with heat treatments, HPP enhances the integrity of starch granules [4], delays starch retrogradation, which causes texture deterioration during storage [5], and decreases the temperature and enthalpy of grain gelatinization [6]. Nevertheless, research has demonstrated that the outcomes anticipated from the implementation of HPP are contingent upon the specific rice cultivar, particularly its amylose content. The impact of HPP on waxy (non-pearled) and *indica* rice cultivars with elevated amylose content has been extensively investigated. This encompasses starch retrogradation [4,5], which has been demonstrated to vary significantly as a function of the leached amylose content. Additionally, alterations in solvent retention capacity exhibited notable distinctions between floury and waxy rice cultivars [7].

The impact of HPP on the microstructure and textural properties of *japonica* rice varieties has been the focus of many studies. However, most of them were carried out on rice grains from oriental *japonica* cultivars [8–11]. To date, only one study has been carried out to evaluate the changes in the microstructure of starches from Western *japonica* rice grains after HPP [12]. Furthermore, textural changes in rice grains of these varieties have not yet been examined.

The hypothesis of this study is that an HPP pretreatment may improve the cooking and textural properties of a *japonica* cultivar with similar or slightly higher amylose content than *Bomba* rice. Therefore, the objective was to evaluate the impact of HPP pretreatment (400 and 600 MPa for 10 min) on the textural properties of *Nuovo Maratelli* rice cooked by two methods: boiling and microwaving.

2. Materials and Methods

2.1. Materials and Experimental Design

A medium-grain *japonica* rice cultivar (*Nuovo Maratelli*) was selected for the present study. Samples were supplied by a local company (Producción Arroz Navarra S.L, Arguedas, Spain) and were stored at ambient temperature in the dark until use.

A two-stage approach was used to fulfill the research objectives. Firstly, the optimal hydration level of the rice grain was determined by examining the soaking time. Secondly, the impact of HPP pretreatment on the rice and on the texture and cooking properties of microwave- and boiling-cooked rice was assessed (Figure 1).

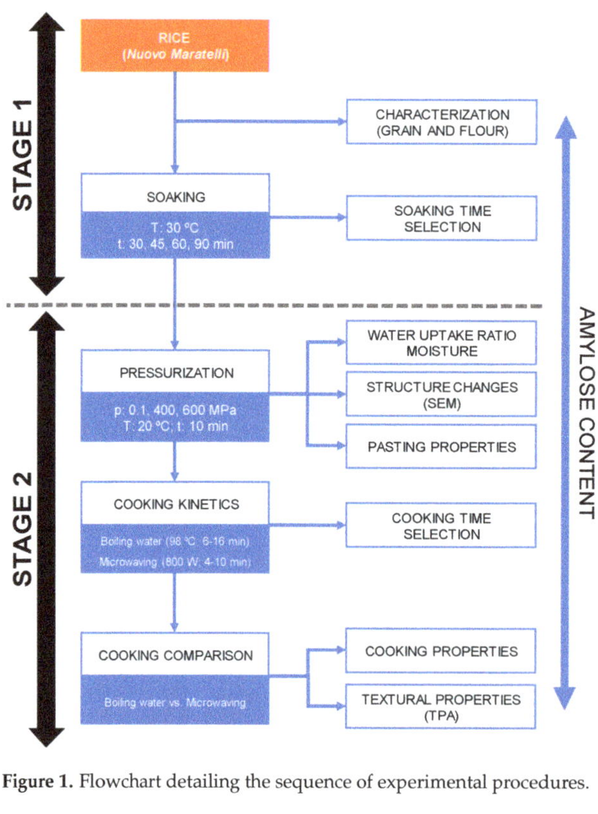

Figure 1. Flowchart detailing the sequence of experimental procedures.

2.2. Soaking Time and Water Uptake Ratio

The optimal soaking time for achieving the appropriate grain moisture was determined through the method of Rattanamechaiskul et al. [13], with some modifications. The samples (100 g of grain rice and 500 g of water) were placed into a polyamide/polyethylene (PA/PE) 20/70 plastic bag and soaked at 30 °C for 30, 45, 60, and 90 min. The samples were then air-dried for 2 min and weighed. The water uptake ratio of the rice samples (soaked and pressurized) was obtained by dividing the weight of the soaked rice by the initial weight. Once the samples reached a steady state, the optimal soaking time was established. Tests were carried out in triplicate.

2.3. High-Hydrostatic Pressure Treatment

A high-pressure unit of 10 L vessel capacity and 800 MPa of maximum pressure (Idus HPP Systems S.L.U., Noain, Spain) was employed for all the experimental assays. Samples of 100 g of rice and 500 mL of osmotized water were placed in a PA/PE 20/70 plastic bag and soaked at 30 °C for the time selected in Section 2.2. The samples were introduced in the HPP equipment at 25 $\pm$ 4 °C, and two different levels of pressure were applied, 400 and 600 MPa, for 10 min in both cases. The 10 min treatment time did not include come-up or depressurization times. The pressure-transmitting medium was water, the pressurization rate was 600 MPa/min, and the depressurization was almost instantaneous. The product temperature was registered during processing with a pressure-resistant data logger (LDT200 and Pt-100 thermometer, Leyro Instruments, Barcelona, Spain; Tempmate®-B5 data button, Heilbronn, Germany). The collected dataset included both pressure and temperature readings. After treatment, the water was removed, and they were left to dry in air for 2 min before being stored for a maximum of 1 day until further analysis.

2.4. Moisture and Amylose Content Determination

The moisture content was estimated following the reference method [14]. For amylose determination, soaked and cooked rice samples were first dried at 40 °C to reduce moisture up to 6–12%. Treated and raw samples were ground for 1 min with an electric grinder (Moulinette, Moulinex, Barcelona, Spain) and passed through a 150-μm sieve to ensure homogeneous particle size.

The apparent amylose content (AAC) was quantified using the method suggested by Juliano et al. [15]. A standard curve for the estimation of AAC (10 to 40 mg/100 mL of potato amylose) was used, and the absorbance was measured at 720 nm using a spectrophotometer (UH5300 UV/VIS Spectrophotometer, HITACHI Corp., Tokyo, Japan). The AAC of rice samples were expressed as mg/100 g dry matter (DM).

2.5. Microstructure by Scanning Electron Microscopy

Changes in the microstructure of pressurized rice samples were examined using scanning electron microscopy (SEM) imaging, according to the method described by Xu et al. [16]. Freeze-dried rice grain samples were mounted onto a SEM plate using conductive carbon tabs and Pt-coated via sputtering of 7 nm. The three-dimensional network microstructure of the rice grain samples was then observed and photographed using a JSM-5610-LV scanning electron microscope (JEOL Ltd., Tokyo, Japan) at a 20-kV acceleration voltage. For image analysis, the ImageJ2 v.1.54f image analysis software was used on the Java 8 platform [17]. Particle circularity was quantified with the method of Jordan et al. [18]. A circularity value of 1 is indicative of a perfect circle, while a value close to zero (0) suggests an elongated shape.

2.6. Pasting Properties

Flour samples (about 10% of moisture) were prepared as for the AAC determination (Section 2.3). Pasting properties of the rice flours were evaluated in a rotational Rheometer HAAKE RotoVisco 1 (Thermo Scientific, Karlsruhe, Germany) equipped with a cup Z43S and a starch blade FL2B paddle-shaped rotor with 2 blades using the method described by

Kaur et al. [19] with slight modifications. Heating and cooling cycles were programmed. Each rice flour suspension (6 g/50 g) was held at 50 °C for 6 min, heated to 95 °C at 2.25 °C/min, held at 95 °C for 400 s, before cooling from 95 to 50 °C at 3 °C/min. The rotating speed of the paddle was set at 160 rpm during the measurements, except 960 rpm during the first 10 s. Device management and data acquisition were performed using HAAKE RheoWin 4 software (Thermo Scientific, Karlsruhe, Germany). Outputs were expressed as Pa·s.

2.7. Cooking Kinetics and Cooking Time Selection

To obtain the rice-cooking kinetics by boiling, 100 g of rice was submerged in 500 mL of osmotizedwater in a beaker and boiled (98 ± 1 °C) using a heating plate (Combiplac, JP Selecta, Barcelona, Spain). Subsamples of 15 g of rice were taken at two-minute intervals between minutes 6 and 14 and were placed on qualitative filter paper for 5 min. This process allowed for the removal of surface moisture from the rice grains.

For microwave-cooking kinetics, 60 g of raw rice was mixed with 300 mL of osmotized water and cooked using a domestic microwave oven (Taurus Luxus Grill 21 L, Taurus Group, Oliana, Spain) at 800 W. At two-minute intervals between 2 and 10 min, samples of 5 g of rice were collected for analysis and placed on filter paper, as previously described. The cooking time and textural properties for both cooking methods were determined by an instrumental analysis of the texture and a visual analysis of the grains. For each extraction time, samples were subjected to the texture profile analysis (TPA) as described in Section 2.9. Additionally, two grains of rice placed between two plates were crushed using a 5 kg weight load. The grains were analyzed under a magnifying glass, and the rice was identified as cooked when approximately less than 10% of the grains showed an opaque center [20].

2.8. Cooking Properties

Once the cooking time was determined for each cooking method, samples of cooked rice were prepared in accordance with the method outlined in Section 2.7, and the cooking properties were evaluated. Elongation ratio, gruel solid loss, water uptake ratio, and expansion volume were quantified according to the methods described by Bhattacharya et al. [21].

2.9. Instrumental Analysis of Texture Profile

The texture profile analysis (TPA) was performed with a texture analyzer TA-XT Plus (Stable Micro System Ltd., Surrey, UK), equipped with a 50 kg load cell and an aluminum compression platen (75 mm Ø, code P/75). The Exponent Lite v.6.1. software (Stable Micro Systems Ltd., Surrey, UK) was used. A double-compression test was programmed for TPA. A rice sample of 2 g (room temperature) was placed carefully on the base platform, under the center of probe, and compressed up to 90% at a test rate of 1 mm/s. Pre-test and post-test rates were 5 mm/s. Before analysis, the analyzer was calibrated using a 1 kg weight. Hardness (N), adhesiveness (N s), cohesiveness, resilience (%), springiness (%), gumminess (N), and chewiness (N) were evaluated applying the method described by Lyon et al. [22]. TPA tests were performed in triplicate.

2.10. Statistical Analysis

A one-way ANOVA was applied to data of amylose content and pasting properties. The HPP pretreatment × cooking time and HPP pretreatment × cooking method effects were evaluated using a two-way ANOVA. Differences between pairwise of means were tested via a Tukey's test (95% significance level). The textural properties that were most affected by the dual treatments (HPP pretreatment and cooking method) applied to the rice in stage 2 were determined through a discriminant analysis. All the statistical analyses were carried out using SPSS Statistics for Windows, vers. 28.0 (IBM Corp., Armonk, NY, USA).

3. Results and Discussion

3.1. HPP

The pressure–time relation and the internal temperature of the sample recorded during rice HPP at 400 and 600 MPa are shown in Figure S1. The temperature of the filling water ranged from 25 to 30 °C during the treatments. The temperature of the product never exceeded 32 °C, with thermal leaps in the range of 8 to 10 °C. This means and ensures that the pressure treatment did not cause any heat-induced structural changes in the rice and starch and that any change in the properties, characteristics, and behavior of the samples should be attributed to the pressure.

3.2. Raw Grain Characterization and Water Uptake

The raw samples of *Nuovo Maratelli* rice have a length of 6.1 ± 0.4 mm and a length/width ratio of 2.12 ± 0.19, belonging to the medium grain classification [23].

The moisture content of *Nuovo Maratelli* samples was found to be 11.78 ± 0.53%, with an AAC of 22.61 ± 0.83% (25.31 ± 0.93 mg/100 g DM). Based on these findings, the samples can be classified as intermediate amylose rice [24]. Commonly, amylose content was positively related with hardness and negatively with stickiness, so cooked-rice hardness and stickiness were strongly and inversely correlated [25]. Although this *japonica* rice cultivar with intermediate content of amylose could be used to prepare *paella*, as HPP could reduce firmness and stickiness, the application of this technology could be of special interest to improve eating quality.

The soaking time at 30 °C necessary to reach the minimum recommended moisture content (from 25 to 35%) was determined to be 45 min.

3.3. Effect of Pressurization on Water Uptake Ratio

The water uptake ratios in the soaked and pressurized samples were 1.27 ± 0.00, 1.33 ± 0.02, and 1.69 ± 0.01 g treated/g initial rice for 0.1, 400, and 600 MPa, respectively. As a consequence, when comparing atmospheric pressure soaking to HPP soaking, the moisture content (g water/100 g) of normal rice grains significantly rose from 39.5 ± 0.45 (unpressurized samples) to 42.13 ± 0.17 (400 MPa) and 39.07 ± 1.51 (600 MPa). This is in accordance with the results described by Wu et al. [26], who found water absorption rates around 1.28 g/100 g in *Wuchang* cultivar, a non-waxy rice, after 400 MPa soaking for 15 min. The rise in moisture levels indicated that water molecules could effectively penetrate the peripheral regions of starch granules in rice grains under conditions of HPP [27]. The increase in weight with rising pressure levels did not correlate with a corresponding increase in moisture content. These findings suggest that solids retention was greater in samples subjected to 600 MPa, which is attributable to gelatinization processes that take place at these pressures [28,29].

3.4. Effect of Pressurization on Rice Grain Microstructure

Figure 2 depicts the alterations in the rice grain microstructure according to pressure treatments, as observed through SEM. The microstructure of rice grains was affected by HPP treatment. Moderate pressurization (400 MPa, 10 min) enhanced hydration of starch granules, making the cavities more homogenous (Figure 2(a1,a2)). The characteristic polyhedral forms of the starch of the raw rice grain (Figure 2(b1)) began to be lost after the application of 400 MPa (Figure 2(b2)). At 600 MPa, the microstructure underwent reorganization as a result of compaction and partial aggregation of the starch grains (Figure 2(b3)). Almeida et al. [12] observed similar changes in *japonica* rice starches treated at 400 and 600 MPa for 30 min. Seo et al. [11] pressurized *japonica* rice flours for 10 min and found that the starch structure was densest at 400 MPa and collapsed at 600 MPa. The pressurization favors the entry of water molecules into the granules, where they form new hydrogen bonds with the hydroxyl groups of the starch that replace the original intramolecular hydrogen bonds, leading to the destruction of the starch structure [30]. The intermediate amount of amylose could contribute to the stabilization of granules and the prevention of disintegrating, as found by Hu et al. [4] in non-waxy starch granules that retained their integrity, which is different from waxy starch granules that disintegrated after HPP at 600 MPa.

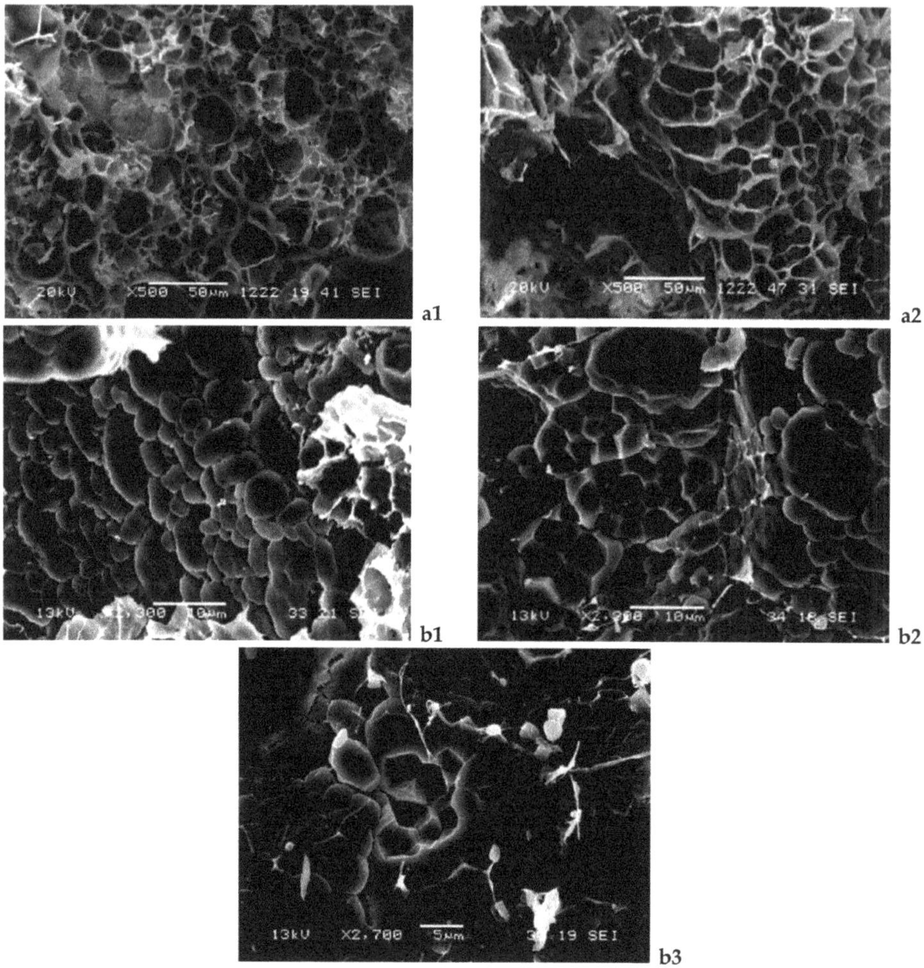

Figure 2. Scanning electron microscopy images of samples subjected to different treatments ((**a1**): raw rice; (**a2**): 400 MPa, ×500; (**b1**): raw rice; (**b2**): 400 MPa; (**b3**): 600 MPa, ×2300–2500).

3.5. Effect of Pressurization on Pasting Properties of Rice Flours

The pasting properties of rice starch are influenced by granule swelling, amylose leaching, starch crystallinity, amylose content, and branch chain-length distribution of amylopectin. The increase in viscosity with rising temperature can be attributed to the removal of water from the exuded amylose by the granules as they undergo swelling [31]. Rice with a high–intermediate amylose content typically exhibits elevated values for pasting temperature (PT), peak viscosity (PV), minimum viscosity (MV), and final viscosity (FV). Additionally, it displays a low breakdown index (BI), indicating stability during the cooking process and a high setback, which means that it retrogrades easily [32,33].

The results of the assays with the *Nuovo Maratelli* rice showed no differences in any of the pasting properties in all the control and HPP-treated samples. However, the HPP-treated samples exhibited a tendency to increase PV at 400 MPa and to decrease this pasting parameter at 600 MPa, in comparison to the control (Table 1). These results are aligned with those previously reported by Li et al. [34], who applied pressures of 120–600 MPa to rice starch.

Table 1. Pasting properties of *Nuovo Maratelli* rice flours as a function of pressurization treatment (10 min). Data are expressed as mean ± standard deviation (n = 3).

Parameter	Control	400 MPa	600 MPa	F Ratio	p-Value
PT	65.60 ± 1.82	66.30 ± 0.37	64.30 ± 1.29	2.752	0.142
PV (η/Pas)	24.10 ± 0.65 ^{ab}	27.00 ± 2.32 ^{a}	23.10 ± 1.11 ^{b}	5.248	0.040
MV (η/Pas)	15.90 ± 0.95	18.30 ± 2.05	15.10 ± 1.17	3.855	0.084
FV (η/Pas)	33.90 ± 0.87	35.90 ± 2.47	34.00 ± 1.27	1.349	0.358
BI	7.90 ± 0.89	7.20 ± 1.04	7.40 ± 0.38	0.580	0.589

PT: pasting temperature; PV: viscosity maximum; MV: viscosity minimum; FV: final viscosity; BI: breakdown index. Different superscripts in each row indicate significant differences between means via Tukey's test.

Modifications were observed in the microstructure of the starch granule at 400 MPa (Figure 2), this pressure being the threshold for rice grain gelatinization for the *japonica* cultivar *Chunyou* 84 [10]. All of this could explain the changes in the pasting properties of the 400 MPa pressurized sample compared to the control. On the contrary, under 600 MPa, the compaction and partial gelatinization of rice starch does not lead to changes in pasting properties respect to control.

In whole rice grains, the physicochemical modifications induced by HPP treatments which affect the pasting properties are less pronounced than in rice flour and rice starch; so, the effect on the pasting properties is therefore more challenging to assess.

3.6. Cooking Behavior: Cooking-Time Selection and Texture Changes

In this study, the cooking time was a prerequisite for evaluating the texture of the cooked rice. The changes in instrumental texture parameters as a function of HPP pretreatment and cooking time are shown in Figures 3 and 4. Data for 600 MPa treated rice boiled for 14 min were not included because the samples were overcooked and were therefore not suitable for instrumental analysis. During cooking, all texture parameters decreased with time, except for adhesiveness, which increased. Overall, HPP treatment increased the cohesiveness and reduced the adhesiveness of the rice grains irrespective of the cooking method. However, the results of the texture kinetics showed that the effect of the other texture parameters depended on the cooking method applied. These effects were generally more pronounced in the 400 MPa than in the 600 MPa treated samples.

Compared to the unpressurized samples, the HPP samples showed a notable reduction in adhesiveness. In addition to changes in the amylose content in the cooked samples, which is negatively related to the stickiness, the structure of the starch affects this parameter [25], so the changes in adhesiveness could be explained by changes in the rice starch during pressurization. As the instrumentally measured adhesiveness is indicative of the stickiness between the rice grain and the TPA probe, it can be deduced that the leached materials (amylose) from the cooking process are a determining factor in the stickiness between the rice grains.

When boil cooking was applied (Figure 3), the pressurized samples had higher values for hardness, cohesiveness, chewiness, gumminess, and resilience than their non-pressurized counterparts at the same cooking time, especially when 400 MPa was used. In microwave-cooked samples, in contrast to boiling-cooked samples, HPP pretreatment reduced the hardness.

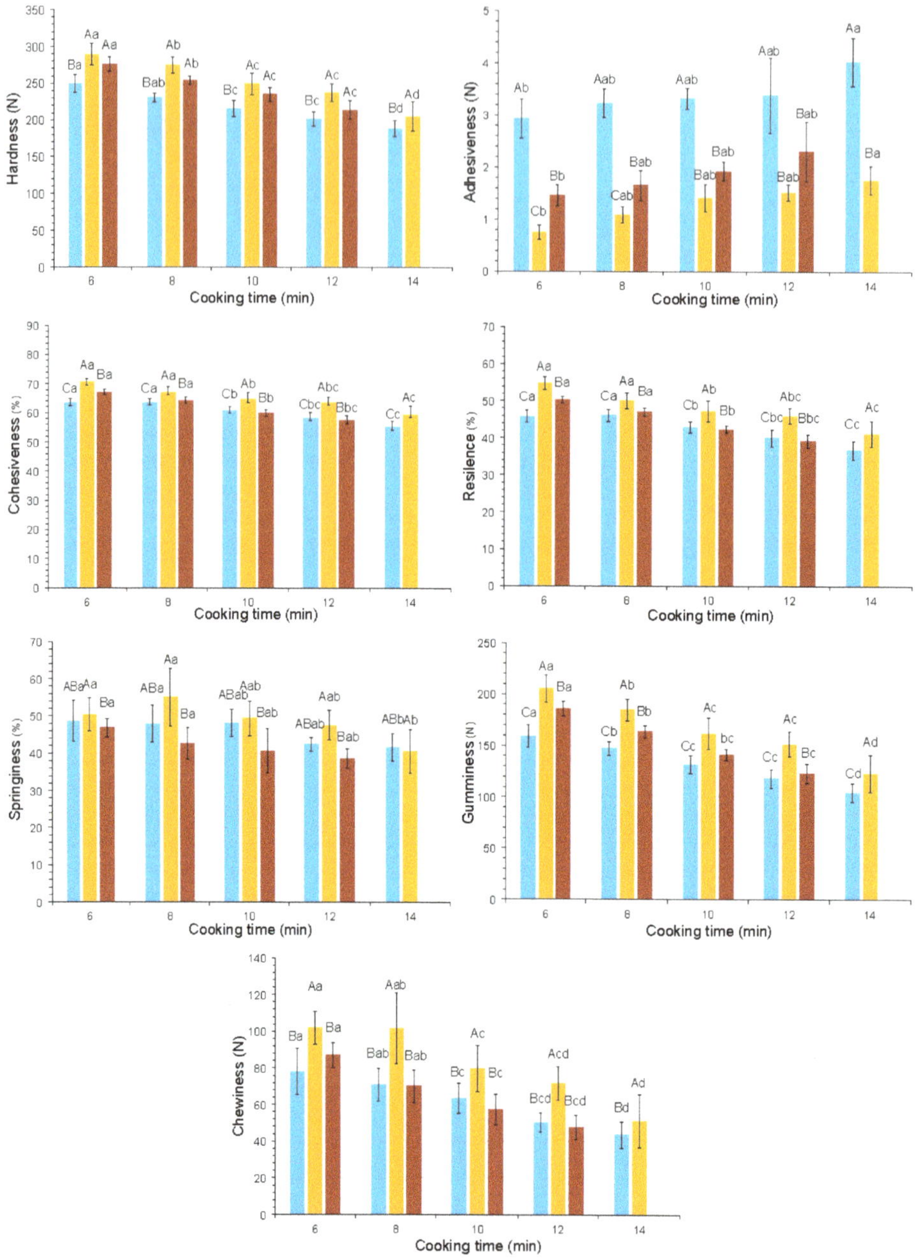

Figure 3. Kinetics of textural parameters measured by instrumental TPA, to establish the cooking time in rice samples treated by boil cooking (■: control; ■: 400 MPa; ■: 600 MPa). Note: Adhesiveness data are expressed as absolute values. The error bars indicate the standard deviation. Uppercase letters compare HPP pretreatments, and lowercase letters compare cooking times. Different letters represent significant differences via Tukey's test (n = 3).

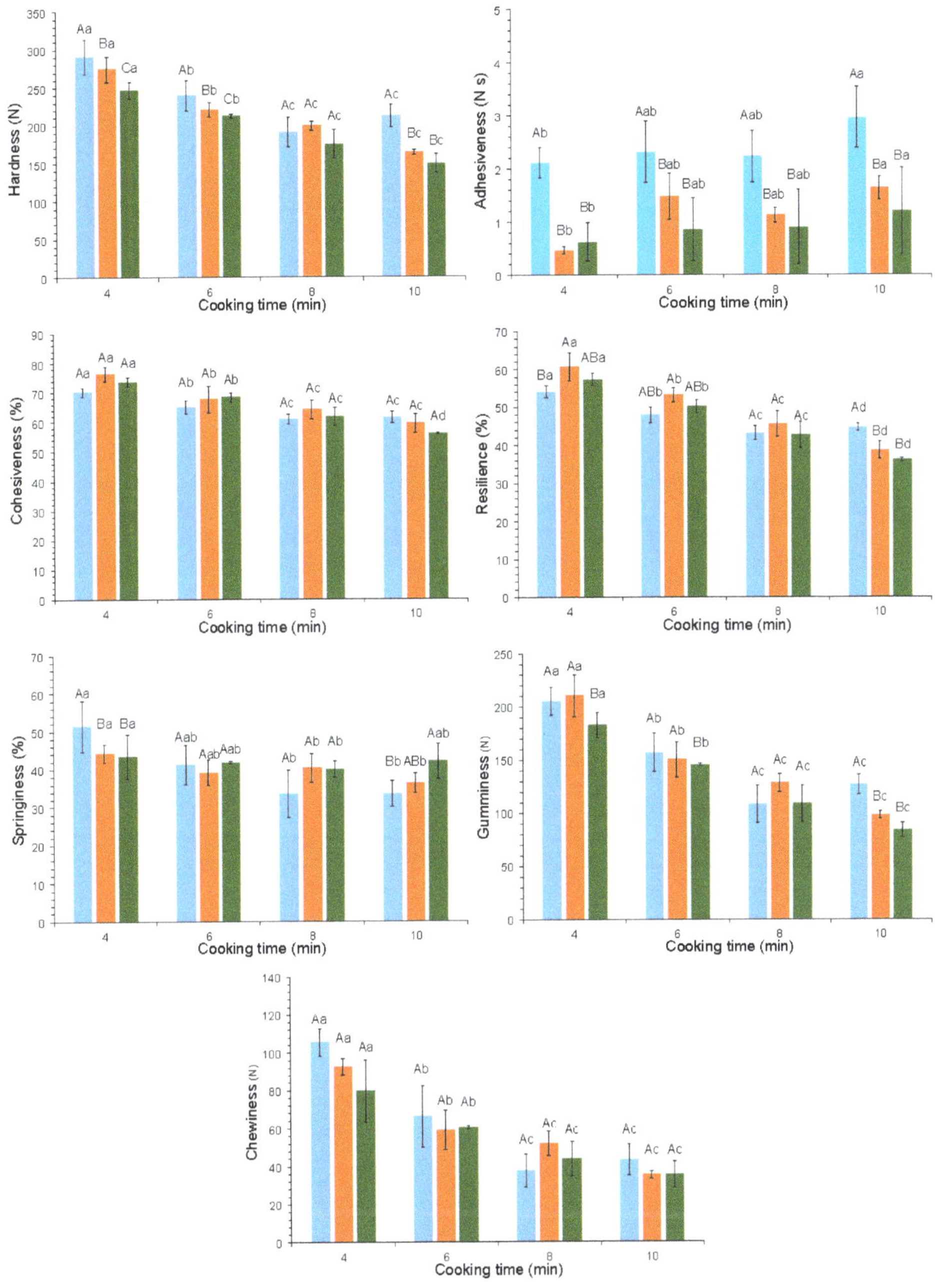

Figure 4. Kinetics of textural parameters measured by instrumental TPA to establish the cooking time in rice samples treated using microwave cooking (■: control; ■: 400 MPa; ■: 600 MPa). Note: Adhesiveness data are expressed as absolute values. The error bars indicate the standard deviation. Uppercase letters compare HPP pretreatments, and lowercase letters compare cooking times. Different letters represent significant differences via Tukey's test (n = 3).

The more uniform structure shown in the 400 MPa pressurized rice than in control (Figure 2(a1,a2)) could be responsible for these changes. Tian et al. [27] also observed a more homogeneous network structure, resulting in lower hardness when normal and waxy rice were soaked at 400 MPa. In addition, the cooked rice also had smaller holes and a more resistant granule structure. Yu et al. [5] also reported that the hardness of HPP brown rice was reduced under pressure of 300 MPa for 10 min.

The cooking time was reduced when a pretreatment of 400 or 600 MPa was applied prior to cooking compared to the unpressurized samples. According to Priestley [35], fast-cooking rice is directly related to the water absorbed in the previous soaking. Since HPP favors water uptake through diffusion, it is expected that cooking time will be reduced. Thus, the application of intermediate pressures (around 400 MPa) induces starch gelatinization and may reduce cooking time by facilitating water diffusion into the rice grain [36,37]. In addition, when the pressure treatment is increased to 600 MPa, rice grain gelatinization occurs depending on the degree of the milling and/or treatment time [10], thus reducing cooking time.

Once the textural behavior of the samples subjected to the different treatments had been determined, it was necessary to fix the cooking times for rice processing in order to continue the research. According to the method described in Section 2.7, these times were therefore set at 14 and 10 min for the control samples cooked by boiling and microwave, respectively, 12 and 8 min for the 400 MPa HPP pretreated samples, and finally, 8 and 6 min for the 600 MPa HPP pre-treated samples. It should be noted that, although Billiris et al. [38] proposed instrumental texture as a method to establish the criteria for "well cooked rice", they only quantified the maximum compression force by means of a one-cycle test.

3.7. *Effect of HPP Pretreatment on Cooking Properties*

To enable a comparison of the effect of the cooking method, the experiment was replicated with the cooking times selected in the previous section. The cooking properties of the raw rice samples (boiled and microwave cooked) and those treated with HPP are shown in Table 2. The soaking and pressurization processes caused water to penetrate the grain, increasing the moisture content of the soaked rice grains pressurized at 400–600 MPa, as indicated in Section 3.1. This pretreatment facilitated subsequent cooking by reducing cooking time, water uptake, and changes in grain physical dimensions. In addition, the cooking properties and impact of the HPP treatment differed depending on the cooking method. Thus, shorter cooking times were required for microwave cooking, although the application of the higher pressures reduced the difference between both methods. The partial gelatinization of starch by pressure at 600 MPa could explain this phenomenon [10].

Table 2. Cooking properties of *Nuovo Maratelli* rice according to pressurization (10 min, 20 °C) and cooking treatments.

	Control		400 MPa		600 MPa	
Parameter	**Boiling**	**Microwaving**	**Boiling**	**Microwaving**	**Boiling**	**Microwaving**
Cooking time (min)	14 ± 0	10 ± 0	12 ± 0	8 ± 0	8 ± 0	6 ± 0
Moisture (g/100 g sample)	67.58 ± 0.06 ^{a}	53.33 ± 0.53 ^{d}	62.02 ± 0.91 ^{b}	58.92 ± 0.35 ^{c}	62.69 ± 0.25 ^{b}	49.43 ± 0.77 ^{e}
Water uptake ratio	3.21 ± 0.21 ^{a}	2.42 ± 0.31 ^{a}	1.90 ± 0.09 ^{a}	1.36 ± 0.05 ^{b}	1.39 ± 0.05 ^{b}	1.28 ± 0.03 ^{c}
Expansion volume	2.75 ± 0.28 ^{a}	1.78 ± 0.31 ^{b}	1.57 ± 0.09 ^{bc}	1.36 ± 0.05 ^{cd}	1.15 ± 0.04 ^{cd}	1.12 ± 0.05 ^{d}
Grain elongation ratio	68.50 ± 10.93 ^{a}	60.48 ± 2.33 ^{ab}	54.64 ± 2.51 ^{bc}	46.99 ± 0.95 ^{cd}	42.62 ± 1.64 ^{cd}	41.59 ± 0.64 ^{d}
Gruel solid loss (%)	3.83 ± 0.67 ^{a}	1.28 ± 0.13 ^{c}	3.81 ± 0.04 ^{a}	1.77 ± 0.12 ^{bc}	2.98 ± 0.85 ^{ab}	2.32 ± 0.20 ^{bc}

Different superscripts in each row indicate significant differences between means via Tukey's test.

As water uptake occurred during HPP and cooking time was reduced, lower ratios of these parameters were detected in the treated samples, especially at the higher pressure. Water absorption facilitates the incorporation of flavors into rice dishes, which is a crucial aspect from a culinary point of view. Similarly, the grain elongation ratio and the volume expansion after cooking in the HPP-treated samples were consequently smaller than those observed in the cooked non-pressurized samples. Furthermore, a lower final moisture content of the HPP-cooked samples was found, as already found in the uncooked pressurized samples, especially when cooked in the microwave. The physical changes in the grain dimensions during cooking were less pronounced at 600 MPa than at 400 MPa, probably due to a shorter cooking time and a partial pressure gelatinization before heat cooking. The compressibility of starch suspensions under pressure showed a reduction in total volume associated with gelatinization [39].

Furthermore, the effect of HPP treatments on the gruel solids loss differed depending on the cooking method. Microwave cooking increased the gruel solids loss, whereas boiling had no effect.

In terms of cooking properties, boiled rice showed superior convenience and quality compared to microwaved rice, which showed lower values for cooking properties, regardless of the pressure applied.

3.8. Effect of HPP Pretreatment on Textural Properties of Cooked Rice

The outputs of the instrumental texture evaluation for boiled and microwave-cooked rice grains are summarized in Table 3. According to the previous section, the cooking time was established as a function of the pressure and cooking method. The statistical analysis revealed that all parameters were significantly different between the unpressurized and pressurized samples, except for springiness (boiling and microwave cooking) and for resilience and chewiness (microwave cooking).

Table 3. Instrumental texture of unpressurized (control) and pressurized rice (10 min, 20 °C), cooked by the boiling or microwaving method (cooking time in Table 2). Data are expressed as mean ± standard deviations (n = 3).

	Control		400 MPa		600 MPa	
Parameter	Boiling	Microwaving	Boiling	Microwaving	Boiling	Microwaving
Hardness (N)	188.8 ± 11.0 ^{c}	227.9 ± 16.1 ^{ab}	237.8 ± 12.3 ^{ab}	189.3 ± 11.6 ^{c}	254.5 ± 5.5 ^{a}	210.4 ± 7.1 ^{bc}
Adhesiveness (N s)	−4.0 ± 0.5 ^{c}	−3.4 ± 0.7 ^{c}	−1.5 ± 0.2 ^{b}	−1.1 ± 0.2 ^{ab}	−1.6 ± 0.3 ^{b}	−0.4 ± 0.1 ^{a}
Cohesiveness (%)	55.2 ± 2.0 ^{c}	62.2 ± 1.4 ^{b}	63.6 ± 2.0 ^{ab}	63.6± 4.1 ^{ab}	64.2 ± 1.2 ^{ab}	69.5 ± 1.0 ^{a}
Resiliency (%)	36.73 ± 2.5 ^{b}	45.0 ± 1.1 ^{a}	45.9 ± 2.1 ^{a}	45.3 ± 4.3 ^{a}	47.0 ± 1.2 ^{a}	51.5 ± 1.5 ^{a}
Springiness (%)	41.7 ± 3.7	36.1 ± 3.4	47.5 ± 4.0	37.8 ± 2.1	42.6 ± 4.3	42.0 ± 6.7
Gumminess (N)	104.4 ± 9.0 ^{c}	127.6 ± 9.1 ^{bc}	151.4 ± 12.2 ^{ab}	126.2 ± 15.2 ^{bc}	163.5 ± 5.8 ^{ab}	146.1 ± 6.7 ^{a}
Chewiness (N)	43.7 ± 7.3 ^{b}	47.7 ± 7.0 ^{b}	72.0 ± 9.3 ^{ab}	46.0 ± 4.8 ^{b}	70.0 ± 8.9 ^{a}	61.2 ± 5.1 ^{ab}

Different superscripts in each row indicate significant differences between means via Tukey's test.

Regarding the hardness, the lowest value was observed in the unpressurized-and boil-cooked sample and the pretreated at 400 MPa plus microwave-cooked sample, whereas the highest was found in the pretreated at 600 MPa plus boiled sample. The lower hardness in pressurized-plus microwaved samples could be explained by a lower amylose leaching during cooking, which would form a film coating the rice grains [40].

Boiled pressurized samples displayed a higher hardness than control samples, while microwave cooked samples showed a reduced hardness, to a higher extent at 400 MPa pressure. A reduction in hardness was also detected for the oriental *japonica* rice treated at up to 300–500 MPa for 10 min [9]. Boluda-Aguilar et al. [41] found a decrease in instrumental hardness of *Jasmine* rice (an *indica* cultivar) heated by microwave in samples subjected to 300–400 MPa for 2–3 min, in comparison to untreated samples.

The pressurized samples, particularly those microwaved, showed lower adhesiveness than the unpressurized ones. Therefore, HPP pretreatment shortened the cooking time and modified the grain microstructure, affecting the adhesiveness. All the pressurized samples had more cohesiveness than the unpressurized and boiled sample, which showed the lowest value. A more compact network might be formed in rice under the high pressure that increases cohesiveness [27,42]. Similarly, differences in resilience only were found between the unpressurized and boiled sample and the other studied samples In the case of gumminess and chewiness, lower values were found in unpressurized and in 400 MPa treated and microwaved samples, whereas the highest value was observed in the 600 MPa pressurized and boiled sample. Moderate HPP treatment (up to 500 MPa) can cause the starch granules to be more firmly bound to the protein, thus decreasing the hardness of the rice and improving the gumminess and chewiness [26].

The discriminant analysis performed to the instrumental texture parameters yielded two functions that collectively explained 99.2% of the sample variance. Discriminant function 1 (87.5%) was closely associated with hardness, gumminess, and resilience. Discriminant function 2 (11.8%) was strongly related to adhesiveness. Figure 5 depicts the samples based on the two discriminant functions. The unpressurized samples (control) are shown in the first quadrant of the graph. The pressurized and boiled samples are grouped in the second quadrant. The pressurized and microwaved samples are collected in the remaining quadrants.

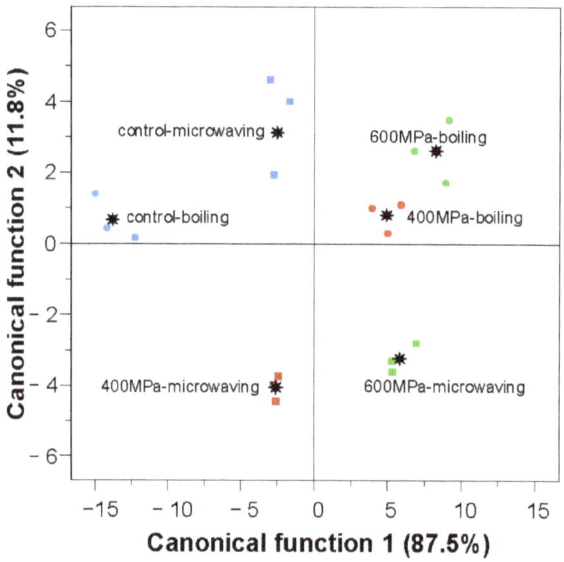

Figure 5. Score plot of the canonical discriminant functions for the instrumental texture evaluated in rice samples (✱: group centroids). blue: unpressurized; red: treated at 400 MPa; green: treated at 600 MPa; circle: boil-cooked; square: microwave-cooked.

In comparison to the unpressurized and boiled samples, the pressurized at 400 MPa and microwave-cooked samples showed comparable hardness values but had decreased adhesiveness. This finding suggests that microwave cooking had the greatest impact on the textural properties of pressurized samples, as hardness and stickiness are the key to improve and manage eating quality for rice [25].

3.9. Effect of Pressurizing and Cooking on Amylose Content

Figure 6 shows the modifications in the AAC of rice samples soaked and subjected to the different treatments (pressurization and cooking). The highest values were recorded in the samples pressurized at 600 MPa, regardless of the cooking method (27.47–28.06 mg/100 g DM). The samples treated at 400 MPa displayed intermediate values (25.84–27.21 mg/100 g DM). Instead, soaked and microwaved samples showed the lowest values (23.49 mg/100 g DM).

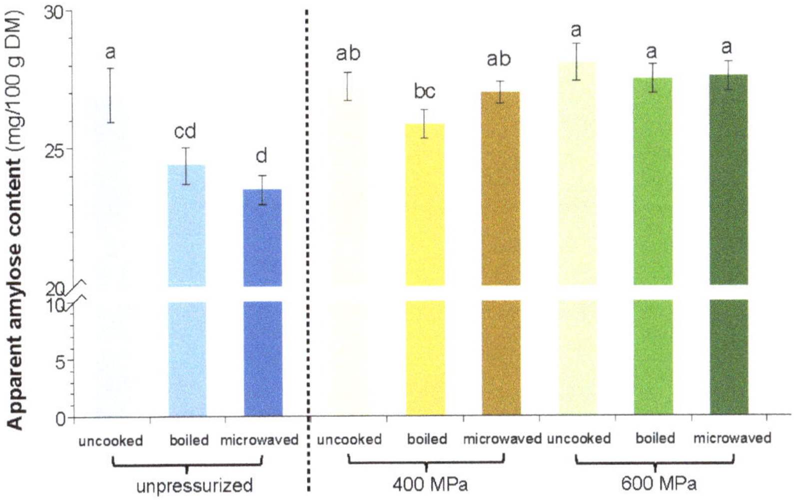

Figure 6. Changes in apparent amylose content in soaked-rice samples as a function of the treatments applied. The dashed vertical line separates unpressurized samples (**left**) from pressurized samples (**right**). Note the broken axis to highlight differences in treatments. Different letters indicate significant differences via Tukey's test ($n = 3$).

Regarding the uncooked sample, the HPP pretreatment did not significantly impact the AAC. On the contrary, the cooked samples exhibited elevated AAC contents when subjected to HPP pretreatment, especially when microwave cooking was used and when the pressure applied was 600 MPa. Therefore, it can be concluded that pressurization would effectively reduce the leaching of amylose during the cooking process. This was also observed in pressurized normal and black rice grains [4,42], which was attributed to the intact starch granules presented in the HPP treatment that make amylose leaching difficult [4].

4. Conclusions

This study evaluates, for the first time, the effect of HPP pretreatment on the properties of *Nuovo Maratelli* rice cooked using two different cooking methods. The optimal cooking time was determined through the use of the TPA test. Based on the results obtained, pretreatments at 400 and 600 MPa for 10 min could be used to modify the cooking properties of the rice. In particular, the HPP treatment reduces cooking times and improves some of the textural properties that are critical for the eating quality of this rice cultivar. However, the shorten of the cooking time and the subsequent modifications depend on both the pressure level and the cooking method used. The application of the highest-pressure level facilitates the incorporation of water into the grain, thereby increasing its availability and reducing the cooking time by up to 6 min when microwaved. Additionally, the HPP pretreatment causes changes in the grain microstructure, which are already observed at 400 MPa and become more pronounced at 600 MPa, when a compact network was formed. Most of the textural parameters were modified as a result, but to improve the eating quality of the cooked rice, 400 MPa before microwave cooking could be selected, as it achieves a similar hardness to that of unpressurized-boiling-cooked rice and reduces adhesiveness. The results obtained in this work suggest that a pretreatment of pressurization could improve the eating quality and convenience of *Nuovo Maratelli* rice to prepare traditional rice dishes such as *paella*, making it more acceptable to consumers.

Supplementary Materials: The following supporting information can be downloaded at: https://www.mdpi.com/article/10.3390/foods13244052/s1, Figure S1: Pressure and temperature profiles recorded in the IDUS HPP 10L unit from rice samples: processing at 400 MPa (a); processing at 600 MPa (b).

Author Contributions: Conceptualization, P.V. and C.A.; methodology, M.J.N. and S.H.; formal analysis, F.C.I.; investigation, S.H. and I.F.-P.; resources, S.H. and I.F.-P.; writing—original draft preparation, M.J.N. and CA.; writing—review and editing, I.F.-P., S.H., and F.C.I.; supervision, P.V. All authors have read and agreed to the published version of the manuscript.

Funding: This research received no external funding.

Data Availability Statement: The original contributions presented in the study are included in the article and Supplementary Materials, further inquiries can be directed to the corresponding author.

Conflicts of Interest: The authors declare no conflicts of interest.

References

1. Mefleh, M. Cereals of the Mediterranean region: Their origin, breeding history and grain quality traits. In *Cereal-Based Foodstuffs: The Backbone of Mediterranean Cuisine*; Boukid, F., Ed.; Springer International Publishing: Cham, Switzerland, 2021; pp. 1–18, ISBN 978-3-030-69228-5.
2. Barea-Ramos, J.D.; Santos, J.P.; Lozano, J.; Rodríguez, M.J.; Montero-Fernández, I.; Martín-Vertedor, D. Detection of aroma profile in Spanish rice paella during socarrat formation by electronic nose and sensory panel. *Chemosensors* **2023**, *11*, 342. [CrossRef]
3. Yu, L.; Turner, M.S.; Fitzgerald, M.; Stokes, J.R.; Witt, T. Review of the effects of different processing technologies on cooked and convenience rice quality. *Trends Food Sci. Technol.* **2017**, *59*, 124–138. [CrossRef]
4. Hu, X.; Xu, X.; Jin, Z.; Tian, Y.; Bai, Y.; Xie, Z. Retrogradation properties of rice starch gelatinized by heat and high hydrostatic pressure (HHP). *J. Food Eng.* **2011**, *106*, 262–266. [CrossRef]
5. Yu, Y.; Pan, F.; Ramaswamy, H.S.; Zhu, S.; Yu, L.; Zhang, Q. Effect of soaking and single/two cycle high pressure treatment on water absorption, color, morphology and cooked texture of brown rice. *J. Food Sci. Technol.* **2017**, *54*, 1655–1664. [CrossRef]
6. Liu, Y.; Selomulyo, V.O.; Zhou, W. Effect of high pressure on some physicochemical properties of several native starches. *J. Food Eng.* **2008**, *88*, 126–136. [CrossRef]
7. Ahmed, J.; Mulla, M.Z.; Arfat, Y.A.; Kumar, V. Effects of high-pressure treatment on functional, rheological, thermal and structural properties of Thai jasmine rice flour dispersion. *J. Food Process. Preserv.* **2017**, *41*, e12964. [CrossRef]
8. Xia, Q.; Wang, L.; Li, Y. Exploring high hydrostatic pressure-mediated germination to enhance functionality and quality attributes of wholegrain brown rice. *Food Chem.* **2018**, *249*, 104–110. [CrossRef]
9. Xia, Q.; Wang, L.; Yu, W.; Li, Y. Investigating the influence of selected texture-improved pretreatment techniques on storage stability of wholegrain brown rice: Involvement of processing-induced mineral changes with lipid degradation. *Food Res. Int.* **2017**, *99*, 510–521. [CrossRef] [PubMed]
10. Zhu, S.M.; Hu, F.F.; Ramaswamy, H.S.; Yu, Y.; Yu, L.; Zhang, Q.T. Effect of high pressure treatment and degree of milling on gelatinization and structural properties of brown rice. *Food Bioprocess Technol.* **2016**, *9*, 1844–1853. [CrossRef]
11. Seo, J.-H.; Jo, Y.-J.; Lee, Y.-R.; Lee, J.; Jeong, H.-S. Physicochemical properties of soft and hard-type rice flour according to moisture content and high hydrostatic pressure treatment. *Foods* **2023**, *12*, 227. [CrossRef] [PubMed]
12. Almeida, R.L.J.; Santos, N.C.; Feitoza, J.V.F.; Ribeiro, V.H.D.A.; Silva, V.M.D.A.; de Figueiredo, M.J.; Ribeiro, C.A.C.; Muniz, C.E.D.S.; Eduardo, R.D.S.; Cavalcante, J.D.A.; et al. Influence of high hydrostatic pressure (HHP) on the ordered structure and on the functional, morphological, and thermal properties of japonica rice starch. *Starch* **2023**, *75*, 2200175. [CrossRef]
13. Rattanamechaiskul, C.; Junka, N.; Prakotmak, P. Modeling and optimization of moisture diffusion of paddy and brown rice during thermal soaking. *J. Food Process Eng.* **2023**, *46*, e14302. [CrossRef]
14. ISO. *Cereals and Cereal Products—Determination of Moisture Content—Reference Method*; International Organization for Standardization: Geneva, Switzerland, 2009.
15. Juliano, B.O.; Perez, C.M.; Blakeney, A.B.; Castillo, T.; Kongseree, N.; Laignelet, B.; Lapis, E.T.; Murty, V.V.S.; Paule, C.M.; Webb, B.D. International cooperative testing on the amylose content of milled rice. *Starch* **1981**, *33*, 157–162. [CrossRef]
16. Xu, X.; Yan, W.; Yang, Z.; Wang, X.; Xiao, Y.; Du, X. Effect of ultra-high pressure on quality characteristics of parboiled rice. *J. Cereal Sci.* **2019**, *87*, 117–123. [CrossRef]
17. Schindelin, J.; Arganda-Carreras, I.; Frise, E.; Kaynig, V.; Longair, M.; Pietzsch, T.; Preibisch, S.; Rueden, C.; Saalfeld, S.; Schmid, B.; et al. Fiji: An open-source platform for biological-image analysis. *Nat. Methods* **2012**, *9*, 676–682. [CrossRef] [PubMed]
18. Jordan, S.C.; Dürig, T.; Cas, R.A.F.; Zimanowski, B. Processes controlling the shape of ash particles: Results of statistical IPA. *J. Volcanol. Geotherm. Res.* **2014**, *288*, 19–27. [CrossRef]
19. Kaur, M.; Kaur, N.; Kaur, M.; Sandhu, K.S. Some properties of rice grains, flour and starches: A comparison of organic and conventional modes of farming. *LWT—Food Sci. Technol.* **2015**, *61*, 152–157. [CrossRef]

20. Juliano, B.O.; Perez, C.M.; Barber, S.; Blakeney, A.B.; Iwasaki, T.A.; Shibuya, N.; Keneaster, K.K.; Chung, S.; Laignelet, B.; Launay, B.; et al. International cooperative comparison of instrument methods for cooked rice texture. *J. Texture Stud.* **1981**, *12*, 17–38. [CrossRef]
21. Bhattacharya, K.R. Some Selected Rice Quality Test Procedures. In *Rice Quality: A Guide to Rice Properties and Analysis*; Woodhead Publishing Series in Food Science, Technology and Nutrition; Woodhead Publishing: Cambridge, UK, 2011; pp. 531–562, ISBN 978-1-84569-485-2.
22. Lyon, B.G.; Champagne, E.T.; Vinyard, B.T.; Windham, W.R. Sensory and instrumental relationships of texture of cooked rice from selected cultivars and postharvest handling practices. *Cereal Chem.* **2000**, *77*, 64–69. [CrossRef]
23. European Union. *European Union Council Regulation (EC) No 1234/2007 of 22 October 2007 Establishing a Common Organisation of Agricultural Markets and on Specific Provisions for Certain Agricultural Products*; European Union: Brussels, Belgium, 2007; Volume 299.
24. Juliano, B.O. Structure, chemistry, and function of the rice grain and its fractions. *Cereal Foods World* **1992**, *37*, 772–779.
25. Li, H.; Gilbert, R.G. Starch molecular structure: The basis for an improved understanding of cooked rice texture. *Carbohydr. Polym.* **2018**, *195*, 9–17. [CrossRef] [PubMed]
26. Wu, Z.; He, Y.; Yan, W.; Zhang, W.; Liu, X.; Hui, A.; Wang, H.; Li, H. Effect of high-pressure pre-soaking on texture and retrogradation properties of parboiled rice. *J. Sci. Food Agric.* **2021**, *101*, 4201–4206. [CrossRef] [PubMed]
27. Tian, Y.; Zhao, J.; Xie, Z.; Wang, J.; Xu, X.; Jin, Z. Effect of different pressure-soaking treatments on color, texture, morphology and retrogradation properties of cooked rice. *LWT—Food Sci. Technol.* **2014**, *55*, 368–373. [CrossRef]
28. Ahromrit, A.; Ledward, D.A.; Niranjan, K. Kinetics of high pressure facilitated starch gelatinisation in Thai glutinous rice. *J. Food Eng.* **2007**, *79*, 834–841. [CrossRef]
29. Huang, S.-L.; Jao, C.-L.; Hsu, K.-C. Effects of hydrostatic pressure/heat combinations on water uptake and gelatinization characteristics of Japonica rice grains: A kinetic study. *J. Food Sci.* **2009**, *74*, E442–E448. [CrossRef] [PubMed]
30. Chen, Z.; Yang, Q.; Yang, Y.; Zhong, H. The effects of high-pressure treatment on the structure, physicochemical properties and digestive property of starch—A review. *Int. J. Biol. Macromol.* **2023**, *244*, 125376. [CrossRef] [PubMed]
31. He, M.; Qiu, C.; Liao, Z.; Sui, Z.; Corke, H. Impact of cooking conditions on the properties of rice: Combined temperature and cooking time. *Int. J. Biol. Macromol.* **2018**, *117*, 87–94. [CrossRef] [PubMed]
32. Tao, K.; Yu, W.; Prakash, S.; Gilbert, R.G. Investigating cooked rice textural properties by instrumental measurements. *Food Sci. Hum. Wellness* **2020**, *9*, 130–135. [CrossRef]
33. Yamakura, M.; Haraguchi, K.; Okadome, H.; Suzuki, K.; Tran, U.T.; Horigane, A.K.; Yoshida, M.; Homma, S.; Sasagawa, A.; Yamazaki, A.; et al. Effects of soaking and high-pressure treatment on the qualities of cooked rice. *J. Appl. Glycosci.* **2005**, *52*, 85–93. [CrossRef]
34. Li, W.; Bai, Y.; Mousaa, S.A.S.; Zhang, Q.; Shen, Q. Effect of high hydrostatic pressure on physicochemical and structural properties of rice starch. *Food Bioprocess Technol.* **2012**, *5*, 2233–2241. [CrossRef]
35. Priestley, R.J. Moisture requirements for gelatinisation of rice. *Starch* **1975**, *27*, 416–419. [CrossRef]
36. Bauer, B.A.; Knorr, D. The impact of pressure, temperature and treatment time on starches: Pressure-induced starch gelatinisation as pressure time temperature indicator for high hydrostatic pressure processing. *J. Food Eng.* **2005**, *68*, 329–334. [CrossRef]
37. Oh, H.E.; Hemar, Y.; Anema, S.G.; Wong, M.; Neil Pinder, D. Effect of high-pressure treatment on normal rice and waxy rice starch-in-water suspensions. *Carbohydr. Polym.* **2008**, *73*, 332–343. [CrossRef]
38. Billiris, M.A.; Siebenmorgen, T.J.; Meullenet, J.-F.; Mauromoustakos, A. Rice degree of milling effects on hydration, texture, sensory and energy characteristics. Part 1. Cooking using excess water. *J. Food Eng.* **2012**, *113*, 559–568. [CrossRef]
39. Douzals, J.P.; Marechal, P.A.; Coquille, J.C.; Gervais, P. Microscopic study of starch gelatinization under high hydrostatic pressure. *J. Agric. Food Chem.* **1996**, *44*, 1403–1408. [CrossRef]
40. Leelayuthsoontorn, P.; Thipayarat, A. Textural and morphological changes of Jasmine rice under various elevated cooking conditions. *Food Chem.* **2006**, *96*, 606–613. [CrossRef]
41. Boluda-Aguilar, M.; Taboada-Rodríguez, A.; López-Gómez, A.; Marín-Iniesta, F.; Barbosa-Cánovas, G.V. Quick cooking rice by high hydrostatic pressure processing. *LWT—Food Sci. Technol.* **2013**, *51*, 196–204. [CrossRef]
42. Meng, L.; Zhang, W.; Wu, Z.; Hui, A.; Gao, H.; Chen, P.; He, Y. Effect of pressure-soaking treatments on texture and retrogradation properties of black rice. *LWT* **2018**, *93*, 485–490. [CrossRef]

foods

MDPI

Article

Optimization of High-Pressure Processing for Microbial Inactivation in Pigmented Rice Grass Juice and Quality Impact Assessment during Refrigerated Storage

Uyen Ha Dao [1], Jitlada Na Lamphun [1], Sitthidat Tongdonyod [1], Sirinya Taya [2], Suphat Phongthai [1,3,4] and Wannaporn Klangpetch [1,3,*]

1 Division of Food Science and Technology, Faculty of Agro-Industry, Chiang Mai University, Chiang Mai 50100, Thailand; daohauyen@gmail.com (U.H.D.); sittidat07@gmail.com (S.T.); suphat.phongthai@cmu.ac.th (S.P.)
2 Functional Food Research Unit, Multidisciplinary Research Institute, Chiang Mai University, Chiang Mai 50200, Thailand; sirinya.t@cmu.ac.th
3 Cluster Research of High Value Products from Thai Rice and Plants for Health, Chiang Mai University, Chiang Mai 50100, Thailand
4 Center of Excellence in Agro Bio-Circular-Green Industry (Agro BCG), Faculty of Agro-Industry, Chiang Mai University, Chiang Mai 50100, Thailand
* Correspondence: wannaporn.u@cmu.ac.th

Abstract: Pigmented rice grass juice (RGJ) is a good source of bioactive compounds, but fresh juice has a relatively short shelf life of only 7 days at 4 °C. The objectives of this study were to determine the optimal growth stage of pigmented rice grass, investigate the optimal condition of high-pressure processing (HPP) for bacterial inactivation in inoculated RGJ using response surface methodology (RSM), and evaluate quality changes in uninoculated HPP-treated juice during storage at 4 °C compared with heat-treated (85 °C/10 min) and untreated samples. Results revealed that the optimal growth stage of rice grass was 9 days with the highest total anthocyanin content of 158.92 mg/L. The optimal condition of HPP was determined to be 612 MPa, 11 min, and 36 °C, and inactivated *Escherichia coli* K12 and *Listeria innocua* with 6.43 and 5.02 log reductions, respectively, meeting FDA regulations. The lethality of bacteria after HPP treatment can be explained by damage to the cell membrane and the leakage of intracellular constituents such as protein and nucleic acid. During 12 weeks of storage at 4 °C, total plate counts and yeast and mold counts in uninoculated HPP-treated juice were not detected. Moreover, HPP did not significantly change phytochemical properties ($p < 0.05$), caused a minor impact on physicochemical properties of RGJ, and maintained the durability of juice samples during storage. Analysis of the phytochemical profile revealed that HPP treatment could preserve most of the phenolic compounds in RGJ and especially increase the contents of protocatechuic acid, 4-hydroxybenzoic acid, syringic acid, transcinnamic acid, isorhamnetin-3-o-glucoside, quercetin, and cyanidin-3-glucoside ($p < 0.05$). Overall, HPP is a potential pasteurization technique for microbial inactivation and nutritional preservation for rice grass juice.

Keywords: high hydrostatic pressure; optimization; *Escherichia coli*; *Listeria innocua*; phytochemicals; cereal grass

1. Introduction

Cereal grass is now considered a functional food [1]. Wheat grass juice is the most common cereal grass juice and contains considerable amounts of bioactive compounds [2]. However, due to the high price and cultivation area of wheat, products from wheat grass are not commonly marketed and consumed in Asian countries. Rice (*Oryza sativa*) belongs to the *Poaceae* family, similar to wheat, and is the most important staple food crop in Asia, which is also the leading continent in rice production, accounting for nearly 90% of the global market share based on the data of FAOSTAT (available online: https://www.fao.org/faostat/en/#data/QCL/visualize) (accessed on 9 September 2024). Kum Doi Saket (KDS) is a black-purple glutinous rice cultivar that is widely grown in the north and northeast of Thailand. Juice squeezed from rice grass harvested at the jointing stage (days 6–15 of the growth stage) exhibits the highest levels of bioactive compounds [1,3]. Therefore, the "growing stage" must be considered to maximize the phytochemical properties of cereal grass juice.

Few studies have addressed the bioactivity of rice grass juice (RGJ), with most revealing superior phytochemical characteristics in pigmented RGJ compared to wheat grass juice [4]. Khanthapoka et al. (2015) found that RGJ from a black-color cultivar exhibited the highest total phenolic content (4.3 mg GAE/g DE) and the highest radical scavenging activity (EC_{50} = 11 mg/mL) compared to other white rice and wheat grass juices [5]. Thepthanee et al. (2021) reported the cancer chemopreventive potential, strong antioxidant activity, and anti-inflammatory capacity of a black rice grass extract which was a rich source of phenolic acids, anthocyanins, and flavonoids [6]. Rice is the main staple food in Asia and black rice cultivars contain high levels of polyphenols and antioxidants. Therefore, pigmented RGJ can be developed competitively as an alternative functional beverage. RGJ is similar to wheat grass juice and is categorized as a low-acid vegetable juice with pH > 4.6 and a short shelf life [7]. Vegetable juice is defined as "the liquid unfermented but fermentable product or lactic acid fermented product intended for direct consumption obtained from the edible part of one or more sound vegetables and preserved exclusively by physical means". The products must be free from microorganisms capable of development under normal conditions of storage [8]. Thus, finding an appropriate preservation technique is necessary to extend the shelf life of RGJ.

Today, minimally processed food products with superior quality and fresh-like characteristics are in high demand, and less harsh preservation techniques with fewer additives are preferred. Due to its advantages over thermal pasteurization, high pressure processing (HPP) has gained popularity. These benefits include (i) immediately inactivating pathogenic and spoilage-causing bacteria, (ii) delaying or deferring the commencement of chemical and enzymatic deteriorative processes, and (iii) maintaining the nutrition and organoleptic properties of food products [9–13].

HPP is usually applied to fruit and vegetable juices as a pasteurization method with pressures ranging from 400 to 600 MPa and treatment duration from 1 to 15 min [9]. Research has shown that HPP preserves the safety and organoleptic quality of low-acid beverages such as carrot juice [14], cucumber juice [15], wheat grass juice [16], pitaya juice [17], and coconut juice [18]. The study by Liu et al. (2016) reported that HPP (500 MPa for 5 min at 20 °C) can inactivate total aerobic bacteria and total yeast and mold counts to less than 1 log (CFU/mL) in a clear cucumber beverage, and prevent the appearance of bacterial colonies for 20 days of refrigerated storage. HPP-treated juice also retained a fresh-like color, demonstrated higher clarity, and preserved more important aroma constituents than high temperature short time (HTST)-treated juice [15].

Therefore, the objectives of this study were to determine the optimal growth stage of black rice grass, optimize the HPP conditions for inactivation of *Escherichia coli* K12 and *Listeria innocua* in Kum Doi Saket RGJ, and investigate the stability of HPP-treated RGJ during storage compared with heat-treated RGJ.

2. Materials and Methods

2.1. Study of Rice Grass Growth Stages

Kum Doi Saket black rice grass was grown by the Lanna Rice Research Center, Chiang Mai University, Chiang Mai, Thailand. The seeds were tested for >95% germination, soaked in water for 12 h and incubated for 48 h (25 °C). Then, 200 g of seeds were spread evenly on a seedling tray (30 × 60 cm) and watered regularly. Germination was conducted under controlled conditions at 25/20 °C and 60 to 70% relative humidity.

The rice grass was grown for 14 days. Harvesting began on day 6 when the grass was 12 to 15 cm high. After harvesting, the fresh grass was weighed, washed with tap water, and pretreated by soaking in sodium bicarbonate solution (0.1%, 15 min) for contaminant removal. The RGJ was then extracted by a single gear horizontal juicer (Oscar DA1200, Ulsan, Republic of Korea) and filtered through cheesecloth three times. The extracted juice was analyzed for some phytochemical properties including total phenolic content, antioxidant activity, and total anthocyanin content. Among them, total anthocyanin content was considered the most important factor. The growth stage that revealed the highest levels of bioactive contents was chosen for further experiments.

2.2. Optimization of Processing Conditions

2.2.1. Bacterial Strain Preparation

Gram-negative bacterium *Escherichia coli* K12 (ATCC 10798) and Gram-positive bacterium *Listeria innocua* JCM 32814 were the two bacteria strains used in this study, which were provided by the American Type Culture Collection and Japan Collection of Microorganisms, respectively.

The bacterium stocks were stored at −20 °C in Tryptic Soy Broth (TSB) supplemented with 50% glycerol. Cell cultures were activated by streaking on Tryptic Soy Agar (TSA) and incubated at 37 °C for 24 h. The strains were then subcultured into 10 mL of TSB at 37 °C until they reached the early stationary phase (10^9 CFU/mL). The cells were then washed with sterile 0.85% (*w*/*v*) NaCl solution and collected by centrifugation (6000× *g*, 10 min). The bacterial pellets were resuspended at 10^9 CFU/mL for further study.

2.2.2. Response Surface Methodology Experimental Design for Microbial Inactivation

RGJ was autoclaved at 12 °C for 15 min to inactivate all naturally occurring microorganisms in the juice. *E. coli* K12 and *L. innocua* suspensions were inoculated individually into sterile RGJ samples in 30 mL PET bottles at a concentration of 10^6 CFU/mL. In this study, we used response surface methodology (RSM) to optimize the processing conditions for microbial inactivation. The process parameters investigated were pressure (400 to 600 MPa), time (5 to 15 min) and temperature (15 to 35 °C). The results of preliminary experiments showed that all single processing factors provided significant effects on microbial inactivation. A Box–Behnken Design (BBD) was applied with three factors including pressure, time, and temperature, coded as A, B, and C, respectively. Each factor had three levels, and the HPP condition at the center point was replicated five times. Log reduction of the two bacteria was defined as the response for the optimization process. Analysis of variance (ANOVA) was employed to evaluate the impacts of the HPP parameters on two responses with $p < 0.05$ indicative of statistically significant parameters.

After treatment, inoculated RGJ samples were investigated for the viability of bacterial cells. Appropriate decimal dilutions of cell suspensions were drop-plated onto TSA. All plates were incubated at 37 °C for 24 to 36 h. Finally, typical colonies were counted and reported as log reduction using Equation (1):

$$\text{Log reduction} = \log(N_0) - \log(N) \quad (1)$$

where N_0 represents the count of colonies in untreated samples and N is the count of colonies in treated samples.

2.3. Bacterial Morphological Studies

The morphological alterations in untreated, heat-treated, and HPP-treated bacteria cells were assessed by scanning electron microscopy (SEM), based on the approach of Sha et al. (2020) with some modifications. Inoculated HPP-treated samples, heat-treated (85 °C, 10 min) samples, and untreated samples were centrifuged (6000× *g*, 10 min) to obtain the bacterial cell pellet [19]. A critical point dryer (Leica EM CPD300, Vienna, Austria) was used to dry the pellet. The dried cell was then coated with gold and subjected to SEM (JEOL, JSM-IT300, Tokyo, Japan) (15 kV) to examine the morphology [20].

2.4. Determination of the Intracellular Constituent Leakage

Preservative treatments like HPP or heat might cause breakage of the bacterial cell membrane resulting in the leakage of intracellular constituents. The release of nucleic acids and proteins from the two bacterial strains after treatment was determined following the method of [19]. After HPP and heat treatment, the bacterial suspension was centrifuged (6000× *g*, 10 min). The supernatant was filtered through sterile syringe filters of 0.22 μm to remove bacterial cells. Then, a UV-Vis spectrophotometer (Genesys 180, Thermo Fisher Scientific, Waltham, MA, USA) was employed to measure the absorbances of the filtered suspensions at wavelengths of 260 nm and 280 nm to examine the leakage of nucleic acid and protein, respectively [19].

2.5. Quality Assessment of the Treated RGJ during Storage

HPP (612 MPa, 11 min, and 36 °C)-treated samples (HPP-RGJ), heat (85 °C, 10 min)-treated samples (HT-RGJ), and control samples (C-RGJ) were evaluated for durability over 12 weeks of storage at refrigerated temperature (4 °C) based on three main attributes, including microbial, physicochemical, and phytochemical qualities.

2.5.1. Microbial Quality

The microbiological quality of control, HPP-treated, and heat-treated RGJ samples during storage was evaluated by analyzing total plate count (TPCt) on plate count agar and yeast and mold count (YMC) on potato dextrose agar. Both TPCt and YMC were measured using the spread plate technique [21,22]. Finally, the samples were stored at 37 °C for 48 h. The amount of bacterial colonies found on agar was represented in log CFU/mL.

2.5.2. Physicochemical Quality

The total soluble solids (TSS) was evaluated by a hand refractometer (ATAGO, Tokyo, Japan). A pH meter (FiveGo F2, Mettler Toledo, Switzerland) was employed to determine the pH value of the rice grass juice. Titratable acidity (TA) was analyzed following the methodology with some modifications [16]. Color Quest XE (Hunter Lab, Tokyo, Japan) Colorimeter was employed to measure L*, a*, and b* color parameters of RGJ.

2.5.3. Phytochemical Quality

Total Phenolic Content (TPC)

The Folin-Ciocalteu method of Khanthapoka et al. (2015), with some modifications, was used to evaluate the TPC of rice grass juice [5]. In summary, 0.5 mL of diluted RGJ was reacted with 2.5 mL of Folin-Ciocalteu reagent 10% (*v*/*v*) (ratio 1:25) and stored in the dark for 5 min. After that, 2 mL of Na_2CO_3 7.5% (*w*/*v*) was added to the reaction mixture followed by incubating in the dark for 60 min. The absorbance of the suspension was detected at 765 nm. TPC of rice grass juice was expressed as milligrams of gallic acid equivalent per liter of juice (mg GAE/L) [5].

Antioxidant Activity by 2,2′ Azinobis (3 Ethylbenzothiazoline 6 Sulfonic Acid) (ABTS) Assay

The ABTS value of RGJ was measured following the methodology described by Re et al. (1999) with certain changes [23]. In short, to produce the $ABTS^+$ radical cation, two solutions of ABTS 7 mM and $K_2S_2O_8$ 2.45 mM (ratio 1:1) were mixed and stored in the dark at room temperature for 16 h. Then, 3 mL of diluted $ABTS^{\cdot+}$ solution (absorbance of 0.7 ± 0.02 at 734 nm) was reacted with 50 μL of diluted RGJ (1:25) and incubated for 30 min in the dark at 25 °C. The absorbance was detected at 734 nm. Antioxidant activity (AOA) was defined as millimoles of Trolox equivalent per liter of juice (mmol TE/L) [23].

Antioxidant Activity by Ferric Reducing Antioxidant Power (FRAP) Assay

The FRAP value of RGJ was evaluated based on the study of Benzie and Strain with some changes. Briefly, FRAP reagent generated by the reaction of 300 mM acetate buffer (pH 3.6), 10 mM TPTZ, and 20 mM $FeCl_3$ was reacted with 0.1 mL diluted juice (1:25). The samples were then stored in the dark for 30 min. After incubation, the absorbance of the solution was detected at 593 nm. $FeSO_4$ was used to contribute the standard curve. The FRAP value of RGJ was defined in µmol Fe^{2+} per liter of juice (µmol Fe^{2+}/L) [24].

Determination of Total Anthocyanin Content (TAC)

Anthocyanin is an important bioactive compound that contributes to the black-purple color of rice grass juice. The TAC of RGJ was investigated following the differential methodology of Lee et al. (2005) with some changes [25]. Briefly, we prepared two buffer solutions including sodium acetate 0.4 M (pH = 4.5) and potassium chloride 0.025 M (pH = 1). Then, the undiluted juice was added into each buffer solution at a ratio of 1:4 (1 part test portion, 4 parts buffer) and stored for 20 min. The absorbance of all samples was analyzed at both 520 and 700 nm. Total anthocyanin content was expressed as milligrams of cyanidin-3-glucoside equivalent per liter of juice (mg/L) and determined using Equation (2) [25]:

$$\text{TAC (mg/L)} = \frac{\text{A} \times 449.2 \times \text{DF} \times 10^3}{26,900 \times 1} \quad (2)$$

where A (absorbance) = $(A_{520} - A_{700})_{pH\ 1.0} - (A_{520} - A_{700})_{pH\ 4.5}$; 449.2 = molecular weight of cyanidin-3-glucoside (g/mol), DF = dilution factor, 1 = pathlength (cm), 26,900 = molar extinction coefficient ($L \cdot mol^{-1} \cdot cm^{-1}$).

Determination of Total Chlorophyll Content (TCC)

Chlorophyll is also an important pigment compound found in cereal grass juice. The TCC of RGJ was investigated based on the methodology of Porra et al. (1989) with minor modifications. In short, 50 µL of RGJ was mixed with 450 µL of 80% acetone. The absorbance (A) of the sample was detected at 663 and 645 nm and TCC was determined using Equation (3) [26]:

$$\text{TCC (mg/L)} = 19.54\text{A}^{645} + 8.29\text{A}^{663} \quad (3)$$

2.6. Phytochemical Profile Analysis

The phytochemical profiles (phenolic acids, flavonoids, and anthocyanin) of fresh juice (C-RGJ), HPP-treated juice (HPP-RGJ), heat-treated juice (HT-RGJ), and commercially available wheat grass juice (WGJ) and barley grass juice (BGJ) were identified and quantitated using a high-performance liquid chromatography (HPLC, Agilent 1290 Infinity II, Santa Clara, CA, USA) system with a Phenomenex C18 column (5 µm, 4.6 × 250 mm), (Torrance, CA, USA) and a diode array detector, following the methodology with some modifications [27]. Two mobile phases of acetonitrile supplemented with 0.1% (v/v) formic acid and water supplemented with 0.1% (v/v) formic acid were employed at room temperature with flow rates of 1 mL/min. The detector wavelengths were set at 280 and 320 nm [27]. There were 16 phenolic compound standards used, including gallic acid, protocatechuic acid, 4-hydroxybenzoic acid, chlorogenic acid, vanillic acid, syringic acid, p-coumaric acid, ferulic acid, transcinnamic acid, catechin, rutin, isorhamnetin-3-o-glucoside, quercetin, naringenin, cyanidin-3-glucoside, and peonidin-3-glucoside.

2.7. Statistical Analyses

Each experiment was conducted in triplicate. Experimental runs and data were generated and analyzed with Design-Expert version 13.0.5.0, consisting of analysis of variance (ANOVA), regression evaluation, creation of response surface plots, and prediction of the optimal conditions for HPP. Response surface methodology (RSM) was employed to optimize the process parameters (pressure, time, and temperature) of HPP. JMP version 12 software (SAS Institute, Cary, NC, USA) was used to conduct two-way ANOVA. Means were compared using Tukey's HSD test. Significance was assigned to comparisons with a p-value lower than 0.05 ($p < 0.05$).

3. Results and Discussion

3.1. Effect of Rice Grass Growth Stage on Phytochemical Properties

TPC, AOA, and TAC levels in rice grass rose at 9 days after planting ($p < 0.05$) and subsequently decreased from days 11–14 (Table 1). Notably, from day 9 to day 10, TPC and ABTS values slightly increased by 7.19% and 7.57%, respectively, while FRAP value did not change significantly, but TAC decreased sharply by 22.34%. The most distinctive feature of Kum Doi Saket grass is its black-purple color, which is created by pigmented compounds such as anthocyanin. According to Khanthapoka et al. (2015), the KDS cultivar had the highest TAC compared to other cultivars of white rice grass and wheatgrass, which may result in superior antioxidant capacity and DNA protection [5]. Therefore, we considered TAC the most important parameter to determine the optimal growth stage of KDS rice grass. As a result, although the TPC and ABTS values of rice grass on day 9 were slightly lower than those on day 10, day 9 was selected as the optimal growth stage of KSD grass because TAC reached a maximum value of 158.92 mg/L. After day 9, the third leaves appeared and the color of the grass turned from dark purple to green (Figure 1), marking a gradual decrease in bioactive compounds ($p < 0.05$).

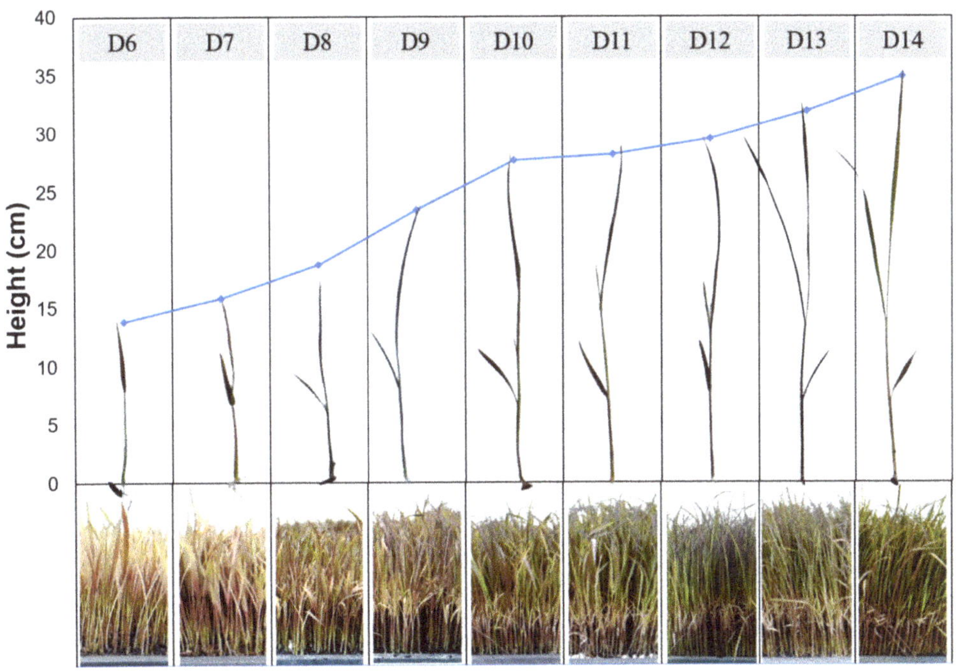

Figure 1. Growth stages of rice grass from day 6 to day 14.

Cereal grasses after 6 to 15 days of germination are at the "jointing stage", when they reveal the highest level of bioactive compounds [1,3]. Young plants need phytonutrients such as polyphenols or antioxidants to protect against threatening external factors [28]. After the jointing stage, bioactive compounds decline sharply as levels of cellulose increase to support the elongation of internodal tissue in the grass leaf, forming a stem [1,4]. In 2005, Randhir and Shetty discovered that polyphenols in immature maize leaves during the early germination stage were soluble. However, in the late growth stages, polyphenol concentration decreased due to lignification, which occurred when phenolic molecules were partitioned in a polymerization process to generate lignans and lignins [29]. The study of Kohler (1944) discovered that fiber content in oat leaves increased as the plant matured, but vitamins, protein, fat, and chlorophyll content declined [30].

Kapkum et al. (2011) investigated changes in TPC and AOA over 18 days of rice grass germination and found that 15 days was the optimal germination age [31]. Kulkarni et al. (2006) also found that wheat grass extract had the highest AOA after 15 days of growth. Different cereal cultivars have diverse optimal growth stages [32]. Rice grass harvested at the optimal growth stage (day 9) was used for further experiments.

Table 1. Phytochemical properties of RGJ at 6 to 14 days of growth.

Attributes	Growth Stage (Days)								
	6	7	8	9	10	11	12	13	14
TPC (mg GAE/L)	3092.99 ± 156.28 [de]	3349.02 ± 81.71 [c]	3568.65 ± 18.57 [b]	3433.11 ± 161.34 [bc]	3680.70 ± 49.14 [a]	3390.44 ± 179.1 [bc]	3263.68 ± 30.43 [cd]	3010.16 ± 104 [e]	3148.22 ± 32.46 [de]
FRAP (mm Fe^{2+}/L)	19.79 ± 0.57 [e]	22.99 ± 0.52 [d]	29.76 ± 1.09 [b]	31.76 ± 1.61 [a]	32.58 ± 0.77 [a]	25.98 ± 1.78 [c]	25.45 ± 0.43 [c]	25.54 ± 1.11 [c]	27.35 ± 1.55 [c]
ABTS (mm TE/L)	26.28 ± 1.87 [de]	27.38 ± 2.5 [de]	32.47 ± 3.17 [b]	31.71 ± 1.86 [bc]	34.11 ± 2.14 [a]	28.21 ± 1.09 [cd]	25.73 ± 2.47 [de]	24.36 ± 1.95 [ef]	20.92 ± 1.56 [f]
TAC (mg/L)	68.97 ± 3.28 [d]	77.65 ± 1.76 [d]	87.11 ± 9.16 [c]	158.92 ± 3.2 [a]	123.41 ± 3.17 [b]	91.4 ± 7.52 [c]	95.13 ± 4.83 [c]	87.84 ± 3.78 [c]	95.46 ± 5.21 [c]

Different superscripts represent significant differences ($p < 0.05$).

3.2. Microbial Inactivation by HPP

3.2.1. Model Fitting

Three factors of HPP including pressure, time, and temperature, along with their codes and levels, are shown in Table 2. Experimental runs generated by BBD and results of the effect of HPP parameters on bacterial inactivation are shown in Table 3. The observed responses were obtained via HPP experiments, whereas the expected responses were generated using a regression equation (Table 4).

Table 4 shows RSM plots demonstrating the effect of process factors on microbial inactivation, as well as coded values for second-order polynomial regression equations. The effects of the independent variables on the log reduction of two bacteria strains were demonstrated by the linear terms (A, B, C), quadratic terms (A^2, B^2, C^2), and interaction terms (AB, AC, BC). The RSM model coefficients (R^2) were 0.9613 and 0.9583, indicating that the generated models explained 96.13% and 95.83% of the variation for *E. coli* K12 and *L. innocua* log reduction, respectively. A model with a relatively good fit usually has a coefficient above 0.08 [33]. The findings showed that experimental and predicted values had good correlation, indicating that the quadratic models were efficacious in this research.

Analysis of variance (ANOVA) was used to examine the influence of process parameters on log reduction of the two bacteria strains, with *p*-values less than 0.05 indicating statistical significance. The data shown in Table 4 reveal that the quadratic models were significant for the optimization of HPP process parameters, with *p*-values < 0.05.

Table 2. Three independent variables (pressure, time, temperature) used for optimization of HPP via Box–Behnken Design (BBD).

Independent Variable (Factor)	Label	Level		
		−1	0	−1
Pressure (MPa)	A	400	500	600
Time (h)	B	5	10	15
Temperature (°C)	C	15	25	3

Table 3. The impact of independent HPP factors (pressure, time, temperature) on the inactivation of *E. coli* K12 and *L. innocua*.

Run	Factors			Response (*E. coli* K12)		Response (*L. innocua*)	
	A: Pressure (MPa)	B: Time (min)	C: Temperature (°C)	Actual	Predicted	Actual	Predicted
1	600	10	35	6.15	6.25	5.65	5.12
2	600	15	25	5.30	5.05	5.09	4.96
3	600	10	15	5.24	5.77	4.25	3.90
4	500	15	35	5.24	5.75	2.99	3.18
5	500	10	25	5.18	5.27	2.64	2.52
6	500	10	25	5.18	5.27	2.64	2.52
7	500	10	25	5.18	5.27	2.64	2.52
8	500	10	25	5.18	5.27	2.64	2.52
9	500	5	35	5.18	5.30	1.97	1.70
10	500	10	25	5.18	5.20	2.64	2.52
11	400	10	35	4.93	4.57	1.88	1.98
12	600	5	25	4.56	4.64	2.92	3.25
13	500	5	15	4.43	4.10	1.71	1.28
14	500	15	15	4.26	4.26	1.70	1.73
15	400	15	25	2.24	2.33	1.94	1.36
16	400	10	15	2.21	2.29	1.05	1.33
17	400	5	25	1.76	2.19	1.26	1.14

Table 4. Second-order polynomial regression equations for microbial inactivation by HPP.

	E. coli K12	*L. innocua*
R^2	0.9613	0.9583
p-value	0.0004 (significant)	0.0005 (significant)
Equation	$Y_1 = 5.18 + 1.27A + 0.1388B + 0.6701C - 0.0680AB - 0.4538AC + 0.0585BC - 0.9316A^2 - 0.7849B^2 + 0.3824C^2$	$Y_2 = 2.64 + 1.47A + 0.4820B + 0.4701C + 0.3727AB + 0.1440AC + 0.2582BC + 0.6345A^2 - 0.4788B^2 - 0.0715C^2$

3.2.2. Effect of HPP Parameters on the Inactivation of *E. coli* and *L. innocua*

The three HPP processing parameters were pressure (A), processing time (B), and processing temperature (C). Figure 2 depicts response surface plots that demonstrate the correlations between the process parameters and *E. coli* K12 and *L. innocua* inactivation. Pressure caused the most significant linear effect on the log reduction of *E. coli* K12 and *L. innocua*, according to the ANOVA (Tables S3 and S4) ($p < 0.05$). At a fixed temperature of 25 °C and processing time of 15 min, an elevation of pressure from 400 to 600 MPa resulted in higher log reduction of both bacterial strains (Table 3). The ANOVA results also showed that treatment temperature was more significant than treatment time for *E. coli* K12 inactivation while the opposite results were obtained for *L. innocua*. Extension of treatment time (5 to 15 min) did not significantly affect *E. coli* K12 log reduction ($p < 0.05$).

Increased pressure (400−600 MPa) and processing temperature (15–35 °C) led to better bacterial elimination (Figure 2B,b). Moreover, the maximum pressure of 600 MPa assisted with an elevated temperature of 35 °C gave the highest inactivation of both *E. coli* K12 and *L. innocua* at 6.15 and 5.65 log reduction, respectively (Table 3). In general, HPP treatment conducted above or below room temperature (20 °C) can enhance microbe inactivation by altering bacterial strain resistance [34]. Treatment with HPP (375 MPa at 35 °C for 5 min) could increase the inactivation of *E. coli* O157:H7 by greater than 5 log reduction compared to the same HPP conditions at 25 °C [35]. Similarly, HPP-treated (400 MPa, 2 min, 20 °C) carrot juice (pH 6.4) resulted in a 3-log reduction of *L. innocua* whereas the inactivation ability was improved to more than 6 log reduction when the mild heat of 35 °C was applied [36]. These phenomena can be explained by the bacterial cells becoming more vulnerable to HPP when exposed to moderate heat due to the membrane phase transition [37].

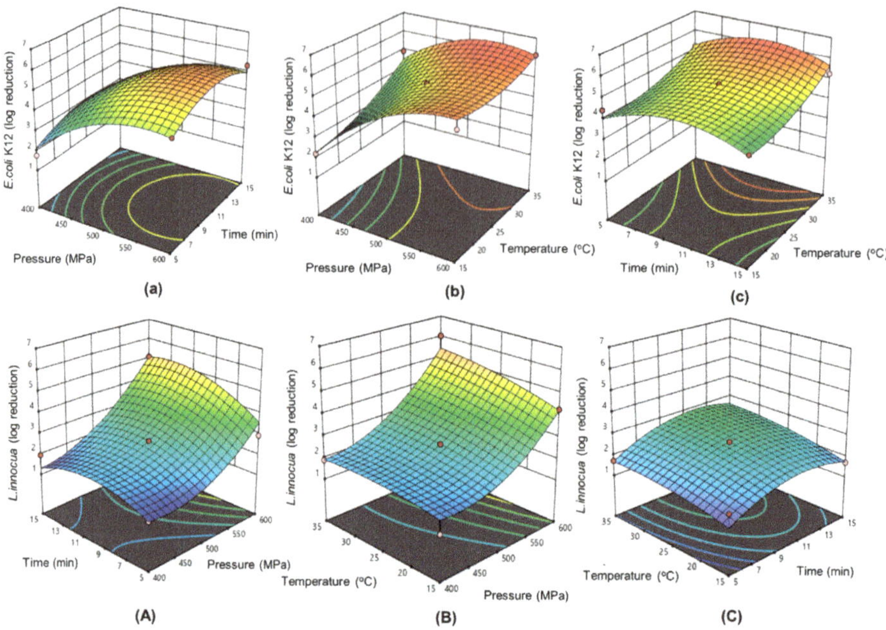

Figure 2. 3D response surface graphs illustrating the effect of three HPP independent factors (pressure, time, temperature) on the log reduction of *E. coli* K12 (**a–c**) and *L. innocua* (**A–C**).

Obviously, the resistance of the Gram-negative *E. coli* K12 and Gram-positive *L. innocua* to HPP variables differed. At all HPP conditions, the log reduction of *E. coli* K12 was higher than the log reduction of *L. innocua* (Table 3). Generally, Gram-negative bacterial cells are more vulnerable to high pressure assisted by mild heat than Gram-positive bacteria because the latter are surrounded by thicker layers of peptidoglycan [38].

Pressure was shown to be the most important parameter for the inactivation of both bacteria strains. Our experiments on sublethal injury of *E. coli* K12 and *L. innocua* after HPP treatment (Table S1) revealed that at 600 MPa and treatment time > 10 min no microbial cells were recovered on the selective medium (tryptic soy broth supplemented with 2% (w/w) sodium chloride). Increases in the injury rate of *E. coli* (89.55–100%) and *L. innocua* (78.93–100%) treated with HPP were detected when pressure increased from 400 MPa to 600 MPa. High pressure impacts microorganisms because of (i) alterations in cell morphology, including disruption of membrane permeability and loss of membrane durability; (ii) the inhibition of physiological functions and denaturation of functional proteins leading to limited proton flow, reduced intracellular pH, inactivation of enzymes responsible for synthesizing ATP, loss of functionality of membrane-bound enzymes, and disintegration of ribosomes; and (iii) the alteration of genetic systems, causing genetic materials to condense and degrade chromosomal DNA [39–43] The combination of high pressure and mild elevated temperature (35 °C) inactivates pathogens more effectively than pressure or temperature alone by damaging the bacterial cell membrane.

3.2.3. Optimization and Verification

The FDA regulations for fruit and vegetable juice pasteurization require that preservative treatment must achieve the inactivation of pathogenic bacteria or surrogate microorganisms with more than 5 log reduction [44]. Therefore, in order to find out the predicted optimal HPP conditions, the responses for numerical optimization of HPP process parameters were set to the range of greater than 5 log reductions of both bacterial strains. Table 5 reveals the predicted optimal conditions for HPP, which was then verified by validated experiments to ensure that HPP could inactivate both bacteria to the target log reduction. As a result, the validated experimental findings varied somewhat from the predicted values. HPP treatment at 612 MPa, 11 min, and 36 °C effectively inactivated *E. coli* K12 and *L. innocua*, resulting in 6.43 and 5.02 log reduction, respectively. Similar results were also observed with pitaya juice, where the application of HPP at optimal conditions of 600 MPa for 12 min could also achieve 5 log reduction of *L. innocua* [17].

Table 5. Predicted and validated values for bacterial inactivation.

Value	Optimized Process Parameters			Responses	
	A: Pressure (P)	B: Time (min)	C: Temperature (°C)	*E. coli* K12 log Reduction	*L. innocua* log Reduction
Predicted	612.00	11.2	36.25	6.25	5.12
Validated	612.00	11.00	36.00	6.43	5.02
% error	0.00	1.78	0.69	1.42	1.95

3.3. Mechanism of HPP on Inactivation of *E. coli K12* and *L. innocua*

The mechanism of HPP on microbial inactivation was elucidated by studying the morphology of untreated, HPP-treated, and heat-treated bacterial cells using SEM and the leakage of intracellular constituents. Figure 3 illustrates SEM micrographs of the surfaces of *E. coli* K12 and *L. innocua* cells. The cells of untreated samples (Figure 3A,D) had smooth and undamaged surfaces. By contrast, the surfaces of both strains of bacteria treated by HPP exhibited severe damage to the cell membrane (Figure 3C,F). Cells treated only with heat revealed mild defects on the surface (Figure 3B,E). Yang et al. (2012) observed cell membrane rupture in *E. coli* caused by HPP (500 MPa/30 min/25 °C) [45]. Ritz et al. (2002) reported that HPP (400 MPa, 10 °C, 20 min) induced scarring on the surface, loss of membrane integrity, and decrease in membrane functionality in *L. monocytogenes* cells [46].

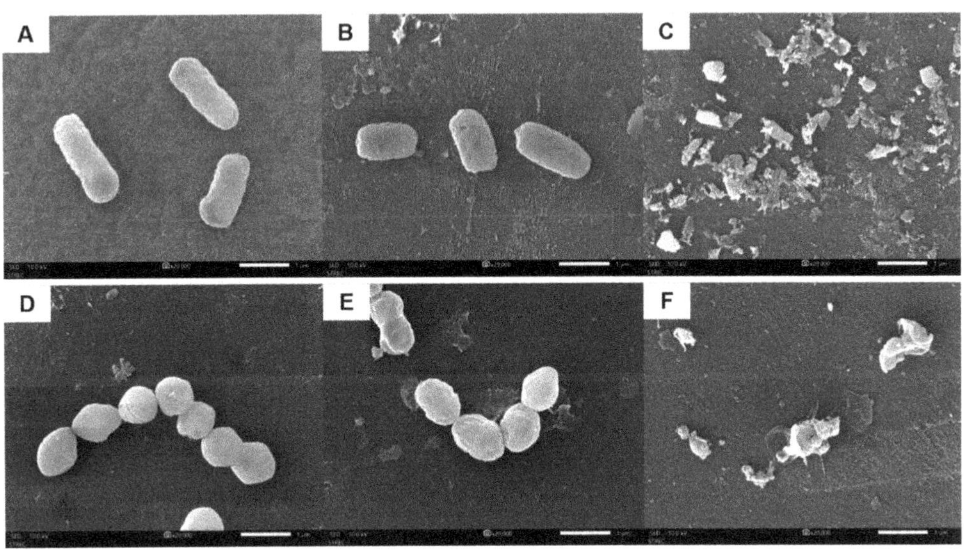

Figure 3. Morphology of *E. coli* K12 (**A–C**) and *L. innocua* (**D–F**) evaluated using SEM. (**A**,**D**): untreated cells; (**B**,**E**): heat-treated samples (85 °C, 10 min); (**C**,**F**): HPP-treated samples (612 MPa, 11 min and 36 °C).

Cell membrane damage after heat and HPP treatment caused leakage of the intracellular constituents. UV–Vis spectrophotometry was used to evaluate the leakage of nucleic acid and protein at 260 and 280 nm, respectively (Figure 4). Differences in absorbance level between untreated and treated samples revealed the release of nucleic acid and protein through the broken cell membrane site into the supernatant. HPP caused significantly higher amounts of leaked nucleic acid compared with the heat-treated sample. These observations were consistent with the results of SEM experiments on morphology changes, which also indicated that HPP caused more severe damage to the cell membranes of both bacteria than heat treatment. Our study on sublethal injured cells of *E. coli* K12 and *L. innocua* after HPP (600 MPa, 10 min) and heat (85 °C, 10 min) treatment (Table S1) revealed that, although both these treatments were capable of achieving lethality of up to 5 log reduction, HPP treatment resulted in a higher sublethal injury ratio (%) of bacterial cells compared to heat treatment.

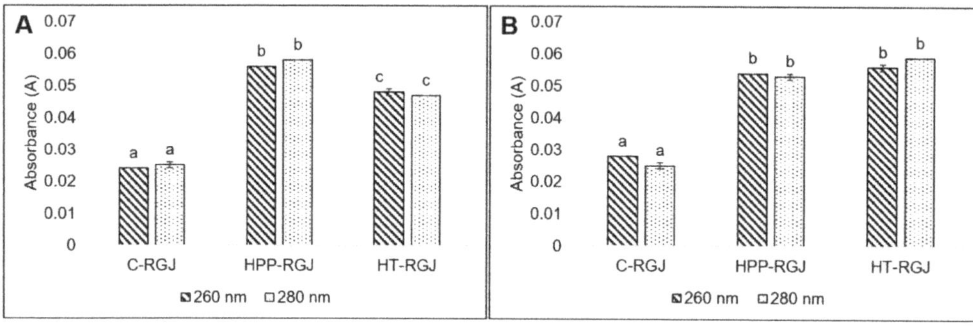

Figure 4. Effect of HPP (HPP-RGJ) and heat treatment (HT-RGJ) on the leakage of nucleic acid (measured at 260 nm) and protein (measured at 280 nm) of *E. coli* (**A**) and *L. innocua* (**B**) compared with the control sample (C-RGJ). Different letters indicate significantly different values among treatments ($p < 0.05$).

The results of bacterial morphology and intracellular component leakage supported the theory of HPP-mediated microbial inactivation. The structure of the bacterial cell wall is constituted of peptidoglycans, which include N-acetylglucosamine, N-acetylmuramic acid, and three amino acids (D-glutamic acid, D-alanine, and meso-diaminopimelic acid) that coat the nuclear membrane. The genetic material is vulnerable to elevated temperature and high pressure [43]. The cell membrane is the primary location of pressure injury in microorganisms [11]. Microorganism inactivation mechanisms under pressure involve modifying noncovalent electrostatic and hydrophobic interactions, resulting in the unfolding of tertiary and quaternary enzyme and protein structures that are crucial for replication, cellular integrity, and metabolism. Moreover, high pressure reduces cell volume, collapses intracellular vacuoles, and disrupts or increases membrane permeability, resulting in the loss of intracellular components like protein and nucleic acids [40].

3.4. *Quality of RGJ during 12 Weeks of Storage*

3.4.1. Microbial Qualities

The mean populations of TPCt and YMC in fresh RGJ were 3.02 ± 0.01 log CFU/mL and 3.21 ± 0.04 log CFU/mL, respectively (Table 6). Immediately after HPP treatment and heat treatment, TPCt and YMC were significantly reduced below the level of detection and remained unchanged during the storage period of 12 weeks. The prolonged microbial shelf life of treated samples in this investigation can be explained by the lag phase period of bacteria increasing as well as the growth rate of injured bacterial cells decreasing [47]. Furthermore, cold storage (4 °C) postponed the commencement of the growth phase of pathogenic bacteria, induced inactivation of sub-lethally wounded cells, impeded bacterial spore germination, and prevented quality deterioration owing to residual enzymatic reactions [48–51].

Table 6. Changes in the microbial population in untreated, HPP-treated, and heat-treated RGJ during storage at 4 °C.

Week	TPCt (log CFU/mL)			YMC (log CFU/mL)		
	C-RGJ	HPP-RGJ	HT-RGJ	C-RGJ	HPP-RGJ	HT-RGJ
0	3.02 ± 0.01 [j]	n.d	n.d	3.21 ± 0.04 [k]	n.d	n.d
1	3.77 ± 0.02 [h]	n.d	n.d	3.28 ± 0.01 [j]	n.d	n.d
2	4.91 ± 0.01 [d]	n.d	n.d	5.50 ± 0.01 [e]	n.d	n.d
3	5.67 ± 0.04 [a]	n.d	n.d	6.50 ± 0.02 [c]	n.d	n.d
4	5.66 ± 0.02 [a]	n.d	n.d	6.78 ± 0.03 [a]	n.d	n.d
5	5.50 ± 0.03 [c]	n.d	n.d	6.77 ± 0.04 [a]	n.d	n.d
6	5.51 ± 0.01 [c]	n.d	n.d	6.70 ± 0.05 [b]	n.d	n.d
7	5.55 ± 0.05 [b]	n.d	n.d	6.71 ± 0.01 [b]	n.d	n.d
8	4.89 ± 0.01 [e]	n.d	n.d	6.02 ± 0.07 [d]	n.d	n.d
9	4.88 ± 0.01 [f]	n.d	n.d	5.22 ± 0.01 [f]	n.d	n.d
10	4.55 ± 0.01 [g]	n.d	n.d	5.03 ± 0.01 [g]	n.d	n.d
11	3.28 ± 0.09 [i]	n.d	n.d	4.32 ± 0.01 [h]	n.d	n.d
12	3.01 ± 0.02 [j]	n.d	n.d	4.21 ± 0.11 [i]	n.d	n.d

C-RGJ: control sample; HPP-RGJ: HPP-treated sample; HT-RGJ: heat-treated sample. n.d.: not detected. Different superscripts represent significant differences ($p < 0.05$).

Treatment with HPP (450 and 600 MPa, 5 min) resulted in the inactivation of mesophiles, yeast and molds, and lactic acid bacteria in apple juice, and the counts remained below the detection limits over 12 weeks of storage [52]. The study of Stinco et al. (2019) reported that carrot juice treated at the pressure of 450 MPa for 5 min resulted in no detection of total aerobic mesophiles and yeast and mold viable cell numbers during 14 weeks [53]. Many studies have shown that treatment by high pressure processing extended shelf life under refrigerated storage for kiwi juice (42 days) [54], grape juice (20 days) [55], cucumber juice (20 days) [15], and cloudy hawthorn berry juice (150 days) [56]. Overall, the results showed that HPP treatment improved the safety and prolonged the shelf life of RGJ at 4 °C.

3.4.2. Physicochemical Qualities

The influences of HPP and heat treatment on the stability of pH and TA values of RGJ during storage are shown in Figure 5A,B. TA (%) and pH values of the control RGJ samples were 26.05 ± 0.07 and 5.68 ± 0.01, respectively. These values did not significantly change after HPP treatment while thermal treatment increased TA and decreased pH values. The research on wheat grass juice and on cucumber juice also reported that pH and TA remained unchanged after HPP treatment [15,57]. HPP maintained the pH of juice samples for 3 weeks of storage. The overall trend of sample pH values was downward, whereas TA (%) increased gradually during the storage period. At week 12, the acidity (%) of untreated, HPP-treated, and heat-treated samples had increased by 12.84%, 6.64%, and 12.78%, respectively. Fresh cucumber juice had a similar pH to RGJ at 5.7. The study of Liu et al. (2016) also reported that HPP (500 MPa, 5 min) and high temperature short time (110 °C, 8.6 s) decreased pH and increased TA in cucumber juice after 20 days of refrigeration [15]. Similar findings on cucumber juice also revealed that the pH values of HPP-treated (400 MPa, 4 min) and heat-treated (85 °C, 15 s) juice samples decreased with refrigeration [58]. Depending on the nature of the juice matrix and pressure levels applied, HPP caused the ionization of small molecules like H_2O and weak acids, resulting in the release of hydrogenic ions and increased juice acidity. Thus, compression of the foodstuff influenced the pH and acidity levels [12,59].

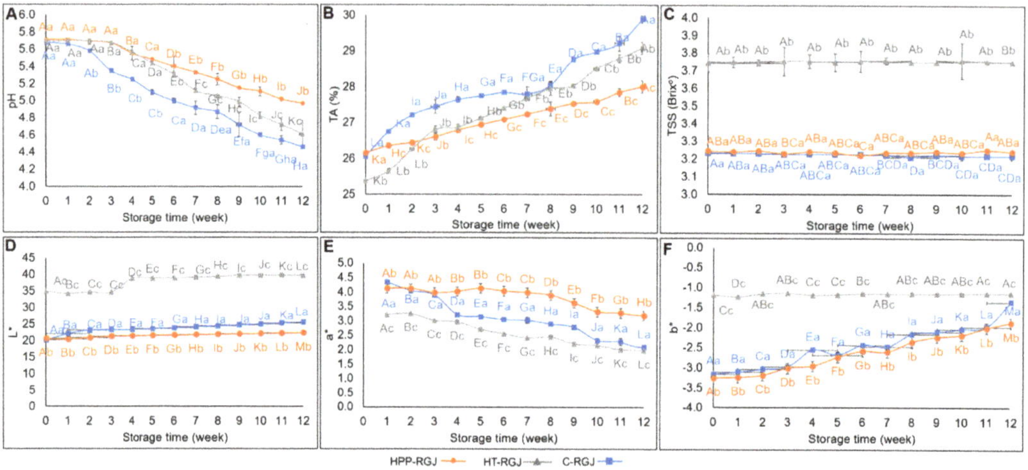

Figure 5. The effect of storage periods at 4 °C on pH (**A**), TA (%) (**B**), TSS (**C**), L* (**D**), a* (**E**), and b* (**F**) of HPP-treated samples (HPP-RGJ), heat-treated samples (HT-RGJ), and the control sample (C-RGJ). Different uppercase letters represent significant differences ($p < 0.05$) between storage times. Different lowercase letters represent significant differences ($p < 0.05$) between samples.

TSS did not vary significantly across samples throughout the 12 weeks of storage ($p > 0.05$) (Figure 5C). Liu et al. (2016) found that HPP and HTST treatment resulted in no significant influence on the TSS value of clear cucumber juice and TSS of all samples remained unchanged during storage (4 °C) [15]. The result is also consistent with research on cloudy cucumber juice [58] and clear apple juice [60].

The color of fruit juices is a key factor in determining consumer sensory acceptability and marketability. The alteration in color parameters (L*, a*, and b*) of HPP and heat-treated RGJ during 12 weeks of storage were evaluated (Figure 5D–F). L*, a*, and b* values of the control RGJ samples were 20.14 ± 0.01, 4.34 ± 0.02, and −3.16 ± 0.02, respectively. After treatment, thermal pasteurization caused changes in RGJ color by elevating the L* and b* values and decreasing the a* value, turning the juice color from black-purple to brown (Figure 6). By contrast, HPP treatment decreased L*, a*, and b* values. Throughout the storage period, the L* and b* values of all samples gradually increased while the a* value decreased. Similar L* value increases were observed in carrot juice during storage (4 °C) [61] and in mulberry juices after thermal and HPP treatment [62]. The increase in L* values was caused by the expulsion of air from the pulp tissue due to cell disruption by pressurization or thermal action, thus making the juice lighter and more opaque [63,64]. All treatments resulted in an increase in the b* value (increase in yellowness) and decrease in a* value (decrease in redness). These results concurred with studies on jabuticaba juice [65] and cucumber juice [15]. Color changes in HPP-treated samples during storage might be induced by the activities of PPO and POD enzymes, which cause discoloration in polyphenols [63]. Overall, HPP preserved the natural color of RGJ after treatment and minimized color changes during storage. A study on pineapple juice reported that HHP had a minor influence on the color of fruit products [66].

Figure 6. Color of HPP-treated sample (HPP-RGJ), heat-treated sample (HT-RGJ), and control sample (C-RGJ).

3.4.3. Phytochemical Qualities

Rice grass juice (RGJ) is a good source of bioactive components. The TPC, FRAP, and ABTS values of initial RGJ samples were 2520.24 ± 67.44 (mgGAE/L), 31.87 ± 0.35 (mmFe^{2+}/L), and 56.28 ± 0.17 (mmTE/L), respectively. As shown in Figure 7, thermal treatment had an intense impact on phenolic compounds and antioxidants in RGJ by decreasing TPC, FRAP, and ABTS values by 36.23%, 41.20%, and 41.93%, respectively. By contrast, treatment with high pressure from 400 to 600 MPa did not significantly affect the ABTS value and slightly increased TPC in RGJ.

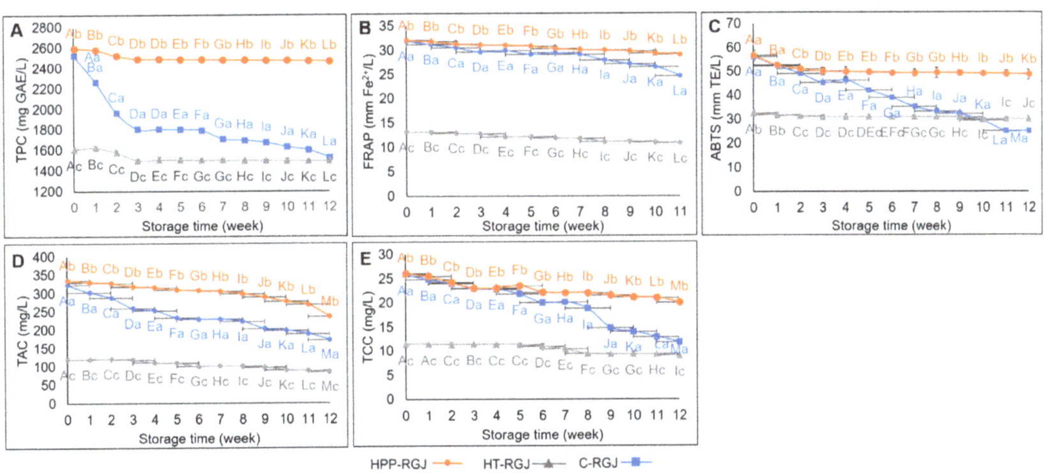

Figure 7. The effect of storage period at 4 °C on TPC (**A**), FRAP (**B**), ABTS (**C**), TAC (**D**), and TCC (**E**) of the HPP-treated sample (HPP-RGJ), the heat-treated sample (HT-RGJ), and the control sample (C-RGJ). Different uppercase letters represent significant differences ($p < 0.05$) between storage times. Different lowercase letters represent significant differences ($p < 0.05$) between samples.

Black rice grass cultivars have high contents of pigmented compounds. The TAC and TCC of fresh RGJ were 322.47 ± 2.76 mg/L and 25.82 ± 0.16 mg/L, respectively (Figure 7D,E). Concurring with color changes in heat-treated RGJ described previously, thermal treatment significantly decreased both TAC and TCC by 62.89% and 55.77%, respectively. The HPP-treated samples recorded an increase in TAC and TCC compared with the control sample, consistent with the increase in TPC after HPP treatment. No significant effects of HPP treatments on cyanidin-3-glucoside content were reported compared with the control, with over 98% of cyanidin-3-glucoside content retained [67].

The increase in TPC, TAC, and TCC was related to the improved extractability of bioactive components. Based on Le Chatelier's theory, a system's volume decreases throughout the pressure-promoting period. HPP promoted solvent penetration that integrated with bioactive components by breaking the cell membranes and accelerating mass transfer by increasing membrane permeability [68,69]. Based on the study of Lou et al. (2022), HPP treatment made the microstructure of the cloudy hawthorn berry juice matrix looser, wrinkled, porous, and fractured, boosting the rate of phenolic compound extraction. They also explained that the phenolic increase following HPP treatment was mostly attributable to an increase in procyanidins and flavonoids [70].

The preservation effect of HPP on bioactive compounds can be explained by two causes. High pressure from 400 to 600 MPa only impacts non-covalent bonds, with little effect on phytochemical constituents that impart flavor, color, and nutritional compounds, while uniform pressure is instantaneously transmitted across all dimensions of the food products, regardless of their size and geometry [9,71].

During storage at 4 °C, the bioactive compound levels of fresh juice samples gradually degraded, while HPP and heat treatment slowed down the degradation of phytochemical constituents and maintained TPC, AOA, TAC, and TCC values due to effective enzyme inactivation. Results of the enzyme inactivation experiments (Table S2) indicated that HPP inactivated peroxidase (POD) and polyphenol oxidase (PPO) in RGJ with decreases of 38.96% and 23.08%, respectively, while heat treatment completely inactivated enzyme activities. A study on cloudy apple juice indicated that the optimal pressure was 600 MPa to retain the content of flavonoids and other polyphenols while reducing the activity of polyphenoloxidase and peroxidase [72].

3.5. Phytochemical Profiles

The phytochemical profiles of the control sample (C-RGJ), HPP-treated sample (HPP-RGJ), and heat-treated sample (HT-RGJ) were compared with commercially available wheat grass juice (WGJ) and barley grass juice (BGJ) samples (Table 7). Results showed that RGJ was a good source of phytochemical compounds. Compared to fresh RGJ, heat treatment caused significant degradation of the 11 phenolic compounds. HPP application preserved most of the contained compounds and also increased the contents of some polyphenols including protocatechuic acid, 4-hydroxybenzoic acid, syringic acid, transcinnamic acid, isorhamnetin-3-o-glucoside, quercetin, and cyanidin-3-glucoside compared with the control samples, consistent with the increases in TPC and TAC in the HPP-treated samples mentioned previously. HPP promoted increased concentrations of some polyphenols in cloudy apple juice such as p-coumaroylquinic acid, caffeoylquinic acid isomer, chlorogenic acid, (-)-epicatechin, phloridzin, phloretin 2-o-xylosylglucoside, procyanidin B2, and procyanidin C1. Improved extraction and determination of bioactive compounds after HPP treatment were achieved because the high pressure broke the cell membrane, leading to better solvent penetration and membrane permeability [56,73].

Table 7. Phytochemical profiles of the control sample (C-RGJ), HPP-treated sample (HPP-RGJ), and heat-treated sample (HT-RGJ) compared with commercially available wheat grass juice (WGJ) and barley grass juice (BGJ).

Phenolic Compound	C-RGJ	HPP-RGJ	HT-RGJ	WGJ	BGJ
Gallic acid	22.07 ± 0.02 [a]	18.94 ± 0.01 [b]	2.97 ± 0.03 [c]	2.13 ± 0.01 [d]	1.84 ± 0.02 [e]
Protocatechuic acid	6.69 ± 0.21 [a]	9.26 ± 0.01 [b]	6.76 ± 0.01 [c]	0.65 ± 0.03 [d]	0.55 ± 0.02 [e]
4-Hydroxybenzoic acid	12.36 ± 0.02 [a]	10.91 ± 0.01 [b]	1.83 ± 0.01 [c]	0.64 ± 0.22 [d]	0.96 ± 0.01 [e]
Chlorogenic acid	26.74 ± 0.03 [a]	25.74 ± 0.03 [b]	14.33 ± 0.04 [c]	13.25 ± 0.01 [d]	12.87 ± 0.07 [e]
Vanillic acid	13.1 ± 0.11 [a]	14.52 ± 0.02 [b]	1.00 ± 0.01 [c]	0.39 ± 0.02 [d]	0.22 ± 0.01 [e]
Syringic acid	10.14 ± 0.02 [a]	146.57 ± 0.01 [b]	146.99 ± 0.07 [c]	68.49 ± 0.06 [d]	39.05 ± 0.01 [e]
p-coumaric acid	686.57 ± 0.01 [a]	673.32 ± 0.01 [b]	12.67 ± 0.01 [c]	n.d	9.61 ± 0.01 [e]
Ferulic acid	254.59 ± 0.02 [a]	254.64 ± 0.01 [a]	35.97 ± 0.02 [b]	4.53 ± 0.03 [c]	6.5 ± 0.01 [d]
Transcinnamic acid	1.3 ± 0.02 [bc]	3.43 ± 0.02 [a]	1.25 ± 0.01 [c]	2.52 ± 0.22 [ab]	2.07 ± 0.04 [ab]
Catechin	7.33 ± 0.02 [a]	12.22 ± 0.01 [b]	51.38 ± 0.02 [c]	n.d	n.d

Table 7. *Cont.*

Phenolic Compound	C-RGJ	HPP-RGJ	HT-RGJ	WGJ	BGJ
Rutin	10.12 ± 0.02 [a]	11.22 ± 0.03 [b]	137.31 ± 0.02 [c]	2.93 ± 0.02 [d]	3.64 ± 0.01 [e]
Isorhamnetin-3-o-glucoside	27.95 ± 0.01 [a]	102.02 ± 0.07 [b]	100.14 ± 0.02 [c]	n.d	n.d
Quercetin	9.85 ± 0.01 [a]	11.43 ± 0.01 [b]	2.03 ± 0.05 [c]	4.12 ± 0.01 [d]	n.d
Naringenin	115.81 ± 0.07 [a]	88.79 ± 0.02 [b]	39.04 ± 0.13 [c]	137.03 ± 0.12 [d]	0.59 ± 0.01 [e]
Cyanidin-3-glucoside	2.26 ± 0.05 [a]	7.01 ± 0.04 [b]	0.73 ± 0.01 [b]	n.d	n.d
Peonidin-3-glucoside	n.d	n.d	n.d	n.d	n.d

Different superscripts represent significant differences ($p < 0.05$). n.d: not detected.

The commercially available wheat grass juice and barley grass juice samples did not contain catechin, isorhamnetin-3-o-glucoside, naringenin, or two anthocyanin compounds including cyanidin-3-glucoside and peonidin-3-glucoside. Black RGJ revealed a 2 to 10-fold higher content of most phytochemicals determined in this study compared to the two commercial products. This result concurred with Khanthapoka et al. (2015) who reported that RGJ from a black-color rice cultivar exhibited higher TPC and AOA compared to a wheat grass juice sample [5].

4. Conclusions

High pressure processing is an innovative non-thermal processing technique which is useful for industrial processing of fruit and vegetable juices. This study reports that HPP at 612 MPa, 11 min and 36 °C was determined to be the optimal condition to inactivate *E. coli* and *L. innocua* for more than 5 log reduction in RGJ, meeting the regulations of the FDA. HPP treatment at the optimal condition could prolong the microbial shelf life of RGJ by at least 12 weeks. This technique also preserved both the physicochemical and phytochemical properties during storage at 4 °C, unlike heat treatment (85 °C/10 min). Moreover, HPP-treated RGJ contained significantly higher phytochemical amounts compared with commercially available wheat grass juice and barley grass juice. The results reveal that high pressure processing is a potential alternative to thermal pasteurization in improving the shelf life and preserving the nutritional quality of low-acidity juice. Therefore, this finding suggests the possibility of developing a new functional cereal beverage preserved by HPP from pigmented rice grass. However, incomplete enzyme inactivation was the limitation of HPP, which caused the deterioration of bioactive compounds during long-term storage. Thus, more in-depth study on this issue is necessary in the future.

Supplementary Materials: The following supporting information can be downloaded at: https://www.mdpi.com/article/10.3390/foods13182995/s1, Table S1: Sublethal injury ratio (%) of *E. coli* and *L. innocua* in HT-treated and HPP-treated RGJ; Table S2: Changes in enzyme activity in untreated, HPP-treated, and heat-treated RGJ during storage at 4 °C; Table S3: ANOVA of the effect of HPP variables on *E. coli* K12 log reduction; Table S4: ANOVA of the effect of HPP variables on *L. innocua* log reduction.

Author Contributions: Investigation, U.H.D., J.N.L. and W.K.; methodology, U.H.D., S.T. (Sirinya Taya), S.P. and W.K.; visualization, U.H.D.; writing, U.H.D., J.N.L., S.T. (Sitthidat Tongdonyod) and W.K.; editing, U.H.D. and W.K.; conceptualization, J.N.L., S.T. (Sitthidat Tongdonyod) and W.K.; data analysis, J.N.L. and S.T. (Sirinya Taya); formal analysis, S.T. (Sitthidat Tongdonyod); resources, S.P. and W.K.; review, S.T. (Sitthidat Tongdonyod), S.T. (Sirinya Taya), S.P. and W.K.; project administration, W.K.; funding acquisition, W.K.; supervision, W.K. All authors have read and agreed to the published version of the manuscript.

Funding: This research was funded by the National Research Council of Thailand and High Pressure Food Products (Thailand) Limited Partnership, the project was supported by Fundamental Fund 2024, Chiang Mai University (Grant No. FF059/2567).

Data Availability Statement: The original contributions presented in the study are included in the article; further enquiries can be directed to the corresponding authors.

Conflicts of Interest: The authors declare no conflict of interest.

References

1. Seibold, R.L. *Cereal Grass: Nature's Greatest Health Gift*; Keats Pub: New Canaan, CT, USA, 1991; pp. 20–100.
2. Rodríguez, F.C.; Gallagher, E.; Rai, D.K.; Burgess, C.M. Nutritional and physiochemical properties of wheatgrass juice and preservation strategies. *Food Chem. Adv.* **2022**, *1*, 100136. [CrossRef]
3. Meyerowitz, S. *Wheatgrass, Natures Finest Medicine*; Book Publishing Company: Summertown, TN, USA, 2006.
4. Chomchan, R.; Siripongvutikorn, S.; Puttarak, P.; Rattanapon, R. Investigation of Phytochemical Constituents, Phenolic Profiles and Antioxidant Activities of Ricegrass Juice compared to Wheatgrass *Juice. Funct. Foods Heal. Dis.* **2016**, *6*, 822. [CrossRef]
5. Khanthapoka, P.; Muangpromb, A.; Sukronga, S. Antioxidant activity and DNA protective properties of rice grass juices. *Sci. Asia* **2015**, *41*, 119. [CrossRef]
6. Thepthanee, C.; Liu, C.; Yu, H.; Huang, H.; Yen, C.; Li, Y.; Liaw, E. Evaluation of Phytochemical Contents and In Vitro Antioxidant, Anti-Inflammatory, and Anticancer Activities of Black Rice Leaf (*Oryza sativa* L.) Extract and Its Fractions. *Foods* **2021**, *10*, 2987. [CrossRef] [PubMed]
7. Suppakul, P.; Kim, D.Y.; Yang, J.H.; Lee, S.B.; Lee, S.J. Practical design of a diffusion-type time-temperature indicator with intrinsic low temperature dependency. *J. Food Eng.* **2018**, *223*, 22–31. [CrossRef]
8. World Health Organization; Food and Agriculture Organization of the United Nations. *Codex Alimentarius: General Requirements (Food Hygiene)*; Food and Agriculture Organization of the United Nations: Rome, Italy, 2001; p. 247.
9. Balasubramaniam, V.; Barbosa-Cánovas, G.V.; Lelieveld, H. *High Pressure Processing of Food: Principles, Technology and Applications*; Springer: Berlin/Heidelberg, Germany, 2016.
10. Koseki, S.; Yamamoto, K. Water activity of bacterial suspension media unable to account for the baroprotective effect of solute concentration on the inactivation of Listeria monocytogenes by high hydrostatic pressure. *Int. J. Food Microbiol.* **2007**, *115*, 43–47. [CrossRef]
11. Neetoo, H.; Haiqiang, C. Application of High Hydrostatic Pressure Technology for Processing and Preservation of Foods. In *Progress in Food Preservation*; Bhat, R., Alias, A.K., Paliyath, G., Eds.; John Wiley & Sons Ltd.: Oxford, UK, 2012; Chapter 12; pp. 247–276.
12. Ravichandran, C.; Jayachandran, L.E.; Kothakota, A.; Pandiselvam, R.; Balasubramaniam, V. Influence of high pressure pasteurization on nutritional, functional and rheological characteristics of fruit and vegetable juices and purees-an updated review. *Food Control* **2023**, *146*, 109516. [CrossRef]
13. Sun, D.W. *Emerging Technologies for Food Processing*; Academic Press: Cambridge, MA, USA; Elsevier Science Ltd.: Dublin, Ireland, 2014.
14. Zhang, Y.; Liu, X.; Wang, Y.; Zhao, F.; Sun, Z.; Liao, X. Quality comparison of carrot juices processed by high-pressure processing and high-temperature short-time processing. *Innov. Food Sci. Emerg. Technol.* **2015**, *33*, 135–144. [CrossRef]
15. Liu, F.; Zhang, X.; Zhao, L.; Wang, Y.; Liao, X. Potential of high-pressure processing and high-temperature/short-time thermal processing on microbial, physicochemical and sensory assurance of clear cucumber juice. *Innov. Food Sci. Emerg. Technol.* **2016**, *34*, 51–58. [CrossRef]
16. Ali, N.; Popović, V.; Koutchma, T.; Warriner, K.; Zhu, Y. Effect of thermal, high hydrostatic pressure, and ultraviolet-C processing on the microbial inactivation, vitamins, chlorophyll, antioxidants, enzyme activity, and color of wheatgrass juice. *J. Food Process Eng.* **2020**, *43*, 13036. [CrossRef]
17. Quiroz-González, B.; Rodríguez-Martínez, V.; García-Mateos, M.d.R.; Torres, J.A.; Welti-Chanes, J. High hydrostatic pressure inactivation and recovery study of Listeria innocua and Saccharomyces cerevisiae in pitaya (*Stenocereus pruinosus*) juice. *Innov. Food Sci. Emerg. Technol.* **2018**, *50*, 169–173. [CrossRef]
18. Raghubeer, E.V.; Phan, B.N.; Onuoha, E.; Diggins, S.; Aguilar, V.; Swanson, S.; Lee, A. The use of High-Pressure Processing (HPP) to improve the safety and quality of raw coconut (*Cocos nucifera* L.) water. *Int. J. Food Microbiol.* **2020**, *331*, 108697. [CrossRef] [PubMed]
19. Sha, X.; Li, P.; Feng, Y.; Xia, D.; Tian, X.; Wang, Z.; Yang, Y.; Mao, X.; Liu, L. Self-Assembled Peptide Nanofibrils Designed to Release Membrane-Lysing Antimicrobial Peptides. *ACS Appl. Bio Mater.* **2020**, *3*, 3648–3655. [CrossRef] [PubMed]
20. Sharma, A.; Yadav, M.; Sharma, N.; Kumari, A.; Kaur, S.; Meenu, M.; Garg, M. Comparison of wheatgrass juices from colored wheat (white, black, blue, and purple) for health promoting phytochemicals. *Food Res. Int.* **2022**, *161*, 111833. [CrossRef]
21. Apang, T.; Xavier, K.M.; Lekshmi, M.; Kannuchamy, N.; Layana, P.; Balange, A.K. *Garcinia* spp. extract incorporated icing medium as a natural preservative for shelf life enhancement of chilled Indian mackerel (*Rastrelliger kanagurta*). *LWT* **2020**, *133*, 110086. [CrossRef]
22. Jafari, S.; Pongsarn, K.; Srestasupana, C.; Wetchasart, N.; Assatarakul, K. Kinetic study of microbial inhibition by dimethyl dicarbonate and quality attributes of pomegranate juice during cold storage. *LWT* **2021**, *152*, 112309. [CrossRef]
23. Re, R.; Pellegrini, N.; Proteggente, A.; Pannala, A.; Yang, M.; Rice-Evans, C. Antioxidant activity applying an im-proved ABTS radical cation decolorization assay. *Free. Radic. Biol. Med.* **1999**, *26*, 1231–1237. [CrossRef]
24. Benzie, I.F.F.; Strain, J.J. The ferric reducing ability of plasma (FRAP) as a measure of "antioxidant power": The FRAP assay. *Anal. Biochem.* **1996**, *239*, 70–76. [CrossRef]

25. Lee, J.; Durst, R.W.; Wrolstad, R.E.; Wrolstad, R.E. Determination of Total Monomeric Anthocyanin Pigment Content of Fruit Juices, Beverages, Natural Colorants, and Wines by the pH Differential Method: Collaborative Study. *J. AOAC Int.* **2005**, *88*, 1269–1278. [CrossRef] [PubMed]
26. Porra, R.J.; Thompson, W.A.; Kriedemann, P.E. Determination of accurate extinction coefficients and simultaneous equations for assaying chlorophylls a and b extracted with four different solvents: Verification of the concentration of chlo-rophyll standards by atomic absorption spectroscopy. *Biochim. Biophys. Acta (BBA)-Bioenerg.* **1989**, *975*, 384–394. [CrossRef]
27. Chen, X.; Chhun, S.; Xiang, J.; Tangjaidee, P.; Peng, Y.; Quek, S.Y. Microencapsulation of *Cyclocarya paliurus* (Batal.) Iljinskaja Extracts: A Promising Technique to Protect Phenolic Compounds and Antioxidant Capacities. *Foods* **2021**, *10*, 2910. [CrossRef] [PubMed]
28. Zeb, A. *Phenolic Antioxidants in Foods: Chemistry, Biochemistry and Analysis*; Springer Nature: Dordrecht, The Netherlands, 2021; p. 247.
29. Randhir, R.; Shetty, K. Developmental stimulation of total phenolics and related antioxidant activity in light- and dark-germinated corn by natural elicitors. *Process Biochem.* **2004**, *40*, 1721–1732. [CrossRef]
30. Kohler, G.O. The effect of stage of growth on the chemistry of the grasses. *J. Biol. Chem.* **1944**, *152*, 215–223. [CrossRef]
31. Kapkum, N.; Phimphilai, S.; Srichairatanakool, S.; Varith, J. Reduction in antioxidant properties lost during processing of a powdered beverage from young organic rice plants. *Asian J. Food Ag.-Ind.* **2011**, *4*, 388–398.
32. Kulkarni, S.D.; Tilak, J.C.; Acharya, R.; Rajurkar, N.S.; Devasagayam, T.P.A.; Reddy, A.V.R. Evaluation of the antioxidant activity of wheatgrass (*Triticum aestivum* L.) as a function of growth under different conditions. *Phytother. Res. Int. J. Devoted Pharmacol. Toxicol. Eval. Nat. Prod. Deriv.* **2006**, *20*, 218–227. [CrossRef]
33. Chakraborty, S.; Rao, P.S.; Mishra, H.N. Response Surface Optimization of Process Parameters and Fuzzy Analysis of Sensory Data of High Pressure–Temperature Treated Pineapple Puree. *J. Food Sci.* **2015**, *80*, 63–75. [CrossRef] [PubMed]
34. Podolak, R.; Whitman, D.; Black, D.G. Factors Affecting Microbial Inactivation during High Pressure Processing in Juices and Beverages: A Review. *J. Food Prot.* **2020**, *83*, 1561–1575. [CrossRef]
35. Alpas, H.; Kalchayanand, N.; Bozoglu, F.; Ray, B. Interactions of high hydrostatic pressure, pressurization temperature and pH on death and injury of pressure-resistant and pressure-sensitive strains of foodborne pathogens. *Int. J. Food Microbiol.* **2000**, *60*, 33–42. [CrossRef]
36. Pokhrel, P.R.; Toniazzo, T.; Boulet, C.; Oner, M.E.; Sablani, S.S.; Tang, J.; Barbosa-Cánovas, G.V. Inactivation of *Listeria innocua* and *Escherichia coli* in carrot juice by combining high pressure processing, nisin, and mild thermal treatments. *Innov. Food Sci. Emerg. Technol.* **2019**, *54*, 93–102. [CrossRef]
37. Kalchayanand, N.; Sikes, A.; Dunne, C.; Ray, B. Interaction of hydrostatic pressure, time and temperature of pres-surization and pediocin AcH on inactivation of foodborne bacteria. *J. Food Prot.* **1998**, *61*, 425–431. [CrossRef]
38. Silhavy, T.J.; Kahne, D.; Walker, S. The bacterial cell envelope. *Cold Spring Harb. Perspect. Biol.* **2010**, *2*, 414. [CrossRef]
39. Fellows, P.J. *Food Processing Technology: Principles and Practice*; Elsevier: Amsterdam, The Netherlands, 2009.
40. Huang, H.-W.; Lung, H.-M.; Yang, B.B.; Wang, C.-Y. Responses of microorganisms to high hydrostatic pressure processing. *Food Control* **2014**, *40*, 250–259. [CrossRef]
41. Rajauria, G.; Tiwari, B.K. *Fruit Juices: Extraction, Composition, Quality and Analysis*; Academic Press: Cambridge, MA, USA, 2017.
42. Roohinejad, S.; Koubaa, M.; Sant'Ana, A.S.; Greiner, R. Mechanisms of microbial inactivation by emerging technologies. In *Innovative Technologies for Food Preservation*; Elsevier: Amsterdam, The Netherlands, 2018; Chapter 4; pp. 111–132.
43. Sehrawat, R.; Kaur, B.P.; Nema, P.K.; Tewari, S.; Kumar, L. Microbial inactivation by high pressure processing: Principle, mechanism and factors responsible. *Food Sci. Biotechnol.* **2020**, *30*, 19–35. [CrossRef] [PubMed]
44. Goodrich, R.M.; Schneider, K.R.; Parish, M. The Juice HACCP Program: An Overview. *Food Saf. Toxicol. Ser.* **2005**, 15. [CrossRef]
45. Yang, B.; Shi, Y.; Xia, X.; Xi, M.; Wang, X.; Ji, B.; Meng, J. Inactivation of foodborne pathogens in raw milk using high hydrostatic pressure. *Food Control.* **2012**, *28*, 273–278. [CrossRef]
46. Ritz, M.; Tholozan, J.; Federighi, M.; Pilet, M. Physiological damages of Listeria monocytogenes treated by high hydrostatic pressure. *Int. J. Food Microbiol.* **2002**, *79*, 47–53. [CrossRef] [PubMed]
47. Chen, D.; Xi, H.; Guo, X.; Qin, Z.; Pang, X.; Hu, X.; Liao, X.; Wu, J. Comparative study of quality of cloudy pome-granate juice treated by high hydrostatic pressure and high temperature short time. *Innov. Food Sci. Emerg. Technol.* **2013**, *19*, 85–94. [CrossRef]
48. Briones, L.S.; Reyes, J.E.; Tabilo-Munizaga, G.E.; Pérez-Won, M.O. Microbial shelf-life extension of chilled Coho salmon (*Oncorhynchus kisutch*) and abalone (*Haliotis rufescens*) by high hydrostatic pressure treatment. *Food Control* **2010**, *21*, 1530–1535. [CrossRef]
49. Cruz-Romero, M.; Kerry, J.; Kelly, A. Changes in the microbiological and physicochemical quality of high-pressure-treated oysters (*Crassostrea gigas*) during chilled storage. *Food Control* **2008**, *19*, 1139–1147. [CrossRef]
50. Scolari, G.; Zacconi, C.; Busconi, M.; Lambri, M. Effect of the combined treatments of high hydrostatic pressure and temperature on *Zygosaccharomyces bailii* and *Listeria monocytogenes* in smoothies. *Food Control* **2015**, *47*, 166–174. [CrossRef]
51. Varela-Santos, E.; Ochoa-Martinez, A.; Tabilo-Munizaga, G.; Reyes, J.E.; Pérez-Won, M.; Briones-Labarca, V.; Morales-Castro, J. Effect of high hydrostatic pressure (HHP) processing on physicochemical properties, bioactive compounds and shelf-life of pomegranate juice. *Innov. Food Sci. Emerg. Technol.* **2011**, *13*, 13–22. [CrossRef]

52. Szczepańska, J.; Pinto, C.A.; Skąpska, S.; Saraiva, J.A.; Marszałek, K. Effect of static and multi-pulsed high pressure processing on the rheological properties, microbial and physicochemical quality, and antioxidant potential of apple juice during refrigerated storage. *LWT* **2021**, *150*, 20–38. [CrossRef]
53. Stinco, C.M.; Szczepańska, J.; Marszałek, K.; Pinto, C.A.; Inácio, R.S.; Mapelli-Brahm, P.; Barba, F.J.; Lorenzo, J.M.; Saraiva, J.A.; Meléndez-Martínez, A.J. Effect of high-pressure processing on carotenoids profile, colour, microbial and enzymatic stability of cloudy carrot juice. *Food Chem.* **2019**, *299*, 35. [CrossRef] [PubMed]
54. Xu, X.; Deng, J.; Luo, D.; Bao, Y.; Liao, X.; Gao, H.; Wu, J. Comparative study of high hydrostatic pressure and high temperature short time processing on quality of clear and cloudy Se-enriched kiwifruit juices. *Innov. Food Sci. Emerg. Technol.* **2018**, *49*, 1–12. [CrossRef]
55. Chang, Y.; Wu, S.; Chen, B.; Huang, H.; Wang, C. Effect of high-pressure processing and thermal pasteurization on overall quality parameters of white grape juice. *J. Sci. Food Agric.* **2016**, *97*, 3166–3172. [CrossRef]
56. Lou, X.; Jin, Y.; Tian, H.; Yu, H.; Chen, C.; Hanna, M.; Lin, Y.; Yuan, L.; Wang, J.; Xu, H. High-pressure and thermal processing of cloudy hawthorn berry (*Crataegus pinnatifida*) juice: Impact on microbial shelf-life, enzyme activity and quality-related attributes. *Food Chem.* **2021**, *372*, 131313. [CrossRef]
57. Ali, N. Comparative Study of the Effects of Ultraviolet Light and High Hydrostatic Pressure on the Quality and Health Related Constituents of Wheatgrass Juice. Doctoral Dissertation, University of Guelph, Guelph, ON, Canada, 2016. Available online: https://atrium.lib.uoguelph.ca/items/36c3b97f-cacb-4cd2-9669-fbf8362ebbee (accessed on 9 September 2024).
58. Zhao, L.; Wang, S.; Liu, F.; Dong, P.; Huang, W.; Xiong, L.; Liao, X. Comparing the effects of high hydrostatic pressure and thermal pasteurization combined with nisin on the quality of cucumber juice drinks. *Innov. Food Sci. Emerg. Technol.* **2013**, *17*, 27–36. [CrossRef]
59. Kaushik, N.; Kaur, B.P.; Rao, P.S.; Mishra, H. Effect of high pressure processing on color, biochemical and mi-cro-biological characteristics of mango pulp (*Mangifera indica* cv. Amrapali). *Innov. Food Sci. Emerg. Technol.* **2014**, *22*, 40–50. [CrossRef]
60. Zhao, L.; Wang, Y.; Qiu, D.; Liao, X. Effect of Ultrafiltration Combined with High-Pressure Processing on Safety and Quality Features of Fresh Apple Juice. *Food Bioprocess Technol.* **2014**, *7*, 3246–3258. [CrossRef]
61. Liao, H.; Sun, Y.; NI, Y.; Liao, X.; Hu, X.; Wu, J.; Chen, F. The Effect of Enzymatic Mash Treatment, Pressing, Centrifugation, Homogenization, Deaeration, Sterilization and Storage on Carrot Juice. *J. Food Process. Eng.* **2007**, *30*, 421–435. [CrossRef]
62. You, Y.; Li, N.; Han, X.; Guo, J.; Zhao, Y.; Liu, G.; Huang, W.; Zhan, J. Influence of different sterilization treatments on the color and anthocyanin contents of mulberry juice during refrigerated storage. *Innov. Food Sci. Emerg. Technol.* **2018**, *48*, 1–10. [CrossRef]
63. Abid, M.; Jabbar, S.; Hu, B.; Hashim, M.M.; Wu, T.; Wu, Z.; Khan, M.A.; Zeng, X. Synergistic impact of sonication and high hydrostatic pressure on microbial and enzymatic inactivation of apple juice. *LWT* **2014**, *59*, 70–76. [CrossRef]
64. Landl, A.; Abadias, M.; Sárraga, C.; Viñas, I.; Picouet, P. Effect of high pressure processing on the quality of acidified Granny Smith apple purée product. *Innov. Food Sci. Emerg. Technol.* **2010**, *11*, 557–564. [CrossRef]
65. Geraldi, M.V.; Cazarin, C.B.B.; Dias-Audibert, F.L.; Pereira, G.A.; Carvalho, G.G.; Kabuki, D.Y.; Catharino, R.R.; Pastore, G.M.; Behrens, J.H.; Cristianini, M.; et al. Influence of high isostatic pressure and thermal pasteurization on chemical composition, color, antioxidant properties and sensory evaluation of jabuticaba juice. *LWT* **2020**, *139*, 110548. [CrossRef]
66. Perera, N.; Gamage, T.; Wakeling, L.; Gamlath, G.; Versteeg, C. Colour and texture of apples high pressure processed in pineapple juice. *Innov. Food Sci. Emerg. Technol.* **2010**, *11*, 39–46. [CrossRef]
67. Yu, Y.; Lin, Y.; Zhan, Y.; He, J.; Zhu, S. Effect of high pressure processing on the stability of anthocyanin, ascorbic acid and color of Chinese bayberry juice during storage. *J. Food Eng.* **2013**, *119*, 701–706. [CrossRef]
68. Shouqin, Z.; Jun, X.; Changzheng, W. High hydrostatic pressure extraction of flavonoids from propolis. *J. Chem. Technol. Biotechnol.* **2005**, *80*, 50–54. [CrossRef]
69. Xi, J.; Shen, D.; Zhao, S.; Lu, B.; Li, Y.; Zhang, R. Characterization of polyphenols from green tea leaves using a high hydrostatic pressure extraction. *Int. J. Pharm.* **2009**, *382*, 139–143. [CrossRef]
70. Lou, X.; Xiong, J.; Tian, H.; Yu, H.; Chen, C.; Huang, J.; Yuan, H.; Hanna, M.; Yuan, L.; Xu, H. Effect of high-pressure processing on the bioaccessibility of phenolic compounds from cloudy hawthorn berry (*Crataegus pinnatifida*) juice. *J. Food Compos. Anal.* **2022**, *110*, 104540. [CrossRef]
71. Hogan, E.; Kelly, A.L.; Sun, D.W. High Pressure Processing of Foods: An Overview. In *Emerging Technologies for Food Processing*; Elsevier BV: Amsterdam, The Netherlands, 2005; pp. 3–32.
72. Yi, J.; Kebede, B.T.; Dang, D.N.H.; Buvé, C.; Grauwet, T.; Van Loey, A.; Hu, X.; Hendrickx, M. Quality change during high pressure processing and thermal processing of cloudy apple juice. *LWT* **2017**, *75*, 85–92. [CrossRef]
73. Szczepańska, J.; Barba, F.J.; Skąpska, S.; Marszałek, K. Changes in the polyphenolic profile and oxidoreductases activity under static and multi-pulsed high-pressure processing of cloudy apple juice. *Food Chem.* **2022**, *384*, 24–39. [CrossRef] [PubMed]

foods

MDPI

Article

Inactivation of Hepatitis A Virus and Feline Calicivirus on Model Food Contact Surfaces by Ultraviolet Light (UV-C) Systems

Breanna Polen [1], Brahmaiah Pendyala [2], Ankit Patras [2] and Doris H. D'Souza [1,*]

[1] Department of Food Science, University of Tennessee, Knoxville, TN 37996, USA; jnn668@vols.utk.edu
[2] Department of Food and Animal Sciences, Tennessee State University, Nashville, TN 37209, USA; bpendyal@tnstate.edu (B.P.); apatras@tnstate.edu (A.P.)
* Correspondence: ddsouza@utk.edu

Abstract: Food contact surfaces can harbor and transmit pathogens leading to outbreaks. Decontamination strategies that are user- and environmentally-friendly without toxic by-product formation are needed. Novel UV-C light-emitting diode (LED) technologies are being explored to deliver the required dose to inactivate viruses in food-processing environments. The objective of this study was to compare the effects of 279 nm UV-C LED to 254 nm UV-C against hepatitis A virus (HAV) and feline calicivirus (FCV, a cultivable human norovirus surrogate) on stainless-steel, ceramic, and glass surfaces. Viruses were surface spread on sterile stainless-steel or ceramic coupons (100 μL on 2×2 cm^2), or glass discs (50 μL on 1×1 cm^2), air-dried, and UV-C-treated for up to 3.75 min (surface dose = 0–49.2 mJ/cm^2 for HAV and 0–24.6 mJ/cm^2 for FCV). Each triplicate treatment was assayed in duplicate, and data were statistically analyzed. The D_{10}-values for HAV treated with UV-C at 254 nm on stainless-steel, ceramic, and glass were 9.48 ± 0.34, 14.53 ± 2.52, and 6.91 ± 1.93 mJ/cm^2, while with UV-C LED at 279 nm were 19.53 ± 2.45, 26.05 ± 0.60, and 8.77 ± 2.08 mJ/cm^2, respectively. The D_{10}-values for FCV treated with UV-C at 254 nm on stainless-steel, ceramic, and glass were 3.65 ± 0.06, 6.25 ± 1.90, and 4.69 ± 0.03 mJ/cm^2, while with UV-C LED at 279 nm were 7.097 ± 2.11, 8.31 ± 2.12, and 7.88 ± 0.86 mJ/cm^2, respectively. Higher 279 nm UV-C doses were needed to inactivate HAV and FCV compared to 254 nm UV-C on the tested surfaces. Novel UV-C LED systems using appropriate doses show promise to inactivate foodborne viruses on food contact surfaces.

Keywords: Hepatitis A virus; human norovirus; feline calicivirus; ultraviolet light systems; inactivation; food contact surfaces

1. Introduction

Environmental sanitation is of major importance to the agricultural industry and public health sector as contamination by bacteria, viruses, parasites, or fungi can cause foodborne illnesses affecting 1 in 6 individuals in the United States annually, with 128,000 hospitalizations out of the estimated 48 million illnesses each year [1]. Though *Salmonella* is most frequently reported with 1.35 million infections and 26,500 hospitalizations in the United States annually, foodborne viruses are also increasingly reported to cause human gastroenteritis illnesses worldwide [1]. Hepatitis A virus (HAV) and human noroviruses (HuNoVs) are the epidemiologically significant foodborne viruses of human health concern [2,3]. Hepatitis A virus (HAV) is a non-enveloped, single-stranded RNA virus of the genus Hepatovirus in the family *Picornaviridae* that is transmitted through the fecal-oral route and affects the liver with symptoms that last more than a month [4]. Contact with an HAV infected individual, contamination of food by an HAV infected food handler, or contaminated surfaces can lead to illness as HAV can survive in the environment for extended periods and under low pH conditions [5]. Although, effective HAV vaccinations were introduced in 1995 with a decrease in number of HAV outbreaks, they continue to occur worldwide [6]. As of August 2023, in the U.S. alone, HAV outbreaks were reported in 37 states totaling 44,903 cases, 27,435 hospitalizations, and 423 deaths linked to person-to-person contact [2].

While HAV surveillance is important due to the severity and length of disease symptoms, HuNoV illnesses are the most frequently reported foodborne illness with 21 million illnesses occurring annually in the U.S. [7]. HuNoV's belong to the *Caliciviridae* family and like HAV are also non-enveloped with a positive-sense, single-stranded RNA enclosed in a capsid, but have a "cup-like" shape and though are transmitted in a similar manner as HAV, have a short incubation period [8]. They are also environmentally stable and possess an infection dose between 18 and 2,800 particles which makes prevention of its spread extremely difficult, especially in the absence of commercially available vaccines [9]. Although HuNoV infection is typically self-limiting in healthy individuals, severe health complications can occur in immunocompromised individuals, the elderly, and young children [8]. Foodborne HuNoV outbreaks continue to occur and a multistate HuNoV outbreak linked to raw oysters was reported between November and December of 2022 in Texas, U.S. [7] and in British Columbia with 192 illnesses as of June 2022 [7]. Since all genogroups of HuNoV's cannot reproducibly be cultivated in the laboratory at high titers for inactivation studies, cultivable surrogates including feline calicivirus (FCV-F9), murine norovirus (MNV-1) and Tulane virus (TV) have been used [10–12].

The importance of sanitation of food contact surfaces in the prevention of foodborne outbreak spread is well-recognized within the food industry. Commonly used chemical disinfectants include chlorine, quaternary ammonium, hydrogen peroxide, and peracetic acid [13–15], but can generate hazardous waste and harmful by-products. An alternate surface disinfection approach is the application of ultraviolet (UV) light-C technology in the range of 200 to 280 nm, traditionally using low-pressure mercury lamps (typically at 254 nm), for inactivation of microorganisms [16,17]. UV-C damages DNA and RNA due to photochemical changes, cross-linking, and oxidative damage and prevents DNA replication [18]. The genetic material within viral capsids are strong absorbers of UV radiation, especially near the 254 nm wavelength, whereas other viral components such as proteins are minor absorbers of UV radiation at this wavelength [19]. Research indicates that UV-C at 254 nm inactivation follows a dose-dependent relationship with varying effects depending on the type of microorganism [20]. Previous studies showed that dried FCV on stainless-steel discs was completely inactivated (>5.0 log) at a 254 nm UV-C dose of 60 mJ/cm^2 [21]. Another study reported a 2.6 log reduction of HAV on stainless-steel surfaces with 300 mJ/cm^2 using 254 nm UV-C [22].

However, more recently, UV-C light emitting diodes (LEDs) have become widely studied as a decontamination approach with energy efficient inactivation of microorganisms [23–25]. LEDs possess intrinsic properties regarding optical features, are more sustainable, portable, inexpensive with low heat emission, have increased wavelength diversity, do not require a warmup time, and without hazardous waste generation due to their lack of mercury as opposed to traditional UV-C mercury containing lamps [26]. UV-C LEDs can be designed to have a peak wavelength close to absorption maxima of DNA and RNA or proteins [23].

There is limited research on the use of UV-C LED at varying wavelengths such as 279 nm for the inactivation of foodborne viruses on contact surfaces. Previously, the HuNoV surrogate, feline calicivirus (FCV-F9) on stainless-steel surfaces was shown to be reduced by 3 log PFU/disc using UV-C LED technology at 269 nm at a dose of 22.5 mJ/cm^2 [27]. Furthermore, the dose required to achieve a 1 log or 90% reduction (D_{10}-value) of FCV on Formica coupons was reported to be 23.37 $\pm$ 0.91 mJ/cm^2 using UV-C LED at 279 nm and 9.97 $\pm$ 2.44 mJ/cm^2 using UV-C at 254 nm [28]. This same study also showed that HAV on Formica coupons had similar D_{10}-values using both traditional UV-C at 254 nm (D_{10} = 12.40 $\pm$ 1.15 mJ/cm^2) and UV-C LED at 279 nm (D_{10} = 12.39 $\pm$ 0.70 mJ/cm^2) [28]. Taken together, UV-C LED at 279 nm shows promise for inactivation of viruses on surfaces with its advantages over traditional UV-C at 254 nm. Therefore, this aimed to evaluate the effectiveness of UV-C LED at 279 nm in comparison to traditional UV-C at 254 nm for the inactivation of HAV and the cultivable human norovirus surrogate, FCV-F9 dried on stainless-steel, ceramic and glass surfaces as model food contact surfaces.

2. Materials and Methods

2.1. Animal Host Cell Lines for HAV and FCV Propagation

Fetal rhesus monkey kidney cells (FRHK-4) were used as host cells for the propagation of HAV-HM175 generously provided by Dr. Kalmia Kniel's laboratory (University of Delaware), and Crandell-Reese Feline Kidney (CRFK) cells were used as host cells for the propagation of FCV-F9, obtained from the American Type Culture Collection (ATCC, Manassas, VA, USA), as reported in our earlier studies [11,29]. Dulbecco's Modified Eagle Medium (DMEM-F12) containing 10% fetal bovine serum (FBS) and 1% Penicillin Streptomycin (Pen-Strep) (PS; Fisher Scientific, Pittsburgh, PA, USA) were used to maintain and propagate both cell lines.

2.2. HAV and FCV Propagation

Previously published protocols were used to propagate HAV and FCV [11,29,30]. Briefly, 2 mL of HAV (strain HM175), generously provided by Dr. Kalmia Kniel's laboratory (University of Delaware), was added to confluent FRHK-4 cells within sterile 175 cm^2 cell-culture flasks along with 8 mL DMEM-F12 containing 2% FBS and 1% Pen-Strep and incubated at 37 °C for three hours in a CO_2 incubator with 5% CO_2. This was followed by the addition of 10 mL DMEM-F12 containing 10% FBS and 1% Pen-Strep. The infected flasks were then incubated for five to seven days, within a water-jacketed incubator under 5% CO_2. Once cytopathic effects were observed, the infected flasks were frozen within a −80 °C freezer and freeze-thawed thrice, centrifuged at 5000 rpm for 10 min and filtered through a 0.2-micron filter to be stored back at −80 °C for subsequent use as additional HAV stock as reported before [12].

Similarly, to the cultivation of the previously mentioned FRHK-4 cells and infection of HAV, confluent CRFK cells within sterile 175 cm^2 cell-culture flasks were infected with 2 mL of FCV-F9 stock (purchased from ATCC) and 8 mL DMEM with 2% FBS and 1% Pen-Strep and incubated at 37 °C for three hours, followed by the addition of 10 mL DMEM-F12 containing 10% FBS and 1% Pen-Strep and incubated for three to five days, within a water-jacketed incubator under 5% CO_2 until cytopathic effects were observed. The infected flasks were then freeze-thawed thrice, centrifuged and filtered through a 0.2-micron filter to be stored back at −80 °C for subsequent use as additional FCV-F9 stock as reported earlier [11].

2.3. UV-C LED (279 nm) Treatment of HAV and FCV Inoculated and Dried on Stainless-Steel Coupons, Ceramic Coupons, and Glass Discs

Stainless-steel coupons (Biosurface Technologies, via Fisher Scientific, Pittsburgh, PA, USA, type 304, finish 2B; 2 × 2 cm^2) or ceramic tiles/coupons (Home Depot; 2 × 2 cm^2) or glass discs (Biosurface Technologies; 1 × 1 cm^2) were surface rinsed with ethanol, dried, wrapped with aluminum foil and sterilized by autoclaving. These sterile coupons or discs were placed within sterile petri dishes within a biosafety hood (254 nm UV-C mercury lamp, Labconco Purifier Class II Biosafety cabinet, 36208 020421542 A, Kansas City, MO, USA) and surface decontaminated using 254 nm UV-C light for 10 min as described in earlier studies [28]. Then, 100 µL of either HAV (~5.5 log PFU/mL) or FCV (~6.0 log PFU/mL) in cell-culture grade phosphate buffered saline (pH 7.2, Fisher Scientific) were aseptically spread onto individual sterile coupons (2 × 2 cm^2 area of stainless- steel or ceramic) or 50 µL were spread on glass discs (1 × 1 cm^2 area) and allowed to dry within the biosafety hood for another 10 min for stainless-steel and ceramic coupons or overnight for glass discs (at an ambient temperature of 23 °C and 43% relative humidity) and exposed to various UV doses/treatment times as described below.

UV-C LED (MD 1016-1, Irtronix, Torrence, CA, USA) emits at a peak wavelength of 279 nm. For treatments, the sum of the 10–12 nm full width at half maximum output irradiance and corresponding wavelength absorption spectra were considered for average delivered UV dose irradiance calculation. The exposures were conducted for up to 2.5 min for HAV on stainless-steel coupons using 30 s intervals (0–150 s), 3.75 min for HAV on ceramic coupons using 45 s intervals (0–225 s), and up to 1.0 min for HAV on glass using 10 s intervals until 40 s, followed by a 20 s interval (0–60 s). Treatments up to 1.25 min for FCV on stainless-steel coupons using 15 s intervals (0–75 s), 2.5 min for FCV on ceramic coupons using 30 s intervals (0–150 s), and up to 1.0 min for FCV on glass using 10 s intervals until 40 s, followed by a 20 s interval (0–60 s) (279 nm, 6.5 cm from sample, Surface irradiance = 0.328 mW/cm^2, HAV surface dose = 0–49.2 mJ/cm^2 on stainless-steel, 0–73.8 mJ/cm^2 on ceramic, 0–19.68 mJ/cm^2 on glass; FCV surface dose = 0–24.6 mJ/cm^2 for stainless-steel, 0–49.2 mJ/cm^2 on ceramic, and 0–19.68 mJ/cm^2 on glass). Each experiment was replicated three times for each treatment condition/exposure time/UV dose.

Viruses from the inoculated control (0–min) ceramic and stainless-steel coupons as well as the UV-C LED exposed treated coupons were then aseptically recovered using 750 µL DMEM-F12 containing 2% FBS and 1% Pen-strep that served as an eluant for HAV or FCV and aseptically added to sterile 2 mL centrifuge tubes. Viruses from the inoculated control (0–min) glass discs as well as the UV-C LED exposed treatments were aseptically recovered by placing the glass disc within a 15 mL sterile centrifuge tube containing 9 mL DMEM-F12 with 2% FBS and 1% Pen-strep that served as an eluant for HAV and FCV. The recovered viruses were ten-fold serially diluted using DMEM-F12 containing 2% FBS and 1% Pen-Strep. The viral infectivity was determined using viral plaque assays in duplicate for each ten-fold dilution for each treatment type and treatment time/dose.

2.4. UV-C (254 nm) Treatment of HAV and FCV Inoculated and Dried on Stainless-Steel Coupons, Ceramic Coupons and Glass Discs

Similar to UV-C LED treatments of viruses, sterile stainless-steel and ceramic coupons or glass discs were placed within sterile petri dishes within a biosafety hood (254 nm, Labconco Purifier Class II Biosafety cabinet, 36208 020421542 A, Kansas City, MO, USA) for surface decontamination using 254 nm UV-C light for 10 min and 100 µL of either HAV (~5.5 log PFU/mL) or FCV (~6.0 log PFU/mL) were spread on stainless-steel or ceramic discs or 50 µL of each virus were aseptically spread on glass discs and allowed to dry within the biosafety hood for 10 min for the coupons and overnight for the discs (at an ambient temperature of 23 °C and 43% relative humidity). The UV-C at 254 nm lamp was turned on for a 10-min warm-up time before treatments were conducted. UV-C hood light treatments at 254 nm (254 nm, 55.88 cm/22 inches from sample, Surface irradiance = 0.217 mW/cm^2, HAV surface dose = 0–32.55 mJ/cm^2 for stainless-steel, 0–48.83 mJ/cm^2 for ceramic, 0–13.02 mJ/cm^2 for glass; FCV surface dose = 0–16.28 mJ/cm^2 for stainless-steel and ceramic, and 0–13.02 mJ/cm^2 for glass) were conducted for up to 2.5 min for HAV on stainless-steel coupons using 30 s intervals (0–150 s), 3.75 min for HAV on ceramic coupons using 45 s intervals (0–225 s), treatments for up to 1.25 min for FCV on stainless-steel coupons and ceramic coupons using 15 s intervals (0–75 s), and 1 min treatment for both HAV and FCV on glass discs with 10 s intervals until 40 s followed by a 20 s interval until 1 min (0–60 s). Using the same methods reported above for UV-C LED in Section 2.3, viruses from both the 0–min control and the UV-C 254 nm treatments were recovered, and viral plaque assay was performed in duplicate for each of the three replicates for each treatment condition.

2.5. Infectious Plaque Assays

Standard plaque assays in 6-well plates containing confluent host cells were used to determine the infectivity of both viruses [28]. Briefly, the confluent host FRhK-4 cells were infected with 500 µL of ten-fold serially diluted recovered HAV from control and treated coupons. After infection, the plates were incubated at 37 °C and 5% CO_2 for 2.5 h. Then, the media was aspirated, and the infected cells were overlaid with 2 mL per well of a 1:1 ratio of 1.5% Noble agar and HAV overlay medium [28]. After the overlay, the plates were stored for 72–120 h at 37 °C and 5% CO_2 until visualization and enumeration of plaques, and recovered viruses were reported as log PFU/mL [12,28].

Similarly, confluent host CRFK cells in 6-well plates were infected with 500 μL of ten-fold serially diluted FCV recovered from control and treated coupons. After infection, the plates were incubated at 37 °C and 5% CO_2 for 2.5 h, and the media was then aspirated. The infected cells were overlaid with 2 mL per well of a 1:1 ratio of 1.5% Noble agar and 2× FCV overlay medium as reported earlier [11,28]. After the overlay, the plates were stored for 72–120 h in the CO_2 incubator at 37 °C and 5% CO_2 until visualization and enumeration of plaques.

2.6. UV-C Dose Calculation

Surface irradiances for the UV-C LED at 279 nm device were measured and calculated using a highly sensitive spectrophotometer (QE Pro series, Ocean Optics, Dunedin, FL, USA) within the exposure area, as discussed previously [31,32]. Surface irradiances for the UV-C at 254 nm system were measured using an International Light Technologies (ILT) research radiometer (ILT5000, SED033/F/W, International Light Technologies, Peabody, MA, USA) which was equipped with a light meter, sensor, filter, optic, and calibration. The surface dose was calculated using the UV intensity and various exposure time (s) as reported in previous studies [24,28,32]. The calculated UV intensity for the UV-C (254 nm) was 0.217 mW/cm^2 and the UV intensity of the UV-C LED was 0.328 mW/cm^2, as calculated using Equation (1) [31].

$$\text{Surface UV-C dosage (mJ/cm}^2\text{)} = \text{UV intensity (mW/cm}^2\text{)} \times \text{exposure time (s)} \quad (1)$$

2.7. Surface Roughness Measurements

For each surface type, the surface roughness was measured and reported using a portable tester SJ-210 by Mitutoyo (Sakado, Takatsu-Ku, Kawasaki, Kanagawa, Japan) at Tennessee State University (Nashville, TN, USA) as described in previous studies, using the SJ-210 V.1.210 software with standard configurations [32]. Briefly, as described earlier, the constant measurement conditions (Lc-0.1 in, Ls-0.000813 cm, Sampling lengths (N)-4, Prelength-OFF, Pitch-0.000150 cm) were used for all the measurements and with a 0.0508 cm/s probe (sensor) speed. Similar to previous reports, the surface roughness (μm) of stainless-steel and ceramic coupons as well as glass discs were measured for an average of 3 sampling lengths/periods [32].

2.8. Statistical Analysis

A two-way ANOVA statistical test with Tukey's adjustments ($p < 0.05$) using JMP v.17 was used to analyze the significant differences between the systems, surfaces, and their interactions for each tested virus. No random effects were introduced. A Shapiro-Wilk W and QQ normality plots were used to evaluate the normality of ANOVA residuals. All statistical assumptions regarding normality were met. Excel® was used to determine the D_{10}-values from linear models as reported earlier [28].

3. Results

3.1. Inactivation of HAV and FCV on Stainless-Steel Coupons Using UV-C at 254 nm

HAV on stainless-steel coupons treated with UV-C at 254 nm showed reduction between 1.26 to 3.63 log PFU after treatment times of 0.5 min to 2.5 min (corresponding to doses of 6.51–32.55 mJ/cm^2), respectively (Table S1 and Figure 1). HAV titers were significantly reduced ($p < 0.05$) by 1.26 ± 0.07 log PFU compared to the control after treatment for 0.5 min and was further reduced by 2.53 ± 0.05 log PFU after 1 min treatment and by 3.50 ± 0.16 log PFU after 2 min treatment, followed by a tailing effect in reduction that occurred past the 2 min treatment with UV-C at 254 nm (Figure 1). The calculated D_{10}-value (mJ/cm^2) for the dose required to achieve a 1 log reduction of HAV using 254 nm UV-C on stainless-steel surfaces was 9.48 ± 0.34 mJ/cm^2 as shown in Table 1.

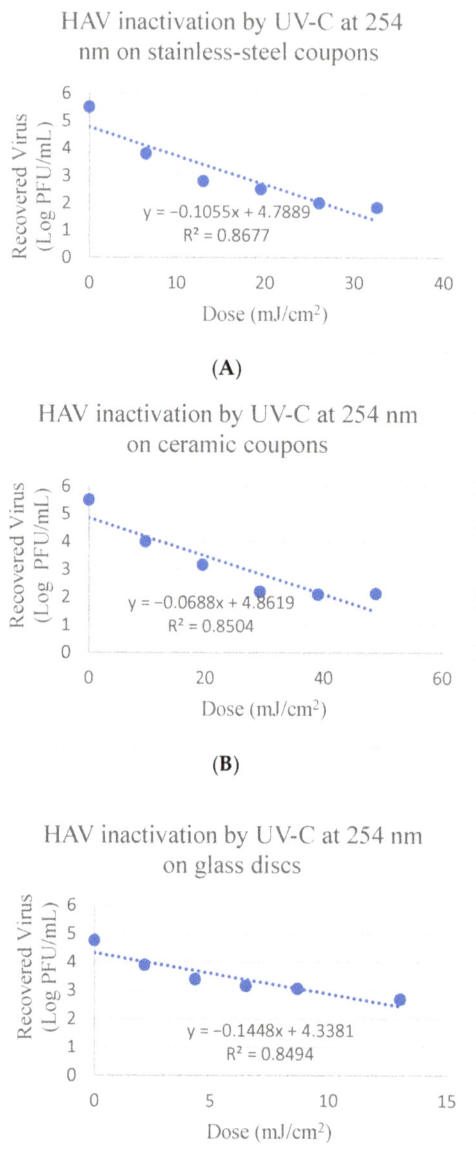

Figure 1. Inactivation of HAV by UV-C at 254 nm on (**A**) Stainless-steel; (**B**) Ceramic; and (**C**) Glass surfaces. Corresponding Linear $1D_{10}$-values for HAV on stainless-steel surfaces = 9.48 mJ/cm^2; Linear $2D_{10}$ = 18.96 mJ/cm^2; Linear $3D_{10}$ = 28.44 mJ/cm^2; Linear $1D_{10}$-values for HAV on ceramic surfaces = 14.53 mJ/cm^2; Linear 2D = 29.07 mJ/cm^2; Linear 3D = 43.60 mJ/cm^2; and on glass discs linear $1D_{10}$-values = 6.91 mJ/cm^2; Linear $2D_{10}$ = 13.81 mJ/cm^2; Linear $3D_{10}$ = 20.72 mJ/cm^2.

FCV on stainless-steel coupons treated with UV-C at 254 nm showed reduction between 1.1 to 4.8 log PFU after 0.25 to 1.25 min (corresponding to doses of 3.26–16.28 mJ/cm^2), respectively (Table S2 and Figure 2). There was significant reduction in FCV titers compared to the control after each treatment time ($p \leq 0.05$). The calculated D_{10}-value (mJ/cm^2) for 1 log reduction of FCV on stainless-steel coupons was 3.65 ± 0.06 mJ/cm^2 as shown in Table 2.

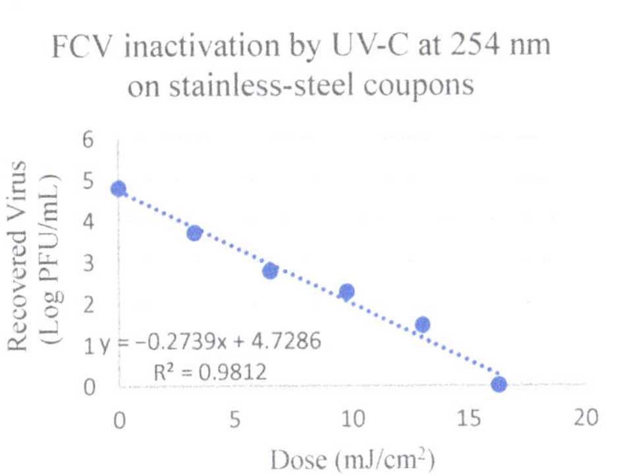

(A)

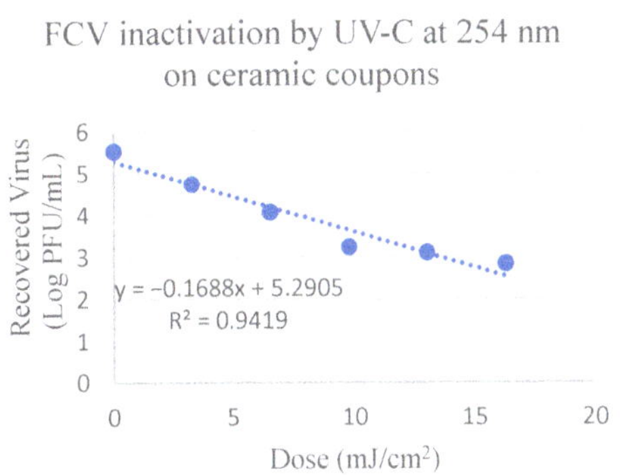

(B)

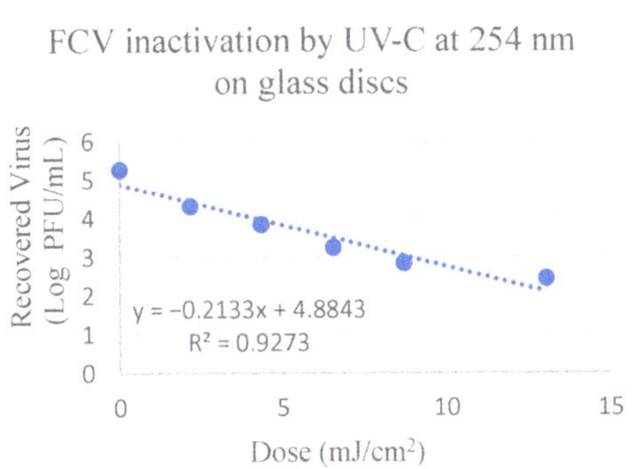

(C)

Figure 2. Inactivation of FCV by UV-C at 254 nm on (**A**) Stainless-steel; (**B**) Ceramic; and (**C**) Glass surfaces. Corresponding Linear $1D_{10}$-values for FCV on stainless-steel = 3.65 mJ/cm^2; Linear $2D_{10}$ = 7.3 mJ/cm^2; Linear $3D_{10}$ = 10.95 mJ/cm^2; Linear 1D-values for FCV on ceramic = 5.92 mJ/cm^2; Linear 2D = 11.85 mJ/cm^2; Linear 3D = 17.78 mJ/cm^2; and Linear $1D_{10}$-values for FCV on glass surfaces = 4.69 mJ/cm^2; Linear 2D = 9.38 mJ/cm^2; Linear 3D = 14.06 mJ/cm^2.

Table 1. D_{10}-values (mJ/cm^2) and D-values (min) of HAV treated with either UV-C (254 nm) or UV-C LED (279 nm) on model food contact surface coupons.

Surface Type	UV-C LED System (279 nm) D_{10}-Value (mJ/cm^2)	UV-C LED System (279 nm) D-Value (min)	UV-C (254 nm) D_{10}-Value (mJ/cm^2)	UV-C (254 nm) D-Value (min)
Stainless-steel	19.53 ± 2.45 ^{Ab}	1.0 ± 0.12 ^{Ab}	9.48 ± 0.34 ^{Bb}	0.73 ± 0.03 ^{Bb}
Ceramic	26.04 ± 0.60 ^{Aa}	1.3 ± 0.03 ^{Aa}	14.53 ± 2.52 ^{Ba}	1.1 ± 0.12 ^{Ba}
Glass	8.77 ± 2.08 ^{Ac}	0.47 ± 0.11 ^{Bc}	6.91 ± 1.93 ^{Bb}	0.56 ± 0.16 ^{Ab}

Capital letters denote statistically significant differences when compared across one row between UV-C systems ($p < 0.05$). Lowercase letters denote statistically significant differences when compared down a treatment medium (one column) ($p < 0.05$). Data are reported as averages of triplicate treatments ± standard deviations; Note: both optical devices have different UV intensities.

Table 2. D_{10}-values (mJ/cm^2) and D-values (min) of FCV treated with either UV-C (254 nm) or UV-C LED (279 nm) on three model food contact surface coupons.

Surface Type	UV-C LED System (279 nm) D_{10}-Value (mJ/cm^2)	UV-C LED System (279 nm) D-Value (min)	UV-C (254 nm) D_{10}-Value (mJ/cm^2)	UV-C (254 nm) D-Value (min)
Stainless-steel	7.097 ± 2.11 ^{Aa}	0.38 ± 0.11 ^{Aa}	3.65 ± 0.06 ^{Ba}	0.28 ± 0.001 ^{Ba}
Ceramic	8.31 ± 2.12 ^{Aa}	1.37 ± 0.34 ^{Ba}	5.92 ± 1.90 ^{Ba}	1.45 ± 0.48 ^{Aa}
Glass	7.82 ± 0.86 ^{Aa}	0.40 ± 0.05 ^{Aa}	4.69 ± 0.03 ^{Ba}	0.36 ± 0.001 ^{Ba}

Capital letters denote statistically significant differences when compared across one row between UV-C systems ($p < 0.05$). Lowercase letters denote statistically significant differences when compared down a treatment medium (one column) ($p < 0.05$). Data represent averages of triplicate treatments ± standard deviations; both optical devices have different UV intensities.

3.2. *Inactivation of HAV and FCV on Ceramic Coupons Using UV-C at 254 nm*

HAV on ceramic coupons treated with UV-C at 254 nm after 0.75 min to 3.75 min (corresponding to doses of 9.77–48.83 mJ/cm^2) showed reduction between 1.52 to 3.35 log PFU, respectively (Table S3 and Figure 1). There was significant reduction in HAV titers after 0.75 min treatment of 1.52 ± 0.12 log PFU, and after 1.5 min and 2.25 min treatment of 2.32 ± 0.11 and 3.30 ± 0.13 log PFU, respectively ($p \leq 0.05$), though tailing effects in reduction were observed after 2.25 min of exposure (Figure 1). The calculated D_{10}-value (mJ/cm^2) for 1 log reduction of HAV on ceramic coupons was 14.53 ± 2.52 mJ/cm^2 as shown in Table 1.

FCV on ceramic coupons treated with UV-C at 254 nm showed reduction between 0.77 to 2.67 log PFU after treatment for 0.25 min to 1.25 min (corresponding to doses of 3.26–16.28 mJ/cm^2), respectively (Table S4 and Figure 2). There was significant reduction in FCV titers of 0.77 ± 0.07 log PFU after 0.25 min treatment, 1.46 ± 0.03 log PFU reduction after 0.5 min treatment, and 2.28 ± 0.07 log PFU reduction after 0.75 min treatment ($p \leq 0.05$), though the reduction tailed after 0.75 min of exposure (Figure 2). The calculated D_{10}-value (mJ/cm^2) for 1 log reduction of FCV on ceramic coupons was 6.25 ± 1.90 mJ/cm^2 using UV-C at 254 nm, as shown in Table 2.

3.3. *Inactivation of HAV and FCV on Glass Discs Using UV-C at 254 nm*

HAV on glass discs treated with UV-C at 254 nm for 0.17 to 1.0 min (corresponding to doses of 2.17–13.02 mJ/cm^2) showed reduction between 0.85 to 2.09 log PFU, respectively (Table S5 and Figure 1). There was significant reduction in HAV titers of 0.85 ± 0.02 log PFU after 0.17 min treatment, with 1.3 ± 0.10 and 2.09 ± 0.10 log PFU reduction after 0.33- and 1-min treatment, respectively ($p \leq 0.05$). The calculated D_{10}-value (mJ/cm^2) to achieve a 1 log reduction of HAV on glass discs was 6.91 ± 1.93 mJ/cm^2 using UV-C at 254 nm, as shown in Table 1.

FCV on glass discs showed reduction between 0.98 to 2.86 log PFU after treatment with UV-C at 254 nm for 0.17 to 1.0 min (corresponding to doses of 2.17–13.02 mJ/cm^2), respectively (Table S6 and Figure 2). There was significant reduction in FCV titers after all times of exposure (0.17, 0.33, 0.5, 0.67 min) until 0.67 min, where tailing reduction began

($p \leq 0.05$) (Figure 2). The calculated D_{10}-value (mJ/cm^2) to achieve a 1 log reduction of FCV on glass discs was 4.69 ± 0.03 mJ/cm^2 using UV-C at 254 nm, as shown in Table 2.

3.4. Inactivation of HAV and FCV on Stainless-Steel Coupons Using UV-C at 279 nm

HAV on stainless-steel coupons showed reduction between 1.2 to 2.75 log PFU after treatment for 0.5 min to 2.5 min (corresponding to doses of 9.84–49.2 mJ/cm^2) with UV-C LED at 279 nm (Table S1 and Figure 3). There was significant reduction in HAV titers of 1.20 ± 0.07 log PFU after 0.5 min treatment from the control, with 1.96 ± 0.02 log PFU reduction after 1.5 min treatment and further reduction of 2.75 ± 0.15 log PFU/mL after 2.5 min treatment. The calculated D_{10}-value (mJ/cm^2) required to achieve a 1 log reduction of HAV on stainless-steel coupons was 19.53 ± 2.45 mJ/cm^2 using UV-C at 279 nm, as shown in Table 1.

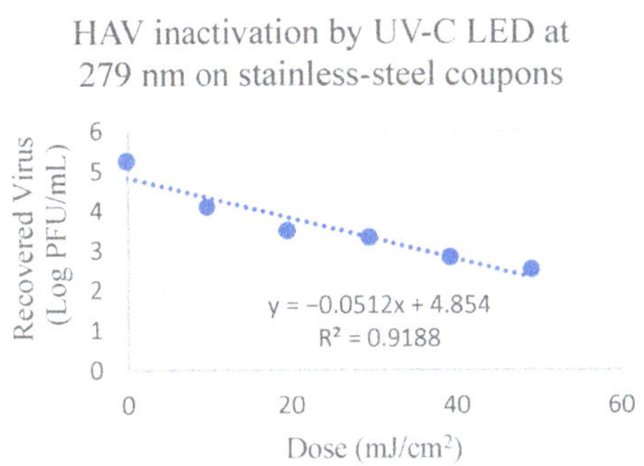

(A)

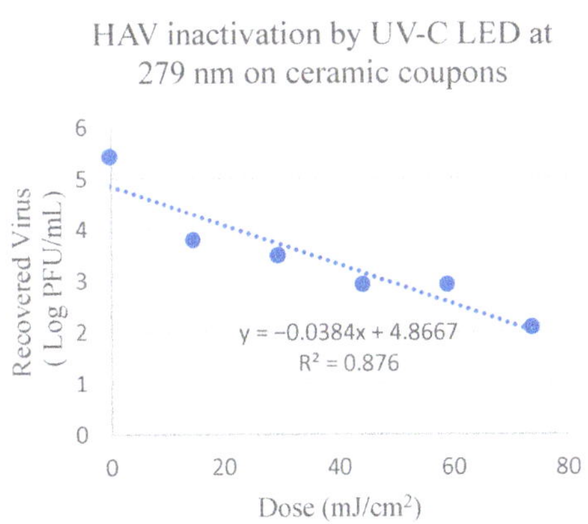

(B)

Figure 3. *Cont.*

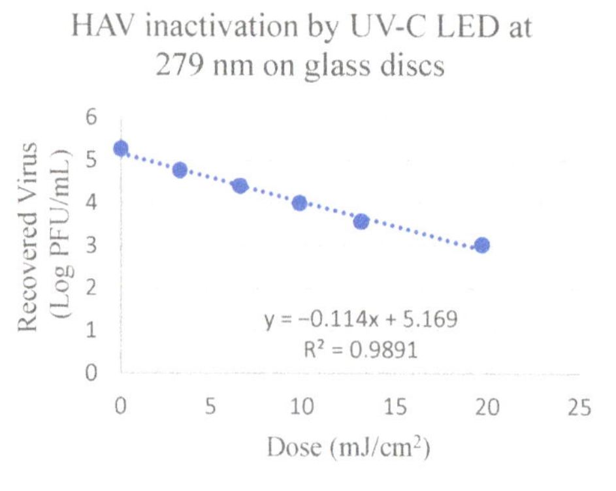

(C)

Figure 3. Inactivation of HAV by UV-C LED at 279 nm on (**A**) Stainless-steel; (**B**) Ceramic; and (**C**) Glass surfaces. Corresponding Linear $1D_{10}$-values for HAV on stainless-steel surfaces 19.53 mJ/cm^2; Linear $2D_{10}$ = 39.06 mJ/cm^2; Linear $3D_{10}$ = 58.59 mJ/cm^2; Linear $1D_{10}$-values for HAV on ceramic surfaces = 26.04 mJ/cm^2; Linear $2D_{10}$ = 52.08 mJ/cm^2; Linear $3D_{10}$ = 78.13 mJ/cm^2; and on glass discs linear $1D_{10}$-values = 8.77 mJ/cm^2; Linear $2D_{10}$ = 17.54 mJ/cm^2; Linear $3D_{10}$ = 26.31 mJ/cm^2.

FCV on stainless-steel coupons showed reduction between 1.93 to 3.89 log PFU/mL after treatment with UV-C LED at 279 nm for 0.25 to 1.25 min (corresponding to doses of 4.92–24.6 mJ/cm^2) (Table S2 and Figure 4). There was significant reduction in FCV titers after 0.25 min of 1.93 ± 0.05 log PFU, with 2.99 ± 0.11 log PFU reduction after 0.75 min, followed by further reduction of 3.89 ± 0.19 log PFU after 1.25 min ($p \leq 0.05$). The calculated D_{10}-value (mJ/cm^2) to achieve a 1 log reduction of FCV on stainless-steel coupons was 7.097 ± 2.11 mJ/cm^2 using UV-C LED at 279 nm, as shown in Table 2.

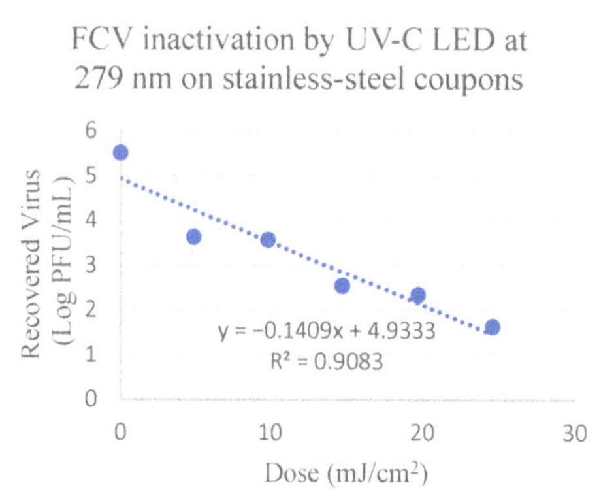

(A)

Figure 4. *Cont.*

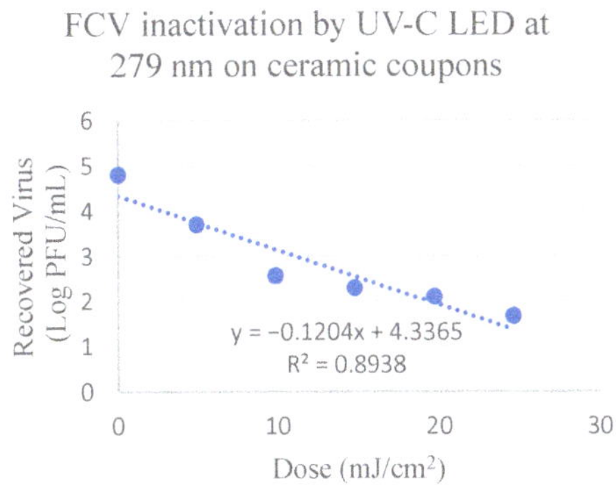

(B)

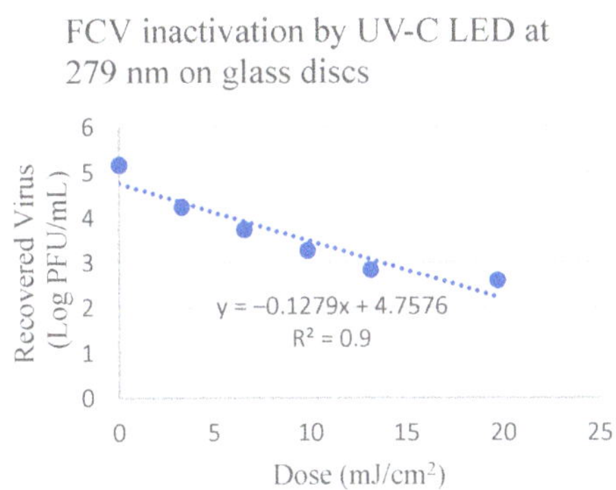

(C)

Figure 4. Inactivation of FCV by UV-C LED at 279 nm on (**A**) Stainless-steel; (**B**) Ceramic; and (**C**) Glass surfaces. Corresponding Linear $1D_{10}$-values for FCV on stainless-steel surfaces = 7.1 mJ/cm^2; Linear $2D_{10}$ = 14.2 mJ/cm^2; Linear $3D_{10}$ = 21.3 mJ/cm^2; Linear $1D_{10}$-values for FCV on ceramic surfaces = 8.31 mJ/cm^2; Linear $2D_{10}$ = 16.61 mJ/cm^2; Linear $3D_{10}$ = 24.92 mJ/cm^2; and on glass discs linear $1D_{10}$-values = 7.82 mJ/cm^2; Linear $2D_{10}$ = 15.64 mJ/cm^2; Linear $3D_{10}$ = 23.46 mJ/cm^2.

3.5. Inactivation of HAV and FCV on Ceramic Coupons Using UV-C at 279 nm

HAV on ceramic coupons showed reduction between 1.63 to 3.37 log PFU after treatment with UV-C LED at 279 nm for 0.75 min to 3.75 min (corresponding to doses of 14.76–73.8 mJ/cm^2), respectively (Table S3 and Figure 4). There was significant reduction after 0.75 min of 1.63 $\pm$ 0.17 log PFU from the control with further significant reduction of 2.52 $\pm$ 0.11 log PFU after 2.25 min and with 3.37 $\pm$ 0.10 log PFU reduction again after 3.75 min treatment. The calculated D_{10}-value (mJ/cm^2) for 1-log inactivation of HAV on ceramic coupons using UV-C at 279 nm was 26.05 $\pm$ 0.60 mJ/cm^2 as shown in Table 1.

FCV on ceramic coupons showed reduction between 1.1 to 3.12 log PFU after treatment treated with UV-C LED at 279 nm for 0.5–2.5 min (corresponding to doses of 9.84–49.2 mJ/cm^2) (Table S4 and Figure 4). There was significant reduction in FCV titers after treatment for 0.5 min of 1.10 $\pm$ 0.04 log PFU, with 2.51 $\pm$ 0.07 log PFU after 1.5 min treatment and 3.12 $\pm$ 0.11 log PFU after 2.5 min treatment ($p \leq 0.05$). The calculated D_{10}-value (mJ/cm^2) for FCV on ceramic coupons was 8.31 $\pm$ 2.12 mJ/cm^2 using UV-C LED at 279 nm, as shown in Table 2.

3.6. Inactivation of HAV and FCV on Glass Discs Using UV-C at 279 nm

HAV on glass discs showed reduction between 0.5 to 2.2 log PFU after treatment with UV-C LED at 279 nm for 0.17 to 1.0 min (corresponding to doses of 3.28–19.68 mJ/cm^2), respectively (Table S5 and Figure 3). There was significant reduction in HAV titers after treatment for 0.17 min of 0.5 ± 0.10 log PFU, with 1.3 ± 0.10 and 1.7 ± 0.10 log PFU reduction after 0.5 and 0.67 min, and 2.2 ± 0.20 log PFU reduction after 1 min treatment ($p \leq 0.05$). The calculated D_{10}-value (mJ/cm^2) for 1 log inactivation of HAV on glass discs was and 8.77 ± 2.08 mJ/cm^2 using UV-C at 279 nm, as shown in Table 1.

FCV on glass discs showed reduction between 0.93 to 2.57 log PFU after treatment with UV-C LED at 279 nm of 0.17 to 1.0 min (corresponding to doses of 3.28–19.68 mJ/cm^2), respectively (Table S6 and Figure 4). There was significant reduction in FCV titers after treatments for 0.17 min, 0.33 min, 0.5 min, and 0.67 min of 0.93 ± 0.02, 1.41 ± 0.05, 1.86 ± 0.11, and 2.31 ± 0.19 log PFU, respectively, while tailing occurred after 0.67 min ($p \leq 0.05$) (Figure 4). The calculated D_{10}-value (mJ/cm^2) for 1-log reduction of FCV on glass discs was 7.88 ± 0.86 mJ/cm^2 using UV-C LED at 279 nm, as shown in Table 2.

3.7. Surface Roughness (Ra) Measurements of Each Model Surface

The average Ra values for stainless-steel coupons, ceramic coupons, and glass discs were calculated to be 7.3, 56.2, and 12.3 μm, respectively.

4. Discussion

Prevention of cross contamination is of utmost concern in the food processing industry as foodborne pathogens can survive on food contact surfaces for extended periods and be transmitted to other items (knives, spoons), food, and surface areas after contact leading to foodborne outbreaks [33]. In addition to surface sanitation with chemical washes, traditional UV-C at 254 nm is commonly used for microbial decontamination of surfaces. Evaluation of the inactivation of *Salmonella* (8–9 log CFU/mL) inoculated on stainless-steel, HDPE, waxed cardboard, and PVC coupons (common surfaces found within the tomato processing industry) using UV-C at 254 nm showed that 2.75, 2.93, 1.39, and 1.91 log CFU reductions, respectively were obtained at doses of 3.3 mJ/cm^2 [34]. Increased UV-C doses of 19.7 mJ/cm^2 were shown to result in reductions of 3.51, 4.32, 1.43, and 3.51 log CFU on stainless-steel, HDPE, waxed cardboard, and PVC, respectively, with waxed cardboard showing the least amount of reduction at both tested dosages, as the wax can provide a shielding effect from UV radiation [34,35]. While traditional UV-C systems have been routinely used, surface disinfection using UV-C LED systems including use of wavelengths at 279 nm continue to be investigated. Stainless-steel discs inoculated by droplets containing 7-log CFU/mL of either *Listeria monocytogenes*, *Escherichia coli*, or *Salmonella enterica* serovar Typhimurium after treatments with UV-C LED systems at 279 nm resulted in D_{10}-values of 3.02 ± 0.1, 2.70 ± 0.1 mJ/cm^2, and 1.90 ± 0.1 mJ/cm^2, respectively (with >3 log CFU/mL inactivation at the highest dosage of 12 mJ/cm^2) for the tested bacteria [24]. Therefore, this research investigated the ability of UV-C at 254 nm and UV-C LED at 279 nm to inactivate HAV and FCV on model food contact surfaces including stainless-steel, ceramic and glass.

HAV is known to be resilient and stable to most environmental stressors. Hence, the high D_{10}-value (mJ/cm^2) of 26.05 ± 0.60 mJ/cm^2 for HAV using UV-C LED at 279 nm and 14.53 ± 2.52 mJ/cm^2 using UV-C at 254 nm on ceramic surfaces, respectively, is not surprising ($p < 0.05$). This difference in dosage between the two systems could be attributed to the surface roughness and inherent properties of HAV that could result in adherence to ceramic and resistance to UV-C LED at 279 nm. These high parameters were followed by D_{10}-values (mJ/cm^2) of HAV on stainless-steel surfaces of 19.53 ± 2.45 mJ/cm^2 for UV-C LED at 279 nm and 9.48 ± 0.34 mJ/cm^2 for UV-C at 254 nm, with HAV on glass discs displaying the lowest D_{10}-values (mJ/cm^2) of 8.77 ± 2.08 mJ/cm^2 for UV-C LED at 279 nm and 6.91 ± 1.93 mJ/cm^2 for UV-C at 254 nm, respectively.

Similarly, FCV on ceramic coupons displayed higher D_{10}-values (mJ/cm^2) of 8.31 ± 2.12 mJ/cm^2 for UV-C LED at 279 nm and 6.25 ± 1.90 mJ/cm^2 for UV-C 254 nm, respectively compared to when inoculated on stainless-steel and glass surfaces. Thus, FCV showed a similar trend to HAV with regards to inactivation on ceramic requiring higher doses using UV-C LED at 279 nm than UV-C at 254 nm. However, unlike HAV, FCV on glass discs showed D_{10}-values of 7.88 ± 0.86 mJ/cm^2 with UV-C LED at 279 nm and 4.69 ± 0.03 mJ/cm^2 with UV-C at 254 nm, and on stainless-steel coupons showed D_{10}-values of 7.097 ± 2.11 mJ/cm^2 for UV-C LED at 279 nm and 3.65 ± 0.06 mJ/cm^2 for UV-C at 254 nm.

Based on the data obtained, HAV and FCV needed the highest dose for inactivation on contaminated ceramic surfaces by both UV-C systems (~9.84–73.8 mJ/cm^2 for UV-C LED at 279 nm and ~3.3–48.8 mJ/cm^2 for UV-C at 254 nm; $p < 0.05$). The high doses of HAV inactivation on ceramic may be associated with surface roughness (as mentioned earlier) and surface properties that may allow for stronger adherence to ceramic, less penetration of UV-C, as well as the resilience of HAV to UV-C inactivation, related to its nucleic acid and capsid protein structure and content. In fact, researchers also state that surface roughness parameters could play a role in decontamination using UV-C systems [24,32]. Therefore, different surfaces may require different doses for microbial inactivation. With ceramic showing a higher surface roughness Ra value of 56.2 μm compared to stainless-steel and glass (7.3 and 12.3 μm, respectively), it is likely that the high Ra values contribute to accumulation of viruses, offering protection against treatments, thus providing higher survival rates and increased dosage requirements, as discussed previously [32]. These researchers compared the inactivation of *E. coli*, *Salmonella enteritidis*, and *Pseudomonas fragi* by UV-C LED at 279 nm on glass, silicone rubber, and stainless-steel surfaces that had surface roughness values 0.020, 0.576, and 1.473 μm with silicone rubber showing the highest resistance to UV-C treatment resulting in only 1.91, 2.91, and 3.08 log CFU/mL reduction for *E. coli*, *Salmonella enteritidis*, and *P. fragi*, respectively. However, the correlation between surface roughness and inactivation abilities is still debatable/unknown, with varying reports among researchers [32].

Other researchers showed that glass displayed the lowest R_a and R_q values of 0.0204 and 0.0492 compared to values of stainless-steel with R_a and R_q values of 0.58 and 0.80, which explained the higher bactericidal effect on glass treated with UV-C at 280 nm [36] and is similar to the findings of this study for HAV inactivation. Kim and Kang [36] showed that the level of inactivation of the tested bacterial pathogens varied dependent on surface type, however glass displayed the highest reduction for all pathogens treated with UV-C LED at 280 nm, followed by PVC, stainless-steel, Teflon, and silicon. Kim and Kang [36] showed that treatments with 280 nm UV-C LED at a dose of 2 mJ/cm^2, caused 0.9–1.44 log reductions of *E. coli*, with highest reduction on glass, while the lowest reduction was obtained on silicon followed by Teflon, stainless-steel and then PVC. These researchers also showed that at a dose of 3 mJ/cm^2, *S. Typhimurium* and *L. monocytogenes* were reduced by 0.5–1.66 log CFU and 0.5–0.91 log CFU, respectively that followed the same surface resistance trend as *E. coli* [36]. Similarly, the current study found that HAV contaminated on glass displayed the lowest D_{10}-value when compared to stainless-steel and ceramic surfaces after treatment with UV-C LED at 279 nm. This could be explained by the ability of the pathogen to adhere to the surfaces as well as a possible back reflection from the stainless-steel coupons that could contribute to the lower D-values of this target microorganism [24].

Currently, literature related to UV-C at 254 nm and 279 nm dose requirements for inactivation of human foodborne viruses including HAV and HuNoV on varied food contact surfaces is limited for comparative purposes to the current study. HAV on Formica coupons was reported to show similar D_{10}-values (mJ/cm^2) using both traditional UV-C at 254 nm (D_{10} = 12.40 ± 1.15 mJ/cm^2) and UV-C LED at 279 nm (D_{10} =12.39 ± 0.70 mJ/cm^2) [28]. In the current study, D_{10}-values using UV-C LED at 279 nm for the inactivation of HAV was shown to be 19.53 ± 2.45, 26.05 ± 0.60, and 8.77 ± 2.08 mJ/cm^2 on stainless-steel, ceramic and glass surfaces, respectively. Thus, HAV showed higher resistance to UV-C LED at 279 nm treatments on ceramic, followed by stainless-steel, Formica, and then glass. Using treatments with UV-C at 254 nm, HAV showed D_{10}-values of 9.48 ± 0.34, 14.53 ± 2.52, and 6.91 ± 1.93 mJ/cm^2 on stainless-steel, ceramic and glass respectively, indicating that HAV had higher resistance to UV-C at 254 nm treatments on Formica, followed by stainless-steel that was similar to ceramic, and then glass. Thus, HAV on glass was found to be most sensitive to UV-C LED at 279 nm compared to ceramic, stainless-steel or Formica surfaces.

Studies with FCV, MNV, echovirus 12 and MS2 on petri dishes showed that doses of 25, 29, 30 and 70 mJ/cm^2 using traditional UV-C at 254 nm were required to achieve 4-log reduction, respectively, while 85 mJ/cm^2 was needed for 4 log reduction of intracellular echovirus 12 [37]. While in the current study, using UV-C at 254 nm, HAV required doses of 26.04 mJ/cm^2 to achieve a 3.50 log PFU/mL reduction on stainless-steel coupons, 29.3 mJ/cm^2 to achieve a 3.3 log PFU/mL reduction on ceramic coupons, and 13.02 mJ/cm^2 to achieve a 2.09 log PFU/mL reduction on glass discs (Tables S1–S3). In addition, FCV required doses of 16.28 mJ/cm^2 to achieve a 4.8 log PFU/mL reduction on stainless-steel surfaces, 9.77 mJ/cm^2 to achieve a 2.28 log PFU/mL reduction on ceramic surfaces, and 13.02 mJ/cm^2 to achieve a 2.86 log PFU/mL reduction on glass surfaces, respectively (Tables S4–S6). Based on these results, MS2 requires the highest dose for inactivation on petri dishes compared to HAV and FCV on stainless-steel, ceramic or glass surfaces. These data also indicate that FCV on Petri dishes was more resistant to UV-C at 254 nm requiring higher doses for similar inactivation than FCV on stainless-steel, ceramic or glass.

When we compare the results of a study with TV, a HuNoV surrogate, significant decreases ($p < 0.05$) in RNA copy number (approx. 1 and 2 log reduction in VP1 gene detection) after 220-nm irradiation treatments at doses between 22.5 and 37.5 mJ/cm^2, or with 254 nm irradiation at doses of 30 or 37.5 mJ/cm^2, respectively were reported [10]. Previous research showed that FCV on Formica surfaces was reduced by 2.26 log PFU/mL at a dose of 21.81 mJ/cm^2, while TV showed reduction of 1.50 log PFU/mL at a dose of 10.91 mJ/cm^2 [28], which were lower than those reported for RNA copy reduction. In the current study, FCV on stainless-steel, ceramic, and glass coupons treated with UV-C at 254 nm showed reduction of 4.8, 2.67, and 2.86 log PFU/mL at dosages of 16.28, 9.77, and 13.02 mJ/cm^2, respectively, while HAV on stainless-steel, ceramic, and glass was reduced by 3.63, 3.35, and 2.09 log PFU/mL at dosages of 32.55, 48.83, and 13.02 mJ/cm^2, respectively. Thus, HAV required higher dosages for inactivation on all tested surfaces compared to FCV on Formica, ceramic, and stainless-steel. The results of this study show that on the tested surfaces, HAV was more resistant to both UV-C systems compared to FCV.

When we compare the inactivation of the pandemic causing SARS-CoV-2 (an enveloped virus at 10^5 PFU/mL) on contaminated synthetic leather and clothing fabric (porous) and glass, stainless-steel, ceramics, and oaks (non-porous) surfaces by UV-C at 254 nm, doses of 132 mJ/cm^2 was needed for 1-log reduction, 264 mJ/cm^2 for a 2-log reduction, and 396 mJ/cm^2 for a 3 log reduction on porous surfaces, while for all non-porous samples, >3 log reduction was achievable at a dose of 8 mJ/cm^2, with exposure times all below 1 min (ordered by oak, glass, ceramics, and stainless-steel for shortest to longest exposure time) [38]. The current study showed that for HAV and FCV on all tested non-porous surfaces (stainless-steel, ceramic, and glass) doses ranging from ~7–9 mJ/cm^2 for HAV and ~3–6 mJ/cm^2 for FCV were required to achieve a 1-log reduction using UV-C at 254 nm. In addition, this current study showed that the tested virus required higher doses using UV-C LED at 279 nm ranging from 7.097 to 26.05 mJ/cm^2 than UV-C at 254 nm to achieve a 1-log reduction. Since, HAV and FCV are both non-enveloped viruses, theoretically higher doses for their inactivation would be needed.

When using traditional UV-C at 254 nm, tailing occurred for the inactivation of HAV on stainless-steel and ceramic coupons with doses > 19.53 and 29.30 mJ/cm^2 (Figure 1), respectively and doses > 6.51 and 8.68 mJ/cm^2 for FCV on ceramic and glass surfaces (Figure 2), respectively proving ineffective at further viral inactivation. In comparison, for treatments with UV-C LED at 279 nm, tailing was not observed, except for the inactivation of FCV on glass discs with doses > 13.12 mJ/cm^2 that were ineffective at further viral inactivation (Figure 4). When observing tailing effects in cases such as UV-C inactivation of HAV on stainless-steel and ceramic surfaces at 254 nm or FCV inactivation on ceramic or glass surfaces by UV-C at 254 nm and also glass with UV-C LED at 279 nm, it is important to recognize that the linear model may not be most suitable for prediction of >2 or 3 log reduction.

As HAV showed the highest D-values using UV-C LED at 279 nm on ceramic coupons (which was significantly higher than on glass surfaces), the differences between the dosage requirements for inactivation of HAV on glass could be associated with the inherent properties of the virus that allow for attachment to the surface (as stated earlier) as well as the properties (structure and components) of the capsid and nucleic acid and the limitation of lower penetration power by UV-C light.

When comparing reductions of FCV and TV on Formica by UV-C at 279 nm with HAV and FCV on stainless-steel, ceramic, and glass surfaces, FCV and TV on Formica showed reductions of 2.45 and 1.83 log PFU/mL at dosages of 54.6 and 27.3 mJ/cm^2 [28]. This study showed that FCV on stainless-steel, ceramic, and glass surfaces using UV-C LED at 279 nm was reduced by 3.89, 3.12, and 2.57 log PFU at dosages of 24.6, 24.6, and 19.68 mJ/cm^2, respectively, while HAV showed reduction of 2.75, 3.37, and 2.2 log PFU/mL at dosages of 49.2, 73.8, and 19.68 mJ/cm^2, respectively. Inactivation of HAV on ceramic coupons by UV-C LED at 279 nm required a higher dose to achieve >3 log reduction, similar to the inactivation of TV on Formica which required a dose of 54.6 mJ/cm^2 to achieve a >1.8 log reduction, compared to inactivation of FCV on all surfaces [28]. As reported above, the resistance could be associated with TV having different binding sites for the surfaces based on its capsid structure making it more resistant to UV-C and preventing damage to the capsid proteins (UV-C at 279 nm) or RNA (UV-C at 254 nm). Also, protein-RNA cross-linking or energy transfer from proteins to RNA is reported to occur at UV-C at 210–240 nm and may also occur at UV-C at 279 nm [28].

Recently, FCV on stainless-steel surfaces was shown to be reduced by 3.3 log PFU/disc at a dose of 27.5 mJ/cm^2 using UV-C LED at 269 nm [27]. In comparison, the current study reported that FCV treated by UV-C LED at 279 nm on stainless-steel, ceramic, and glass surfaces could be decreased 3.89, 3.12, and 2.57 log PFU with doses of 24.6, 49.2, and 19.68 mJ/cm^2, respectively (Tables S4–S6). Thus, the results of the current study are comparable to that reported by Mariita et al. [27] for stainless-steel surfaces given the differences in the wavelengths used. These UV-C LED systems that potentially target the viral capsid structure can cause damage to proteins due to absorption of UV-C by amino acids around 279 nm, with further oxidation and eventual damage to the nucleic acid, while typically 254 nm UV-C is known to cause dimerization of nucleic acids, nucleic acid damage and thus prevention of viral replication [16,24].

Furthermore, UV-C LED systems with 266, 270, 275, and 279 nm were shown to cause 5 to 6 log reduction of *E. coli* O157:H7, *Salmonella enterica* serovar Typhimurium, and *L. monocytogenes* (~10^8–10^9 CFU/mL) with low doses of 0.7 mJ/cm^2 in media within Petri dishes using 279 nm UV-C LED, while inoculated sliced cheese required higher doses of 1, 2, and 3 mJ/cm^2 for inactivation [39]. *Salmonella* cocktails consisting of *Salmonella* Typhimurium, *S.* Newport, *S.* Enteritidis, *S.* Senftenberg, and *S.* Heidelberg (6.5 log CFU/mL) were shown to be reduced by 1.97 and 3.48 log CFU on stainless-steel surfaces using UV-C LED at 260 to 280 nm at a dose of 2 and 4 mJ/cm^2, while reductions of 4.74 and 5.2 log CFU were observed on HD polyethylene for irradiances of 2 and 4 mJ/cm^2 [40]. Though, the *Salmonella* cocktail on stainless-steel surfaces was rapidly reduced by 1.3 log CFU/mL after the initial dose of 30 mJ/cm^2, a maximum reduction of 1.97 log CFU/mL after the final dose of 120 mJ/cm^2 was obtained, similar to the tailing observed in the current study with viruses [40]. Thus, the bacterial pathogens required lower doses for inactivation compared to the higher doses required for inactivation of the HAV and FCV used in the current study on food contact surfaces.

Overall, this study provided the target doses needed for 1-log inactivation of HAV and FCV when dried on three food contact surfaces using UV-C at 254 nm and UV-C LED at 279 nm based on the linear model. This data lays the foundation for the food industry to design systems needed to deliver the desired doses to inactivate the target pathogens on food contact surfaces.

5. Conclusions

This research shows the promising application of UV-C LED at 279 nm for the inactivation of foodborne viruses, HAV and the tested HuNoV surrogate, FCV, on dried model food contact surfaces (stainless-steel, ceramic, and glass). This study helped bridge the gaps in knowledge related to the dose requirements for inactivation of HAV and FCV using both traditional UV-C at 254 nm and UV-C LED at 279 nm. The results also showed notable increased resistance of HAV compared to FCV to UV-C inactivation on the tested surfaces based on the linear model D_{10}-values. While successful at achieving similar levels of inactivation, UV-C LED at 279 nm requires higher doses for inactivation of both HAV and FCV regardless of surface type. This UV-C dose data will be useful in laying the foundation for the design of UV-C LED systems to deliver the optimal doses needed to inactivate foodborne viruses on surfaces in order to decrease the risk of viral contamination. However, it is important to note that only linear models were used in this study that may not be the best fit if and when tailing effects are observed. Future studies are focused on determination of the UV-C dose requirements of aerosolized viral deposits and aerosolized viruses, as well as understanding the effect of temperature, relative humidity, organic load, and varied wavelengths of UV-C LED systems for viral inactivation on surfaces for practical applications to protect public health.

Supplementary Materials: The following supporting information can be downloaded at: https://www.mdpi.com/article/10.3390/foods13182892/s1, Table S1: Inactivation of HAV treated with either UV-C (254 nm) or UV-C LED (279 nm) on Stainless-steel coupons; Table S2: Inactivation of FCV treated with either UV-C (254 nm) or UV-C LED (279 nm) on Stainless-steel coupons; Table S3: Inactivation of HAV treated with either UV-C (254 nm) or UV-C LED (279 nm) on Ceramic coupons; Table S4: Inactivation of FCV treated with either UV-C (254 nm) or UV-C LED (279 nm) on Ceramic coupons; Table S5: Inactivation of HAV treated with either UV-C (254 nm) or UV-C LED (279 nm) on glass discs; Table S6: Inactivation of FCV treated with either UV-C (254 nm) or UV-C LED (279 nm) on glass discs.

Author Contributions: Conceptualization, A.P. and D.H.D.; funding acquisition, A.P., B.P. (Brahmaiah Pendyala) and D.H.D.; supervision, D.H.D.; methodology, B.P. (Breanna Polen) and D.H.D.; lab experiments, B.P. (Breanna Polen); data curation, B.P. (Breanna Polen); formal analysis, B.P. (Breanna Polen) and D.H.D.; writing—original draft preparation, B.P. (Breanna Polen) and D.H.D.; writing—review and editing, B.P. (Breanna Polen), B.P. (Brahmaiah Pendyala), A.P. and D.H.D. All authors have read and agreed to the published version of the manuscript.

Funding: The funding for this research was provided by the USDA AFRI Program Grant #2019-69015-29233 and #2022-67018-36277. This research is part of the M.S. thesis of Breanna Polen in partial fulfillment of the degree requirements at the University of Tennessee.

Data Availability Statement: The data obtained for this study are included in the article, and any further information can be obtained from the corresponding author.

Conflicts of Interest: All authors do not have any conflicts of interest to declare. The names of any vendors, manufacturers, or products are for solely for information and data reproducing purposes and do not imply any endorsement by any of the authors.

References

1. Centers for Disease Control and Prevention (CDC). Reports of Salmonella in the United States. Available online: https://www.cdc.gov/salmonella/index.html (accessed on 20 August 2024).
2. Centers for Disease Control and Prevention (CDC). Person-to-person outbreaks of hepatitis A across the United States. Available online: https://www.cdc.gov/hepatitis/outbreaks/ongoing-hepatitis-a/index.html (accessed on 20 August 2024).
3. Stuart, D.I.; Ren, J.; Wang, X.; Rao, Z.; Fry, E.E. Hepatitis A Virus Capsid Structure. *Cold Spring Harb. Perspect. Med.* **2019**, *9*, a031807. [CrossRef] [PubMed]
4. Sattar, S.A.; Jason, T.; Bidawid, S.; Farber, J. Foodborne spread of hepatitis A: Recent studies on virus survival, transfer and inactivation. *Can. J. Infect. Dis.* **2000**, *11*, 159–163. [CrossRef] [PubMed]

5. Shin, E.C.; Jeong, S.H. Natural History, Clinical Manifestations, and Pathogenesis of Hepatitis A. *Cold Spring Harb. Perspect. Med.* **2018**, *8*, a031708. [CrossRef] [PubMed]
6. Foster, M.A.; Hofmeister, M.G.; Kupronis, B.A.; Lin, Y.; Xia, G.L.; Yin, S.; Teshale, E. Increase in Hepatitis A Virus Infections—United States, 2013-2018. *MMWR Morb. Mortal. Wkly. Rep.* **2019**, *68*, 413–415. [CrossRef]
7. Centers for Disease Control and Prevention (CDC). Burden of Norovirus in the United States. Available online: https://www.cdc.gov/norovirus/data-research/index.html (accessed on 20 August 2024).
8. Graaf, M.; Beek, J.; Koopmans, M.P. Human norovirus transmission and evolution in a changing world. *Nat. Rev. Microbiol.* **2016**, *14*, 421–433. [CrossRef]
9. Debbink, K.; Lindesmith, L.C.; Ferris, M.T.; Swanstrom, J.; Beltramello, M.; Corti, D.; Lanzavecchia, A.; Baric, R.S. Within-host evolution results in antigenically distinct GII.4 noroviruses. *J. Virol.* **2014**, *88*, 7244–7255. [CrossRef]
10. Araud, E.; Fuzawa, M.; Shisler, J.L.; Li, J.; Nguyen, T.H. UV Inactivation of Rotavirus and Tulane Virus Targets Different components of the virions. *Appl. Environ. Microbiol.* **2020**, *86*, e02436-19. [CrossRef] [PubMed]
11. Joshi, S.S.; Dice, L.; Ailavadi, S.; D'Souza, D.H. Antiviral Effects of Quillaja saponaria Extracts Against Human Noroviral Surrogates. *Food Environ. Virol.* **2023**, *15*, 167–175. [CrossRef]
12. Patwardhan, M.; Morgan, M.T.; Dia, V.; D'Souza, D.H. Heat sensitization of hepatitis A virus and Tulane virus using grape seed extract, gingerol and curcumin. *Food Microbiol.* **2020**, *90*, 103461. [CrossRef]
13. Dubuis, M.E.; Dumont-Leblond, N.; Laliberté, C.; Veillette, M.; Turgeon, N.; Jean, J.; Duchaine, C. Ozone efficacy for the control of airborne viruses: Bacteriophage and norovirus models. *PLoS ONE* **2020**, *15*, 0231164.
14. Huang, J.; Park, G.W.; Jones, R.M.; Fraser, A.M.; Vinjé, J.; Jiang, X. Efficacy of EPA-registered disinfectants against two human norovirus surrogates and Clostridioides difficile endospores. *J. Appl. Microbiol.* **2022**, *132*, 4289–4299. [CrossRef] [PubMed]
15. Ao, X.W.; Eloranta, J.; Huang, C.H.; Santoro, D.; Sun, W.J.; Lu, Z.D.; Li, C. Peracetic acid-based advanced oxidation processes for decontamination and disinfection of water: A review. *Water Res.* **2021**, *188*, 116479. [CrossRef] [PubMed]
16. Balamurugan, S.; Zaidi, M.; Arvaj, L.; Pendyala, B.; Gabriel, M.; Farber, J.M.; Sasges, M.; Patras, A. Modeling the UV-C Inactivation Kinetics and Determination of Fluence Required for Incremental Inactivation of *Cronobacter* spp. *J. Food Prot.* **2022**, *85*, 1625–1634. [CrossRef] [PubMed]
17. Vashisht, P.; Pendyala, B.; Gopisetty, V.V.S.; Patras, A. Modeling and validation of delivered fluence of a continuous Dean flow pilot scale UV system: Monitoring fluence by biodosimetry approach. *Food Res. Int.* **2021**, *148*, 110625. [CrossRef]
18. Sadraeian, M.; Zhang, L.; Aavani, F. Viral inactivation by light. *eLight* **2022**, *2*, 18. [CrossRef]
19. Gómez-López, V.M.; Jubinville, E.; Rodríguez-López, M.I.; Trudel-Ferland, M.; Bouchard, S.; Jean, J. Inactivation of Foodborne Viruses by UV Light: A Review. *Foods* **2021**, *10*, 3141. [CrossRef]
20. Nuanualsuwan, S.; Mariam, T.; Himathongkham, S.; Cliver, D.O. Ultraviolet Inactivation of Feline Calicivirus, Human EntericViruses and Coliphages. *Photochem. Photobiol.* **2002**, *76*, 406–410. [CrossRef]
21. Moldgy, A. Comparative Evaluation of the Virucidal Effect of Remote and Direct Cold Air Plasmas with UV-C. *Plasma Process. Polym.* **2020**, *17*, 1900234. [CrossRef]
22. Park, S.Y.; Kim, A.N.; Lee, K.H.; Ha, S.D. Ultraviolet-C efficacy against a norovirus surrogate and hepatitis A virus on a stainless-steel surface. *Int. J. Food Microbiol.* **2015**, *211*, 73–78. [CrossRef]
23. Schöbel, H.; Diem, G.; Kiechl, J.; Chistè, D.; Bertacchi, G.; Mayr, A.; Posch, W. Antimicrobial efficacy and inactivation kinetics of a novel LED-based UV-irradiation technology. *J. Hosp. Infect.* **2023**, *135*, 11–17. [CrossRef]
24. Sharma, A.; Mahmoud, H.; Pendyala, B.; Balamurugan, S.; Patras, A. UV-C inactivation of microorganisms in droplets on food contact surfaces using UV-C light-emitting diode devices. *Front. Food Sci. Technol.* **2023**, *3*, 1182765. [CrossRef]
25. Pendyala, B.; Patras, A.; Pokharel, B.; D'Souza, D. Genomic Modeling as an Approach to Identify Surrogates for Use in Experimental Validation of SARS-CoV-2 and HuNoV Inactivation by UV-C Treatment. *Front. Microbiol.* **2020**, *29*, 572331. [CrossRef] [PubMed]
26. Kheyrandish, A.; Mohseni, M.; Taghipour, F. Protocol for Determining Ultraviolet Light Emitting Diode (UV-LED) Fluence for Microbial Inactivation Studies. *Environ. Sci. Technol.* **2018**, *52*, 7390–7398. [CrossRef] [PubMed]
27. Mariita, R.M.; Wilson Miller, A.C.; Randive, R.V. Evaluation of the virucidal efficacy of Klaran UVC LEDs against surface-dried norovirus. *Access Microbiol.* **2022**, *4*, 000323. [CrossRef] [PubMed]
28. Corson, E.; Pendyala, B.; Patras, A.; D'Souza, D.H. Inactivation of hepatitis A virus, feline calicivirus, and Tulane virus on Formica coupons using ultraviolet light technologies. *Heliyon* **2024**, *10*, e25201. [CrossRef]
29. Choi, J.M.; D'Souza, D.H. Inactivation of Tulane virus and feline calicivirus by aqueous ozone. *J. Food Sci.* **2023**, *88*, 4218–4229. [CrossRef]
30. Bozkurt, H.; D'Souza, D.H.; Davidson, P.M. Determination of thermal inactivation kinetics of hepatitis A virus in blue mussel (Mytilus edulis) homogenate. *Appl. Environ. Microbiol.* **2014**, *80*, 3191–3197. [CrossRef]
31. Baldelli, G.; Aliano, M.P.; Amagliani, G.; Magnani, M.; Brandi, G.; Pennino, C.; Schiavano, G.F. Airborne microorganism inactivation by a UV-C LED and ionizer-based continuous sanitation air (CSA) system in train environments. *Int. J. Environ.* **2022**, *19*, 1559. [CrossRef]
32. Sharma, A.; Singh, A.; Pendyala, B.; Balamurugan, S.; Patras, A. Inactivation of Deposited Bioaerosols on Food Contact Surfaces with UV-C Light Emitting Diode Devices. 2024, to be submitted. [CrossRef]

33. Stein, R.A.; Chirilă, M. Routes of Transmission in the Food Chain. In *Foodborne Diseases*; Academic Press: Cambridge, MA, USA, 2017; pp. 65–103. [CrossRef]
34. Lim, W.; Harrison, M.A. Effectiveness of UV light as a means to reduce Salmonella contamination on tomatoes and food contact surfaces. *Food Control* **2016**, *66*, 166–173. [CrossRef]
35. Yaun, B.R.; Sumner, S.S.; Eifert, J.D.; Marcy, J.E. Inhibition of pathogens on fresh produce by ultraviolet energy. *Int. J. Food Microbiol.* **2004**, *90*, 1–8. [CrossRef]
36. Kim, D.K.; Kang, D.H. Effect of surface characteristics on the bactericidal efficacy of UVC LEDs. *Food Control* **2020**, *108*, 106869. [CrossRef]
37. Park, G.W.; Linden, K.G.; Sobsey, M.D. Inactivation of murine norovirus, feline calicivirus and echovirus 12 as surrogates for human norovirus (NoV) and coliphage (F+) MS2 by ultraviolet light (254 nm) and the effect of cell association on UV inactivation. *Lett. Appl. Microbiol.* **2011**, *52*, 162–167. [CrossRef] [PubMed]
38. Tomás, A.L.; Reichel, A.; Silva, P.M.; Silva, P.G.; Pinto, J.; Calado, I.; Santos, N.C. UV-C irradiation-based inactivation of SARS-CoV-2 in contaminated porous and non-porous surfaces. *J. Photochem. Photobiol. B Biol.* **2022**, *234*, 112531. [CrossRef] [PubMed]
39. Kim, S.J.; Kim, D.K.; Kang, D.H. Using UVC light-emitting diodes at wavelengths of 266 to 279 nanometers to inactivate foodborne pathogens and pasteurize sliced cheese. *Appl. Environ. Microbiol.* **2016**, *82*, 11–17. [CrossRef]
40. Calle, A.; Fernandez, M.; Montoya, B.; Schmidt, M.; Thompson, J. UV-C LED Irradiation Reduces *Salmonella* on Chicken and Food Contact Surfaces. *Foods* **2021**, *10*, 1459. [CrossRef]

foods

MDPI

Article

Impact of Nanoparticle-Based TiO_2 Surfaces on Norovirus Capsids and Genome Integrity

Philippe Raymond [1,*], François St-Germain [2], Sylvianne Paul [1], Denise Chabot [3] and Louise Deschênes [2]

[1] Canadian Food Inspection Agency (CFIA), St-Hyacinthe Laboratory—Food Virology National Reference Centre, St-Hyacinthe, QC J2S 8E3, Canada
[2] Agriculture and Agri-Food Canada (AAFC), St-Hyacinthe Food Research and Development Centre, 3600 Casavant W, St-Hyacinthe, QC J2S 8E3, Canada
[3] Agriculture and Agri-Food Canada (AAFC), Ottawa Food Research and Development Centre, 960 Carling Ave, Ottawa, ON K1A 0C6, Canada
* Correspondence: philippe.raymond@inspection.gc.ca

Abstract: Human noroviruses (HuNoVs) are among the main causes of acute gastroenteritis worldwide. HuNoVs can survive for several days up to weeks at room temperature in the environment, on food, and on food handling and processing surfaces. As a result, this could lead to viral spread through the ingestion of food in contact with contaminated surfaces. The development of stable surface materials with antiviral activity might be useful to reduce viral outbreaks. Metal-based compounds, including photoactivated titanium nanoparticles (TiO_2 NPs), are known for their antiviral activity. In this study, we tested the impact of 2000 µg/mL TiO_2 NPs, with or without UV activation, on HuNoV GII and murine norovirus. Their recovery rates were reduced by 99.6%. We also evaluated a new TiO_2 NP-coating process on a polystyrene surface. This process provided a homogenous coated surface with TiO_2 NPs ranging between 5 nm and 15 nm. Without photoactivation, this TiO_2 NP-coated polystyrene surface reduced the recovery rates of intact HuNoV GII by more than 94%. When a capsid integrity treatment with $PtCl_4$ or a longer reverse transcription polymerase chain detection approach was used to evaluate virus integrity following contact with the TiO_2 NP-coated polystyrene, the HuNoV GII recovery yield reduction varied between 97 and 100%. These results support the hypothesis that TiO_2 NP-coated surfaces have the potential to prevent viral transmission associated with contaminated food surfaces.

Keywords: norovirus; TiO_2; nanoparticles; $PtCl_4$; RNA extraction; capsid integrity

1. Introduction

Human noroviruses (HuNoVs) and other human enteric viruses, including hepatitis A virus (HAV) and rotaviruses (RoV), are considered the leading cause of foodborne illnesses and outbreaks [1–3]. In 2016, the global economic burden of HuNoVs was estimated at 4.2 billion in direct health system costs and USD 60.3 billion in societal costs per year [4]. In the U.S., there was an estimated 20 million annual cases, with less than 1% associated with reported outbreaks [4]. These HuNoV infections represent an annual healthcare cost of USD 430 to USD 740 million in the U.S. [5]. HuNoVs are shed in the vomit and faeces of people who are infected, including asymptomatic carriers [6]. HuNoVs can remain infectious for several days to weeks at room temperature in the environment and on stainless steel, Formica® ceramic, polyethylene, and polyvinyl chloride [7–14]. This leads to the spread of the virus either directly through contaminated hands and surfaces or indirectly through the ingestion of contaminated food and water [2]. This, along with the increased awareness associated with reducing COVID-19 propagation, has generated a renewed interest in the identification of antiviral surface coatings [15,16].

Noroviruses are small (27–40 nm), non-enveloped, single-stranded RNA viruses that belong to a genetically diverse group of viruses of the *Caliciviridae* family [17]. Noroviruses are divided into 10 distinct norovirus genetic groups, which are undergoing constant evolution [18]. While norovirus genogroups I, II, IV, VIII, and IX infect humans, genogroup II (GII) is the most prevalent in humans [17]. According to Teunis et al., the HuNoV 50% human infectious dose is very low (<100 genomic equivalents (gEq)) in secretor-positive subjects [19]. There is currently no culture method established to confirm the presence of infectious HuNoV at the levels found on surfaces, in water, or on food commodities. Current human intestinal enteroïd (HIE) culture systems require inoculum equivalent to 10^5 gEq [20]. The detection of HuNoVs relies on viral recovery from the contaminated matrix, extraction of RNA, and reverse transcription polymerase chain reaction (RT-PCR) amplification methodologies [3]. However, RNA detection does not always correlate with the virus's integrity and infectivity, as non-infectious and fragmented RNA could also be detected [21,22]. Viability RT-qPCRs, such as capsid integrity treatments based on $PtCl_4$ and genomic integrity tests based on the amplification of a long RNA strand, have been used to provide additional information on virus integrity [23–25]. Consequently, virus integrity measurements could provide a more accurate estimation of surface antiviral activity than standard RT-qPCRs.

Metal-based compounds, such as silver, zinc, iron, copper, and titanium, are among the most commonly studied materials as novel antiviral treatments [16]. Their virucidal activity as metal nanoparticles (NPs) could be significantly increased by their high specific surface area and high surface reactivity. These characteristics lead to higher levels of reactive oxygen species (ROS) and ROS-induced oxidative stress responsible for nanotoxicity [26]. The main virucidal mechanisms observed with metal nanoparticles include the binding or disruption of viral surface structures to prevent receptor binding; the production of metal ions and ROS that degrade the viral protein capsid and nucleic acids; the direct interaction with viral surfaces, proteins, and genetic material to damage viral integrity and inhibit protein synthesis and genome replication; and the cleavage of disulfide bonds to denature viral glycoproteins (reviewed in [27]). However some of these compounds, including copper and zinc oxide NPs, are highly cytotoxic when consumed, which limits their potential application as food contact surfaces [28].

Although there is some current debate on its safety as a food additive, the low toxicity of titanium oxide (TiO_2) presents a major advantage for a food surface contact application. Titanium oxide nanoparticles (TiO_2 NPs) were reported to have potential virucidal activity against noroviruses, mainly via a photocatalytic reaction [29,30]. TiO_2 occurs as a rutile, anatase, and, more rarely, brookite crystalline structure with different photocatalytic activities and applications (reviewed in [31]). TiO_2 NPs (<100 nm) are commonly used as food additives (E 171) in candies, salad dressing, creamers, icing, and marshmallows. Approximatively 40% of this additive is found in the nanosize range [32]. A safety margin of 2.25 mg of TiO_2 NPs per kilogram of body weight per day was established by the European Food Safety Authority prior to the year 2021 [33]. However, due mainly to uncertainties concerning the genotoxicity of TiO_2 nanoparticles, an EU expert panel recently concluded that TiO_2 could no longer be considered safe as a food additive [34]. On the other hand, following an extensive review, Health Canada's Food Directorate concluded that there is no scientific evidence to support that the food additive TiO_2 is a concern for human health [35]. Food-grade TiO_2 is still considered safe by the WHO and several other countries [36]. In the U.S., the typical exposure for an adult is around 1 mg of titanium per kilogram of body weight per day [37].

In this paper, we explored the potential of TiO_2 NPs to degrade HuNoV GII using capsid and genomic integrity approaches to evaluate its impact on intact HuNoV GII. The potential antiviral activity of TiO_2 NP-coated surfaces was tested as well, and a new approach to preparing TiO_2 NP-coated polystyrene surfaces was developed for this purpose. The development of stable surface materials integrating TiO_2 NPs might be a safe and useful way to prevent viral transmission associated with contaminated food surfaces.

2. Materials and Methods

2.1. Virus Stocks

Murine norovirus-1 (MNV) was provided by Dr. H. Virgin from Washington University (St. Louis, MO, USA). Aliquots of HuNoV GII.4 (CFIA-FVR-020) were obtained as stool specimens from the British Columbia Centre for Disease Control (Vancouver, BC, Canada). A 10% suspension of HuNoV faeces was prepared as previously described [38]. To prepare the virus inoculum, MNV and HuNoV were mixed in phosphate-buffered saline (PBS), pH 7.4 (ThermoFisher, Asheville, NC, USA). The HuNoV faeces suspension was diluted 1/50 and 1/10 in the inoculum used for the TiO_2 NP suspension and TiO_2 NP-coated surface experiments, respectively. For each assay, the inoculum viral concentrations, in genomic equivalent per microliters (gEq/μL), were assessed by RT-qPCR, as described below.

2.2. Nanoparticle Preparations

Small, anatase TiO_2 NP aqueous stock dispersions (US Research Nanomaterial Inc., Houston, TX, USA) were selected based on their high specific surface areas and supplied with the characteristics described in Table 1.

Table 1. Anatase TiO_2 NP dispersion characteristics.

Appearance	Crystal	pH	Particle Size nm	TiO_2 wt %	Solvent wt %	Surface Treating	Na ppm	Al ppm	Cl ppm	TiO_2 %
Translucent liquid	Anatase	1–4	5–15	≥15.3	Water~85	No	≤98	≤95	≤93	99.9

Prior to the experiment (24–48 h), the stock dispersion was vortexed for 30 s and sonicated for 15 min in an ultrasonic bath (Branson 5510, Branson Ultrasonics, Danbury, CT, USA). Then, the suspension was diluted in PBS, with the pH adjusted to 7.8 using NaOH or HCl, and sonicated again for 15 min. Before each test, the diluted suspension was sonicated for 5 min, mixed vigorously, sonicated again for 10 min, and mixed.

2.3. Plate Surface Preparation

Corning Costar™, flat-bottom polystyrene, 12 and 24 not-treated multiple-well plates (Sigma-Aldrich, Oakville, ON, Canada) were filled with anhydrous ethanol USP (Greenfield, Tiverton, ON, Canada), covered, and incubated for 24 h at room temperature. Meanwhile, the anatase TiO_2 NP stock dispersion was diluted in ethanol to 25 mg/mL and sonicated for 15 min. The plate wells were emptied, and 50 μL of the TiO_2 NP-diluted dispersion was added to each well to cover the bottom (622 μg/cm^2). The plate was incubated for 4 h at room temperature, then placed under vacuum for 4 h at room temperature (Vacuum Oven model #1430, Sheldon MFG Inc., Cornelius, OR, USA), followed by an additional 16 h at 80 °C under vacuum, and then maintained under vacuum at room temperature for a minimum of 8 h before being used.

2.4. Atomic Force Microscopy (AFM)

The morphology and topography of the surface of the bottom of the microplate wells were obtained using a Multimode 8 (Bruker Nano Surfaces, Santa Barbara, CA, USA) equipped with a J-type scanner operating in the PeakForce QNM™ mode. Bruker FMV probes with a nominal resonant frequency of 75 kHz were used with a scan speed of 0.5–0.8 Hz.

2.5. Scanning Electron Microscopy (SEM)

Surface samples from the bottom of the TiO_2 NP-coated plate wells were collected with a razor blade and mounted on aluminium stubs using a double sticky carbon tab. They were cleaned for a duration of 5 min using UV in a ZONESEMII sample cleaner (Hitachi High-Tech Canada, Inc., Toronto, ON, Canada). SEM images were captured using a Hitachi SU7000 (Hitachi, Tokyo, Japan) field emission scanning electron microscope in the variable pressure condition mode from 50 to 100 Pa. The working distance was 6 mm, at 20 kV, and the spot size was set at 3. The secondary electron and backscattered electron signals were collected using an ultra-variable detector (UVD) and a photo diode-type backscattered electron detector (PD-BSED). An image representation was also made using both the UVD and PD1 detectors.

2.6. Inductively Coupled Plasma (ICP)

To recover the TiO_2 for the ICP analysis, 2 mL of deionised water was added to each of the TiO_2 NP-coated polystyrene plate wells, and the plate were sonicated for 15 min on the surface of a Branson 5510 sonicator (Bransonic, Danbury, CT, USA). The extracts were collected in digestion quartz tubes, which were calibrated at 50 mL and sealed with safety pressure caps (SCP Science, Montréal, QC, Canada). The 2 mL extraction was repeated, following which the extracts were combined and dry-evaporated at room temperature in a vacuum oven. A 5 mL mix of sulfuric and nitric acid (4:1) (Trace metal grade, Fisher Scientific, Fair Lawn, NJ, USA) was added to each tube, and the extracts were digested in a NovaWAVE SA microwave digestion system (SCP Science, Montréal, QC, Canada). The temperature was raised to 220 °C for 15 min, maintained for 60 min, and cooled to room temperature for 15 min. The cooled extracts were diluted in 50 mL of deionised water and filtered on a 0.22 µm PVDF membrane (Chromatographic Specialties Inc., Brockville, ON, Canada). The digested sample analysis was carried out using an ICP OES Prodigy (Teledyne Leeman Labs, Hudson, NH, USA), with the peristaltic pump set at 1.4 mL/min. Tests were performed at a wavelength of 336.122 nm using a plasma radial view and integrated for 10 s over 3 replicas. The sample uptake delay time was set at 60 s.

2.7. Norovirus Treatment

2.7.1. Virus Treatment with TiO_2 NPs

In separate and removable Corning Costar™ 12- and 24-well plates (Sigma-Aldrich), 100 µL of PBS or diluted TiO_2 NP suspension was added. An equivalent volume of PBS or virus preparation was added to these same wells, and this mixture was incubated for 60 min at room temperature in a Thermo Scientific 1300 Series A2 biological safety cabinet (Fisher Scientific, Nepean, ON, Canada). Selected well strips were treated with UVA (UVP™ UVG-54) or UVC (UVP™ UVL-56) (Fisher Scientific) or remained untreated for 5 min. The UV intensity was controlled using the Solarmeter Model 4.0 for UVA and Model 8.0 for UVC (Solar Light Company Inc., Glenside, PA, USA). The UVA intensity was between 2.0 and 2.2 mW/cm^2 (500 and 550 mJ/cm^2), while the UVC intensity was between 710 and 770 µw/cm^2 (177 and 195 mJ/cm^2).

2.7.2. Virus Treatment with TiO_2 NP Surface Plate

In each well of the TiO_2 NP-coated surface plate or control plate, 10 µL of virus preparation were added to 200 µL of RNase-free phosphate buffered saline (PBS) (Wisent, St-Bruno, ON, Canada). The plates were incubated for 2 h in a biological safety cabinet at room temperature without cover.

2.8. Capsid Integrity Treatment

To evaluate the integrity of HuNoV GII.4 and MNV capsids, the treated viruses were incubated for 10 min with 2.5 mM $PtCl_4$ (Sigma-Aldrich) at 4 °C and 300 rpm on a Thermomixer C (Eppendorf, Mississauga, ON, Canada). Control-treated virus samples without $PtCl_4$ were processed in parallel. A total of 2 µL of 500 mM EDTA (Sigma-Aldrich), 5 µL of RNA carrier (Qiagen, Mississauga, ON, Canada) and 2 µL of 1% Tween 20 (Sigma-Aldrich) were added to samples with or without $PtCl_4$ treatment prior to the RNA extraction.

2.9. RNA Extraction

A total of 500 µL of RNeasy lysis buffer RLT (Qiagen) and 0.02 M Dithiothreitol (DTT) (ThermoFisher) were added to the treated virus with or without $PtCl_4$ by pipetting up and down. The lysed virus was transferred to a RNeasy column and extracted using the QIAcube supplemented with DNase I, as previously described by the manufacturer (Qiagen). The purified RNA was eluted in 50 μL of RNase-free water, and 40 units of RNasin Plus RNase Inhibitor (Promega, Madison, WI, USA) was added before its storage at −80 °C.

2.10. Short RT-qPCR

Both MNV and HuNoV GII.4 were quantified by RT-qPCR using the TaqMan Fast Virus 1-Step Master Mix (ThermoFisher), as described previously [38]. Briefly, the primer and probe sets developed by Baert et al. [39], Kageyama et al. [40], and Loisy et al. [41] targeting the ORF1 and ORF2 junction regions were used to generate HuNoV and MNV amplicons of 86 bp and 108 bp, respectively.

2.11. Long-Range RNA Quantification (Long RT-qPCR)

A long-range, two-step reverse transcription (RT) process, followed by a qPCR detection protocol, was performed, as described previously [25]. Briefly, 2.5 kb of complementary DNA (cDNA) was synthesized using the Tx30SxN primer and Maxima H Minus reverse transcriptase (ThermoFisher). A real-time PCR (qPCR) was performed on a QuantStudio 6 (ThermoFisher) using the Taq Platinum PCR kit and the primers listed in the short RT-qPCR section, following the manufacturer's recommendations (ThermoFisher).

2.12. Recovery Yields

The impact of the TiO_2 NP suspension and coated surfaces on the norovirus recovery yields were estimated using the cycle threshold (Ct) variation compared to the inoculum reference level either by RT-qPCR or long RT-qPCR. The virus recovery yield = $10^{(\Delta Ct/m)} \times 100\%$, where $\Delta Ct = Ct_{treated} - Ct_{inoculum}$ is the $Ct_{treated}$ value of extracted viral RNA from the treated samples minus the $Ct_{inoculum}$ value of viral RNA extracted from the inoculum, and m is the slope of the virus RNA transcript standard curve, or the virus dilution, for the RT-qPCR, or the long RT-qPCR, respectively.

2.13. Statistical Analysis

The non-parametric Friedman test, with a significance level of $p < 0.05$, was used to identify statistically significant differences in terms of recovery yields following exposition to TiO_2 NPs or TiO_2 NP-coated surfaces using MedCalc (v 19.3.1) (MedCalc Software Ltd., Ostend, Belgium). The Conover test ($p < 0.05$) was used as a post-hoc test to perform sub-group pairwise comparisons. Outlier values were identified and omitted using the Tukey procedure using the MedCalc application.

3. Results

3.1. TiO_2 Nanoparticle Antiviral Activity and UV Impact in PBS

We first explored the antiviral activity of TiO_2 NPs in suspension (Figure 1). The average HuNoV and MNV inocula were $4.4 \pm 0.3 \times 10^4$ gEq and $2.3 \pm 0.3 \times 10^6$ gEq, respectively. The HuNoV and MNV RT-qPCR efficiency ranged from 0.94 to 1.00, while the long RT-qPCR efficiency ranged from 0.9 to 0.92. Without the UV treatment, the $PtCl_4$ capsid integrity treatment reduced the HuNoV recovery rate to 47 ± 2% ($n = 15$). When HuNoV GII were incubated with 20 μg/mL of TiO_2 NPs, the viral recovery rate was reduced to 30 ± 2% ($n = 12$) with or without the $PtCl_4$ treatment ($p < 0.05$). When the concentration of TiO_2 NPs was increased to 2000 μg/mL, the HuNoV recovery rates, with or without the $PtCl_4$ treatment, were further decreased to 0.3 ± 3% ($n = 6$) and 0.2 ± 3% ($n = 6$), respectively ($p < 0.05$). A slight increase in recovery rate at 108 ± 2% ($n = 9$) was noticed when the HuNoV was treated with UVA alone. This recovery rate was reduced to 60 ± 2% ($n = 9$) with the $PtCl_4$ treatment. However, the incubation of HuNoV GII with 20 μg/mL TiO_2 NPs under UVA reduced the recovery rate to 28 ± 2% ($n = 9$) with or without the $PtCl_4$ treatment ($p < 0.05$). In the presence of 2000 μg/mL TiO_2 NPs and UVA, the recovery rates with or without the $PtCl_4$ treatment were reduced to 0.4 ± 4% ($n = 3$) and

0.1 ± 4% (n = 3), respectively (p < 0.05). Without the TiO_2 NPs, UVC reduced the HuNoV recovery rates, with or without the addition of $PtCl_4$, to 7 ± 3% (n = 6) and 41 ± 3% (n = 15), respectively (p < 0.05). The incubation of HuNoV GII with 20 µg/mL TiO_2 NPs under UVC similarly reduced the recovery rate, with or without $PtCl_4$, to 39 ± 4% (n = 3) (p < 0.05). The HuNoV recovery rates following its incubation with 2000 µg/mL TiO_2 NPs and UVC was also reduced to 0.2 ± 4% (n = 3) with or without the $PtCl_4$ capsid integrity treatment.

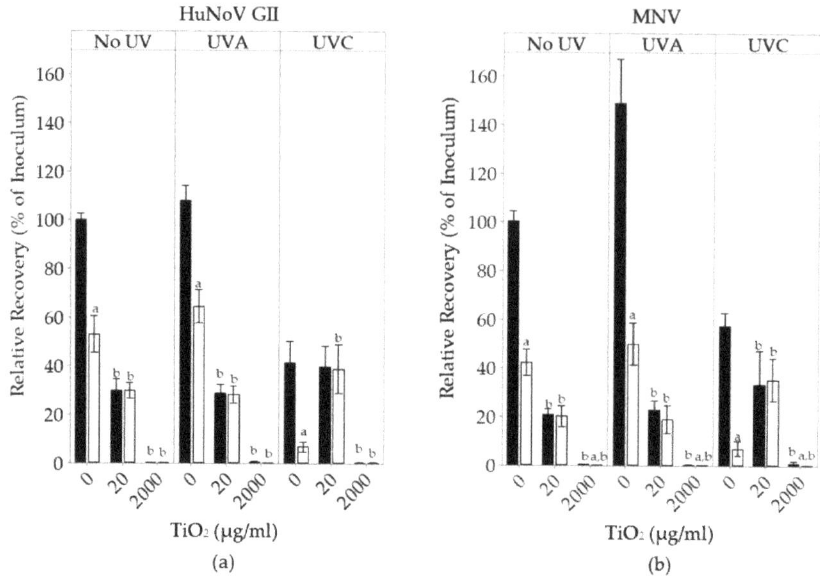

Figure 1. The impact of UV and TiO_2 NPs on HuNoV GII and MNV recovery with or without the PtCl4 capsid integrity treatment, as detected by short RT-qPCR. HuNoV GII were incubated with different concentrations of TiO_2 NPs with or without the UV (UVA; UVC) treatments. After this treatment, HuNoV GII (**a**) and MNV (**b**) were either incubated with PBS (■) or the capsid integrity agent $PtCl_4$ (□). The RNA was extracted and detected by short RT-qPCR. The results are expressed as the percentage of HuNoV GII inoculum recovery (mean ± 95% CI). The Friedman test, followed by the post-hoc Conover test (p < 0.05), was used to identify differences between the PtCl4 treatments ([a]) or between the TiO_2 treatments ([b]) and their respective controls.

The impact of UV and TiO_2 NPs on MNV was similar to that on HuNoV. Without the UV treatment, the recovery rates of MNV incubated with 20 µg/mL of TiO_2 NPs were decreased with or without the $PtCl_4$ treatment to 21 ± 3% (n = 9) and 20 ± 6% (n = 9), respectively (p < 0.05). They were further reduced in the presence of 2000 µg/mL of TiO_2 NPs, with or without the $PtCl_4$ treatment, to 0.3 ± 0.2 (n = 6) and 0 ± 0 (n = 6), respectively (p < 0.05).

3.2. *TiO_2-Coated Plates*

The average amount of TiO_2 recovered from the TiO_2 NP-coated polystyrene of the 24-well multiple plates was estimated at 892 ± 31 µg/well (469 ± 16 µg/cm^2) from the ICP measurements, which corresponds to a recovery of 71.7 ± 2.5%. This could result from the solvent-swelling and heating procedures of the material. This approach could lead a proportion of nanoparticles to be integrated into the polystyrene matrix, and thus to be not accessible for dissolution and TiO_2 quantification. The SEM analysis indicated that TiO_2 NPs were covering the entire bottom of the wells without apparent aggregates (Figure 2A). Some cracks of around 25 to 54 nm were observed. The AFM scans provided a more precise analysis of the TiO_2 NP-coated surfaces (Figure 2B). We noticed a low polydispersity and a relatively homogenous distribution on the surface. The average particle size range (5 to 15 nm) was in agreement with the estimated TiO_2 NP size provided by the supplier.

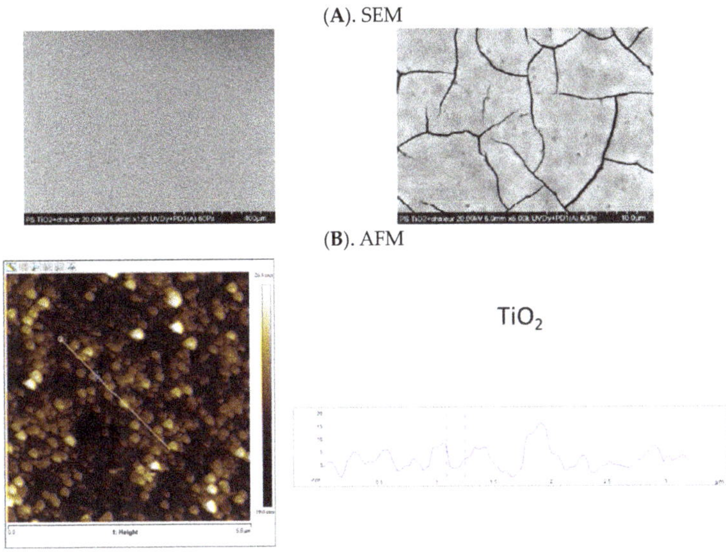

Figure 2. Surface analyses of the TiO_2 NP-coated polystyrene wells. (**A**) SEM scanning images of the TiO_2 NP-coated polystyrene plate well bottoms using a scale bar at 400 μM (**left panel**) and at 10 μm (**right panel**). (**B**) AFM scanning image of the TiO_2 NP-coated polystyrene plate bottom: a 2D image of a typical topography observed (**left panel**) and the height profile of the domains located under the white line of the 2D scan (**right panel**).

Since the TiO_2 NPs were effective against the HuNoV GII without additional UV photoactivation and food is prepared in ambient light, we evaluated the antiviral activity of the TiO_2 NP-coated polystyrene plates without UV. The average HuNoV and MNV inoculum added to the coated surfaces, as estimated by RT-qPCR, were $2.7 \pm 0.5 \times 10^6$ gEq and $5.9 \pm 0.9 \times 10^3$ gEq, respectively. The impact of the TiO_2 NP-coated surfaces on HuNoV GII and MNV was similar (Figure 3). The recovery of HuNoV GII and MNV from the non-coated wells were 13 ± 3% (n = 12) and 16 ± 4% (n = 12), the equivalent of $3.5 \pm 0.8 \times 10^6$ gEq and $9 \pm 2 \times 10^3$ gEq, respectively. The virus recoveries in the non-coated wells were not altered by the $PtCl_4$ treatment. The slight average increase (12 to 19%) observed with both viruses was not statistically significant. Both HuNoV GII and MNV recoveries were significantly reduced, when exposed to the TiO_2 NP-coated wells, to 5 ± 4% (n = 12) and 6 ± 5% (n = 12), respectively ($p < 0.05$). Following the $PtCl_4$ capsid integrity treatment, the HuNoV GII and MNV recovery yields were further reduced to 1.4 ± 1.4% (n = 12) and 1.1 ± 1.6% (n = 12), respectively ($p < 0.05$). Compared to the non-coated wells, this was equivalent to an average HuNoV and MNV log reduction of −2.18 ± 0.654 (n = 12) and <−2.61 (n = 12), respectively.

Long RT-qPCR is another approach that has been employed to estimate the extent of genome degradation (Figure 4). Long-strand RNA RT is more sensitive to $PtCl_4$ degradation than shorter RT. The inoculum recovery rates of HuNoV GII and MNV from the non-coated wells, as estimated by long RT-qPCR, were 14 ± 4% (n = 9) and 24 ± 3% (n = 9), respectively. The HuNoV GII and MNV recovery yields were not modified by the addition of the $PtCl_4$ capsid integrity treatment and were estimated at 17 ± 4% (n = 9) and 25 ± 8% (n = 9), respectively. However, when exposed to the TiO_2 NP-coated surfaces, both the HuNoV and MNV recovery yields were significantly reduced to 1.9 ± 1.9% (n = 9) and 0 ± 0% (n = 9), respectively. The addition of the $PtCl_4$ treatment did not further reduce the recovery yields of both HuNoV GII and MNV exposed to TiO_2 NP-coated surfaces, which were 1.3 ± 2.0% (n = 9) and 3 ± 4% (n = 9), respectively ($p < 0.05$).

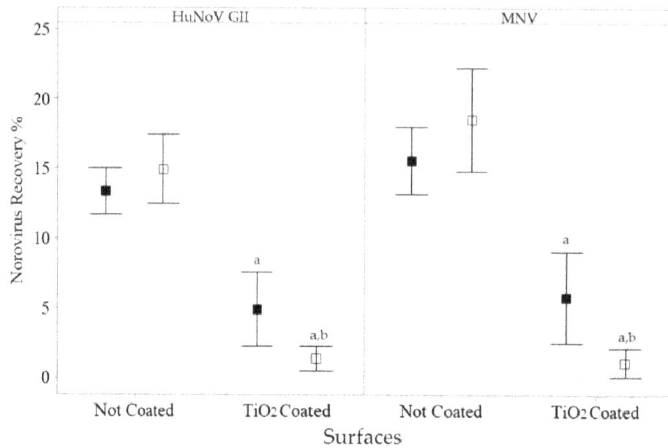

Figure 3. The impact of TiO_2 NP-coated surface on HuNoV GII and MNV recovery with or without the capsid integrity treatment, as detected by short RT-qPCR. HuNoV GII (**left panel**) and MNV (**right panel**) were incubated for 120 min on TiO_2 NP-coated polystyrene surfaces. Then, 200 µL of PBS (■) or the $PtCl_4$ (□) based capsid integrity treatment was added. The RNA was extracted and detected by short RT-qPCR. The results are expressed as percentages of inoculum recovery (mean ± 95% CI). The Friedman test, followed by the post-hoc Conover test ($p < 0.05$), was used to identify differences between the TiO_2 NP surface coatings ([a]) or between the $PtCl_4$ treatments ([b]) and their respective controls.

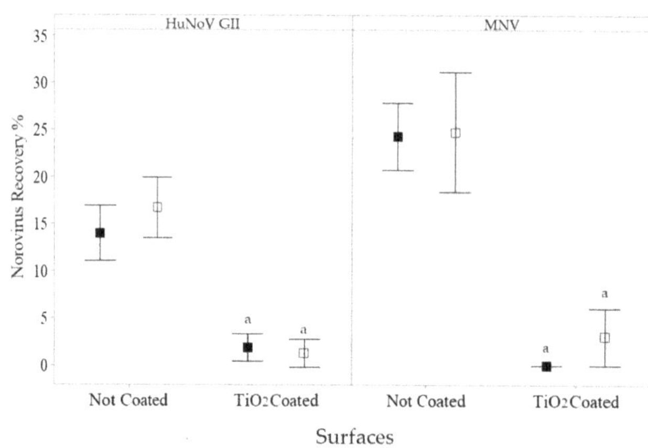

Figure 4. The impact of TiO_2 NP-coated surfaces on HuNoV GII and MNV recovery with or without the capsid integrity treatment, as detected by long RT-qPCR. HuNoV GII (**left panel**) and MNV (**right panel**) were incubated for 120 min on TiO_2 NP-coated polystyrene surfaces. Then, 200 µL of PBS (■) or the $PtCl_4$ (□) based capsid integrity treatment was added. The RNA was extracted and detected by long RT-qPCR. The results are expressed as percentages of inoculum recovery (mean ± 95% CI). The Friedman test, followed by the post-hoc Conover test ($p < 0.05$), was used to identify differences between the TiO_2 NP surface coatings ([a]) and their respective controls.

4. Discussion

The germicidal activity of UV is well known. In this study, UVC alone was active against HuNoV GII, while UVA alone did not decrease HuNoV recovery. UVC is known to be absorbed by nucleic acids, causing the formation of photoproducts such as pyrimidine dimers, and its activity also correlates strongly with the length of the genome [42]. It can also degrade virion capsids. Transmission electron microscopy images have shown that the structure of MNV capsids can be completely disrupted with UVC alone [43]. UV-inactivated viruses could still be detected by RT-qPCR since only a small region of the genome is tested. Moreover, the damage to the capsid might not correlate with damage to the genome [44]. The $PtCl_4$ capsid treatment could be used to differentiate between intact and damaged virions by interacting with free nucleic acids [23]. In our study, $PtCl_4$ further reduced the recovery rates of UVC-treated HuNoV GII, which is indicative of capsid damage. The group of Park and al. [43], using another capsid integrity treatment based on the monoazide dye PMA, also detected a significant decrease (2 log) in MNV capsid integrity following UVC treatment. On the other hand, in our study, the PtCl4 treatment did not further reduce the viral recovery rates following the TiO_2 treatment, which could indicate that there was no remaining non encapsidated RNA in solution.

In the current study, the antiviral activity of UVA was not significant. A previous study suggested that increased antiviral activity could be the result of hydroxyl radical generation from a photocatalytic reaction on TiO_2-coated surfaces [43]. The group of Lee and Ko [29] reported not only that TiO_2 particles were required for UVA to decrease MNV plaque-forming units (less than one log with 500 mJ/cm^2) but also that their combination did not impact MNV RT-qPCR recovery rates. The group of Park et al. [45] reported that after 60 min of UVA exposition, TiO_2 NP-coated glass and surface-fluorinated TiO_2 NP-coated glass decreased MNV plaque-forming units by 0.53 log and 2.6 log, respectively. They reported a small reduction in their MNV RT-qPCR recovery rates (<1 log) following a 10 min treatment with either UVA alone or with TiO_2-coated quartz tubes [43]. Still, these authors also reported no reduction in HuNoV GI.1 viral RNA within the experimental time range. Some of these discrepancies suggest that the RNA concentrations, as estimated by RT-qPCR, might not sufficiently reflect the extent of virus degradation, as well as some reactivity variation between the viruses.

Since most reports on the antiviral activity of TiO_2 are associated with its photocatalytic activity, we did not expect TiO_2 NP antiviral activity without UV against HuNoV GII to be comparable to the combinations of TiO_2 NPs and UVA or UVC. Indeed, a previous group reported that no MNV reduction was detected when they were exposed for 4 min to TiO_2 particles without UV or with up to 1500 mJ/cm^2 UVA [29,46]. Using a UV-cutoff filter, fluorinated TiO_2 NP-coated glass was ineffective against the bacteriophage MS2 without UV excitation [45]. The enhanced surface area of TiO_2 NPs in nanodimensional structures compared to bulk TiO_2 materials could explain some of the discrepancy in reactivity. Anatase TiO_2 NPs have been reported to have higher photocatalytic properties than other TiO_2 crystallites [47]. Nevertheless, others have reported some level of virucidal activity of TiO_2 NPs when not UV-illuminated. TiO_2 NPs in the dark or under daylight could decrease influenza titres and degrade this RNA virus's envelope [48]. Anastase TiO_2 NPs were also reported to reduce human adenovirus, a non-enveloped DNA virus, under dark and ambient light via sorption and inactivation [49]. Nanostructured TiO_2 thin films composed of anatase, rutile crystallites, and carbon were recently reported to exhibit antimicrobial activity against a range of pathogens, such as *E. coli*, *S. aureus*, *P. aeruginosa*, and *S. cerevisiae*, without UV activation, although with the UV photocatalyst, the TiO_2 thin film was more efficient by an order close to one log [50]. Similarly, TiO_2-coated fluoroplastic mediated the photocatalytic inactivation of astrovirus, rotavirus, and feline calicivirus using visible light (0,18 to 0,88 log reduction) but at a lower extent than with UV light photoactivation (1.75 to 2.78 log reduction) [51]. In the current study, the TiO_2 NP activity was dose-dependent. The TiO_2 NP antiviral activity was increased at 2000 μg/mL, while the group of Lee, Zoh, and Ko [46] reported a lower antiviral activity with 1000 μg/mL of TiO_2 particles. They hypothesized that increasing TiO_2 particles to this level blocks UV penetration and virus exposure to UV light. Differences in TiO_2 crystallite structures (Rutile:Anatase/85:15 mix) and preparation, UV intensity, and target viruses might explain the observed variations between the two studies. In our setting, we could not exclude that there could be some interference from high TiO_2 concentrations that could explain the absence of increased activity associated with TiO_2 UV photoactivation.

Based on these results, we explored the antiviral activity of TiO_2 NP-coated surfaces. The SEM and AFM analyses of the TiO_2 NP-coated polystyrene surface indicated that the selected coating process developed and applied in the present study allows for a relatively homogenous coating. The selected RNA extraction process involved the use of chaotrophic agents to denature proteins, inhibit RNAse, lyse viral capsids, and extract viral RNA. RNA greater than 1000 nucleotides in length have been shown to have low affinity to TiO_2 materials when using chaotrophic agents at pH 8 [52]. However, the extraction process did not allow us to differentiate some of these mechanisms, including photocatalytic degradation and adsorption, when the lysis is incomplete, for instance, owing to aggregation. A detailed understanding of the virus removal mechanism by TiO_2 NPs could be important for the development of potential applications [49].

In this study, the TiO_2 NP-coated plate had a similar level of antiviral activity compared to the TiO_2 NP suspension against both noroviruses without direct UV photoactivation. However, since the incubation time was 2 h, we could not exclude that some level of the photoactivation process was occurring under ambient light and that reactive oxygen species were generated and altering the viruses. As could be observed with the controls without the coating, some antiviral activity, ~0.9 log reduction, could be associated with the 2 h incubation time and attributed to desiccation as well as binding to the polystyrene surface combined with limited RNA extraction efficiency. Highly hydrophobic plastics such as polystyrene are known to bind viruses. The faeces and virus culture samples tested contained infectious, intact viruses as well as incomplete, non-infectious viruses that could be more susceptible to viral RNA degradation by the remaining enzymes, RNases, or PtCl4 treatment in the tested conditions. Still, the level of HuNoV extracted from the control wells represent more than 3000 times the HID_{50}. The remaining antiviral activity, an additional ~1 log reduction from the controls levels, appears to be the result of the TiO_2 NP coating itself. The TiO_2 NP coating increases the surface area, and virus binding could explain part of this additional antiviral activity. While some traces of the HuNoV long-strand genome could be recovered, there was no remaining long-strand MNV after the 2 h incubation period on the TiO_2 NP-coated plates. In terms of the difference in inoculum level, the MNV inoculum was lower by 2.7 log compared to HuNoV, and the assay sensitivity limits could explain, in part, these differences. A higher inoculum concentration could provide a more precise estimation of antiviral activity, as well as experimental conditions, to evaluate HuNoV infectivity reductions in HIE culture systems. However, when there is no or few viral RNA genome remaining, there are not enough viral particles to trigger viral replication and infection. Recently, the HIE has been shown to be a more precise model than PMAxx-based capsid integrity assays to assess HuNoV infectivity following heat or high pressure treatments since viral replication was more susceptible to treatments than its RNA [53]. In this study, there was almost no intact viral RNA remaining, as we observed with the TiO_2 NP suspension and T_IO_2 NP-coated surfaces, indicating that there should only be a low level of remaining infectious virus, if any. Exploring alternative extraction processes, or further analysing virus degradation byproducts, for instance, by mass spectrometry, might also provide additional insights into the antiviral mechanisms of the TiO_2 NP surfaces [54]. Overall, HuNoV GII were spiked at more than 20,000 times the HID_{50}. Compared to the inoculum, the TiO_2 NP-coated polystyrene resulted in a total signal reduction for HuNoV-intact virus genomes of 98.7 ± 2%.

Several parameters remain to be investigated to better define the potential applications of such TiO_2 NP-coated polystyrene. These included testing the impact of various incubation times, temperatures, and following incubation in the dark. The target pathogens should also be investigated to compare TiO_2 NP surface virucidal activity towards enveloped vs. non-enveloped viruses, as well as DNA vs. RNA viruses. For instance, attachment, aggregation, or photoactivation mechanisms could be impacted by the selected pH and control conditions used in this study and be specific to viruses with similar capsid isoelectric points and structures. In addition, organic matter and protein residues may reduce the level of binding as well as the photocatalytic activity, if any, of TiO_2 NPs by radical scavenging. For instance, the impact of the visible light-catalytic TiO_2 film on feline calicivirus titres was reduced by one log with the addition of 1 mg/mL bovine serum albumin [51]. In the current study, the TiO_2 NP-coated surfaces were efficient even when the tested samples contained 1% faeces (~ 50 µg proteins). The food residue's impact on TiO_2 NP-coated polystyrene virucidal activity remains to be investigated as well.

Another aspect that remains to be evaluated is the safety of the TiO_2 NP-coated polystyrene, despite its relative broad use in food products [31]. Part of the concern is the lack of available data on TiO_2 nanoparticle genotoxicity, which could be difficult to test owing to their low solubility [34,36]. For a coated surface, the migration of the integrated TiO_2 NPs to the food matrices would also be an important aspect to be verified. For instance, TiO_2 NPs could migrate from the polylactic/TiO_2 nanocomposite films to the cheese surface but far below the migration limit of 10 mg/kg defined by the European Food Safety Agency for food contact materials [55].

The potential application of TiO_2 NPs in food packaging materials and self-sterilizing surfaces is promising [56]. In addition to their antiviral activity and increased UV protection, antimicrobial TiO_2 NPs could increase the shelf-life of food and inhibit biofilm development. Several groups have already explored TiO_2 NP applications, mainly as a nanocomposite with chitosan, nanofibers, or polylactic acid. These applications include antiviral TiO_2 NP-coated cotton fibre, prolonging the storage life of tomato fruits, and antimicrobial cottage cheese packaging [55,57,58]. Since HuNoVs are associated with 58 to 65% of known causes of foodborne illnesses in North America, reducing the food transmission pathway even slightly could have an impact.

5. Conclusions

HuNoVs are the leading cause of foodborne illness in several countries. We reported the impact, with and without UV activation, of TiO_2 NPs on HuNoV GII and MNV. A new process of preparing TiO_2 NP-coated microplate polystyrene surfaces using a solvent and soft heating was also developed. Under ambient light, these surfaces significantly reduce the level of HuNoV GII and MNV detected. The extent of virus integrity impact was better reflected using the capsid and genomic integrity approaches. Our results support other observations on the antiviral activity of TiO_2 NP materials. However, the mechanisms involved remain to be investigated. TiO_2 NP-coated surfaces could have potential practical applications to reduce norovirus transmission associated with contaminated food surfaces. Additional work is required to develop such an application.

Author Contributions: Conceptualization, P.R. and L.D.; funding acquisition, P.R. and L.D.; supervision, P.R.; methodology, P.R., S.P., F.S.-G., L.D. and D.C.; lab experiments, S.P., F.S.-G. and D.C.; data curation, P.R., S.P. and F.S.-G.; formal analysis, P.R. and L.D.; writing-original draft preparation P.R., S.P., F.S.-G. and L.D.; writing-review and editing, P.R. and L.D. All authors have read and agreed to the published version of the manuscript.

Funding: This research was supported by the Canadian Food Inspection Agency, Research Partnership Strategy (STH-F-1607Av2).

Institutional Review Board Statement: Not applicable.

Informed Consent Statement: Not applicable.

Data Availability Statement: The original contributions presented in the study are included in the article, further inquiries can be directed to the corresponding author.

Acknowledgments: The authors acknowledge the work of analysts from the CFIA Food Virology National Reference Centre (FVNRC).

Conflicts of Interest: The authors declare no conflicts of interest. The names of specific vendors, manufacturers, or products are included for informational purposes only and do not imply endorsement by the authors.

References

1. Scallan, E.; Hoekstra, R.M.; Angulo, F.J.; Tauxe, R.V.; Widdowson, M.A.; Roy, S.L.; Jones, J.L.; Griffin, P.M. Foodborne illness acquired in the United States--major pathogens. *Emerg. Infect. Dis.* **2011**, *17*, 7–15. [CrossRef]
2. Banyai, K.; Estes, M.K.; Martella, V.; Parashar, U.D. Viral gastroenteritis. *Lancet* **2018**, *392*, 175–186. [CrossRef] [PubMed]
3. Ludwig-Begall, L.F.; Mauroy, A.; Thiry, E. Noroviruses-The State of the Art, Nearly Fifty Years after Their Initial Discovery. *Viruses* **2021**, *13*, 1541. [CrossRef] [PubMed]
4. Bartsch, S.M.; Lopman, B.A.; Ozawa, S.; Hall, A.J.; Lee, B.Y. Global Economic Burden of Norovirus Gastroenteritis. *PLoS ONE* **2016**, *11*, e0151219. [CrossRef] [PubMed]
5. Burke, R.M.; Mattison, C.P.; Pindyck, T.; Dahl, R.M.; Rudd, J.; Bi, D.; Curns, A.T.; Parashar, U.; Hall, A.J. Burden of Norovirus in the United States, as Estimated Based on Administrative Data: Updates for Medically Attended Illness and Mortality, 2001–2015. *Clin. Infect. Dis.* **2021**, *73*, e1–e8. [CrossRef] [PubMed]
6. Wang, J.; Gao, Z.; Yang, Z.R.; Liu, K.; Zhang, H. Global prevalence of asymptomatic norovirus infection in outbreaks: A systematic review and meta-analysis. *BMC Infect. Dis.* **2023**, *23*, 595. [CrossRef]
7. Cook, N.; Knight, A.; Richards, G.P. Persistence and Elimination of Human Norovirus in Food and on Food Contact Surfaces: A Critical Review. *J. Food Prot.* **2016**, *79*, 1273–1294. [CrossRef]
8. Mattison, K.; Karthikeyan, K.; Abebe, M.; Malik, N.; Sattar, S.A.; Farber, J.M.; Bidawid, S. Survival of calicivirus in foods and on surfaces: Experiments with feline calicivirus as a surrogate for norovirus. *J. Food Prot.* **2007**, *70*, 500–503. [CrossRef] [PubMed]
9. Lamhoujeb, S.; Fliss, I.; Ngazoa, S.E. Molecular Study of the Persistence of Infectious Human Norovirus on Food-Contact Surfaces. *Food Environ. Virol.* **2009**, *1*, 51–56. [CrossRef]
10. Liu, P.; Jaykus, L.A.; Wong, E.; Moe, C. Persistence of Norwalk virus, male-specific coliphage, and Escherichia coli on stainless steel coupons and in phosphate-buffered saline. *J. Food Prot.* **2012**, *75*, 2151–2157. [CrossRef]
11. Lui, P.; Chien, Y.W.; Papafragkou, E.; Hsiab, H.M.; Jayjus, L.A.; Moe, C. Persistence of human noroviruses on food preparation surfaces and human hands. *Food Environ. Virol.* **2009**, *1*, 141–147.
12. Park, G.W.; Boston, D.M.; Kase, J.A.; Sampson, M.N.; Sobsey, M.D. Evaluation of liquid- and fog-based application of Sterilox hypochlorous acid solution for surface inactivation of human norovirus. *Appl. Environ. Microbiol.* **2007**, *73*, 4463–4468. [CrossRef] [PubMed]
13. Warnes, S.L.; Summersgill, E.N.; Keevil, C.W. Inactivation of murine norovirus on a range of copper alloy surfaces is accompanied by loss of capsid integrity. *Appl. Environ. Microbiol.* **2015**, *81*, 1085–1091. [CrossRef] [PubMed]
14. Mormann, S.; Heissenberg, C.; Pfannebecker, J.; Becker, B. Tenacity of human norovirus and the surrogates feline calicivirus and murine norovirus during long-term storage on common nonporous food contact surfaces. *J. Food Prot.* **2015**, *78*, 224–229. [CrossRef] [PubMed]
15. Bregnocchi, A.; Jafari, R.; Momen, G. Design strategies for antiviral coatings and surfaces: A review. *Appl. Surf. Sci. Adv.* **2022**, *8*, 100224. [CrossRef]
16. Balasubramaniam, B.; Prateek; Ranjan, S.; Saraf, M.; Kar, P.; Singh, S.P.; Thakur, V.K.; Singh, A.; Gupta, R.K. Antibacterial and Antiviral Functional Materials: Chemistry and Biological Activity toward Tackling COVID-19-like Pandemics. *ACS Pharmacol. Transl. Sci.* **2021**, *4*, 8–54. [CrossRef] [PubMed]
17. Vinje, J. Advances in laboratory methods for detection and typing of norovirus. *J. Clin. Microbiol.* **2015**, *53*, 373–381. [CrossRef]
18. Chhabra, P.; de Graaf, M.; Parra, G.I.; Chan, M.C.; Green, K.; Martella, V.; Wang, Q.; White, P.A.; Katayama, K.; Vennema, H.; et al. Updated classification of norovirus genogroups and genotypes. *J. Gen. Virol.* **2019**, *100*, 1393–1406. [CrossRef] [PubMed]
19. Teunis, P.F.M.; Le Guyader, F.S.; Liu, P.; Ollivier, J.; Moe, C.L. Noroviruses are highly infectious but there is strong variation in host susceptibility and virus pathogenicity. *Epidemics* **2020**, *32*, 100401. [CrossRef]
20. Estes, M.K.; Ettayebi, K.; Tenge, V.R.; Murakami, K.; Karandikar, U.; Lin, S.C.; Ayyar, B.V.; Cortes-Penfield, N.W.; Haga, K.; Neill, F.H.; et al. Human Norovirus Cultivation in Nontransformed Stem Cell-Derived Human Intestinal Enteroid Cultures: Success and Challenges. *Viruses* **2019**, *11*, 638. [CrossRef]
21. Leifels, M.; Cheng, D.; Sozzi, E.; Shoults, D.C.; Wuertz, S.; Mongkolsuk, S.; Sirikanchana, K. Capsid integrity quantitative PCR to determine virus infectivity in environmental and food applications—A systematic review. *Water Res. X* **2021**, *11*, 100080. [CrossRef] [PubMed]
22. Knight, A.; Li, D.; Uyttendaele, M.; Jaykus, L.A. A critical review of methods for detecting human noroviruses and predicting their infectivity. *Crit. Rev. Microbiol.* **2013**, *39*, 295–309. [CrossRef] [PubMed]
23. Fraisse, A.; Niveau, F.; Hennechart-Collette, C.; Coudray-Meunier, C.; Martin-Latil, S.; Perelle, S. Discrimination of infectious and heat-treated norovirus by combining platinum compounds and real-time RT-PCR. *Int. J. Food Microbiol.* **2018**, *269*, 64–74. [CrossRef] [PubMed]
24. Li, D.; De Keuckelaere, A.; Uyttendaele, M. Application of long-range and binding reverse transcription-quantitative PCR to indicate the viral integrities of noroviruses. *Appl. Environ. Microbiol.* **2014**, *80*, 6473–6479. [CrossRef]

25. Raymond, P.; Paul, S.; Guy, R.A. Impact of Capsid and Genomic Integrity Tests on Norovirus Extraction Recovery Rates. *Foods* **2023**, *12*, 826. [CrossRef] [PubMed]
26. Fu, P.P.; Xia, Q.; Hwang, H.M.; Ray, P.C.; Yu, H. Mechanisms of nanotoxicity: Generation of reactive oxygen species. *J. Food Drug. Anal.* **2014**, *22*, 64–75. [CrossRef]
27. Lin, N.; Verma, D.; Saini, N.; Arbi, R.; Munir, M.; Jovic, M.; Turak, A. Antiviral nanoparticles for sanitizing surfaces: A roadmap to self-sterilizing against COVID-19. *Nano Today* **2021**, *40*, 101267. [CrossRef] [PubMed]
28. Perera, K.Y.; Jaiswal, S.; Jaiswal, A.K. A review on nanomaterials and nanohybrids based bio-nanocomposites for food packaging. *Food Chem.* **2021**, *376*, 131912. [CrossRef] [PubMed]
29. Lee, J.E.; Ko, G. Norovirus and MS2 inactivation kinetics of UV-A and UV-B with and without TiO_2. *Water Res.* **2013**, *47*, 5607–5613. [CrossRef]
30. Prakash, J.; Cho, J.; Mishra, Y.K. Photocatalytic TiO_2 nanomaterials as potential antimicrobial and antiviral agents: Scope against blocking the SARS-COV-2 spread. *Micro Nano Eng.* **2022**, *14*, 100100. [CrossRef]
31. Baranowska-Wojcik, E.; Szwajgier, D.; Oleszczuk, P.; Winiarska-Mieczan, A. Effects of Titanium Dioxide Nanoparticles Exposure on Human Health-a Review. *Biol. Trace Elem. Res.* **2020**, *193*, 118–129. [CrossRef] [PubMed]
32. Blaznik, U.; Krusic, S.; Hribar, M.; Kusar, A.; Zmitek, K.; Pravst, I. Use of Food Additive Titanium Dioxide (E171) before the Introduction of Regulatory Restrictions Due to Concern for Genotoxicity. *Foods* **2021**, *10*, 1910. [CrossRef] [PubMed]
33. EFSA. ANS Panel, Scientific opinion on the re-evaluation of titanium dioxide (E 171) as a food additive. *EFSA J.* **2016**, *14*, 83.
34. Efsa Panel on Food Additives Flavourings; Younes, M.; Aquilina, G.; Castle, L.; Engel, K.H.; Fowler, P.; Frutos Fernandez, M.J.; Furst, P.; Gundert-Remy, U.; Gurtler, R.; et al. Scientific Opinion on the safety assessment of titanium dioxide (E171) as a food additive. *EFSA J.* **2021**, *19*, e06585. [PubMed]
35. Health Canada. *State of the Science of Titanium Dioxide (TiO2) as a Food Additive*; Health Canada: Ottawa, ON, Canada, 2022; ISBN 978-0-660-44121-4.
36. JECFA Joint FAO/WHO Expert Committee on Food Additives Risk Assessment of Titanium Dioxide Risk Released—Background Information. Available online: https://www.who.int/publications/m/item/jecfa-risk-assessment-of-titanium-dioxide-risk-released-background-information (accessed on 31 January 2024).
37. Weir, A.; Westerhoff, P.; Fabricius, L.; Hristovski, K.; von Goetz, N. Titanium dioxide nanoparticles in food and personal care products. *Environ. Sci. Technol.* **2012**, *46*, 2242–2250. [CrossRef] [PubMed]
38. Raymond, P.; Paul, S.; Perron, A.; Deschenes, L. Norovirus Extraction from Frozen Raspberries Using Magnetic Silica Beads. *Food Environ. Virol.* **2021**, *13*, 248–258. [CrossRef] [PubMed]
39. Baert, L.; Wobus, C.E.; Van Coillie, E.; Thackray, L.B.; Debevere, J.; Uyttendaele, M. Detection of murine norovirus 1 by using plaque assay, transfection assay, and real-time reverse transcription-PCR before and after heat exposure. *Appl. Environ. Microbiol.* **2008**, *74*, 543–546. [CrossRef] [PubMed]
40. Kageyama, T.; Kojima, S.; Shinohara, M.; Uchida, K.; Fukushi, S.; Hoshino, F.B.; Takeda, N.; Katayama, K. Broadly reactive and highly sensitive assay for Norwalk-like viruses based on real-time quantitative reverse transcription-PCR. *J. Clin. Microbiol.* **2003**, *41*, 1548–1557. [CrossRef]
41. Loisy, F.; Atmar, R.L.; Guillon, P.; Le Cann, P.; Pommepuy, M.; Le Guyader, F.S. Real-time RT-PCR for norovirus screening in shellfish. *J. Virol. Methods* **2005**, *123*, 1–7. [CrossRef]
42. Gomez-Lopez, V.M.; Jubinville, E.; Rodriguez-Lopez, M.I.; Trudel-Ferland, M.; Bouchard, S.; Jean, J. Inactivation of Foodborne Viruses by UV Light: A Review. *Foods* **2021**, *10*, 3141. [CrossRef]
43. Park, D.; Shahbaz, H.M.; Kim, S.H.; Lee, M.; Lee, W.; Oh, J.W.; Lee, D.U.; Park, J. Inactivation efficiency and mechanism of UV-TiO_2 photocatalysis against murine norovirus using a solidified agar matrix. *Int. J. Food Microbiol.* **2016**, *238*, 256–264. [CrossRef] [PubMed]
44. Rockey, N.; Young, S.; Kohn, T.; Pecson, B.; Wobus, C.E.; Raskin, L.; Wigginton, K.R. UV Disinfection of Human Norovirus: Evaluating Infectivity Using a Genome-Wide PCR-Based Approach. *Environ. Sci. Technol.* **2020**, *54*, 2851–2858. [CrossRef] [PubMed]
45. Park, G.W.; Cho, M.; Cates, E.L.; Lee, D.; Oh, B.T.; Vinje, J.; Kim, J.H. Fluorinated TiO_2 as an ambient light-activated virucidal surface coating material for the control of human norovirus. *J. Photochem. Photobiol. B* **2014**, *140*, 315–320. [CrossRef] [PubMed]
46. Lee, J.; Zoh, K.; Ko, G. Inactivation and UV disinfection of murine norovirus with TiO_2 under various environmental conditions. *Appl. Environ. Microbiol.* **2008**, *74*, 2111–2117. [CrossRef] [PubMed]
47. Zhang, J.; Zhou, P.; Liu, J.; Yu, J. New understanding of the difference of photocatalytic activity among anatase, rutile and brookite TiO_2. *Phys. Chem. Chem. Phys.* **2014**, *16*, 20382–20386. [CrossRef] [PubMed]
48. Mazurkovaa, N.A.; Spitsynaa, Y.E.; Shikinac, N.V.; Ismagilovc, Z.R.; Zagrebel'nyia, S.N.; Ryabchikovaa, E.I. Interaction of Titanium Dioxide Nanoparticles with Influenza Virus. *Nanotechnologies Russ.* **2010**, *5*, 417–420. [CrossRef]
49. Syngouna, V.I.; Georgopoulou, M.P.; Bellou, M.I.; Vantarakis, A. Effect of Human Adenovirus Type 35 Concentration on Its Inactivation and Sorption on Titanium Dioxide Nanoparticles. *Food Environ. Virol.* **2024**. [CrossRef] [PubMed]
50. Wasa, A.; Land, J.G.; Gorthy, R.; Krumdieck, S.; Bishop, C.; Godsoe, W.; Heinemann, J.A. Antimicrobial and biofilm-disrupting nanostructured TiO_2 coating demonstrating photoactivity and dark activity. *FEMS Microbiol Lett.* **2021**, *368*, fnab039. [CrossRef]
51. Sang, y.; Phan, T.G.; Sugihara, S.; Yagyu, F. Photocatalytic inactivation of diarrheal viruses by visible-light-catalytic titanium dioxide. *Clin. Lab.* **2007**, *53*, 413–421.

52. Jimenez, L.A.; Gionet-Gonzales, M.A.; Sedano, S.; Carballo, J.G.; Mendez, Y.; Zhong, W. Extraction of microRNAs from biological matrices with titanium dioxide nanofibers. *Anal. Bioanal. Chem.* **2018**, *410*, 1053–1060. [CrossRef]
53. Wales, S.Q.; Pandiscia, A.; Kulka, M.; Sanchez, G.; Randazzo, W. Challenges for estimating human norovirus infectivity by viability RT-qPCR as compared to replication in human intestinal enteroids. *Int. J. Food Microbiol.* **2024**, *411*, 110507. [CrossRef] [PubMed]
54. Shoemaker, G.K.; van Duijn, E.; Crawford, S.E.; Uetrecht, C.; Baclayon, M.; Roos, W.H.; Wuite, G.J.; Estes, M.K.; Prasad, B.V.; Heck, A.J. Norwalk virus assembly and stability monitored by mass spectrometry. *Mol. Cell. Proteom.* **2010**, *9*, 1742–1751. [CrossRef] [PubMed]
55. LI, W.; LI, L.; Zhang, H.; Yuan, M.; Qin, Y. Evaluation of PLA nanocomposite films on physicochemical and microbiological properties of refrigerated cottage cheese. *J. Food Process. Preserv.* **2017**, *42*, e13362. [CrossRef]
56. Couto, C.; Almeida, A. Metallic Nanoparticles in the Food Sector: A Mini-Review. *Foods* **2022**, *11*, 402. [CrossRef]
57. Kim, M.G.; Kang, J.M.; Lee, J.E.; Kim, K.S.; Kim, K.H.; Cho, M.; Lee, S.G. Effects of Calcination Temperature on the Phase Composition, Photocatalytic Degradation, and Virucidal Activities of TiO_2 Nanoparticles. *ACS Omega* **2021**, *6*, 10668–10678. [CrossRef]
58. Kaewklin, P.; Siripatrawan, U.; Suwanagul, A.; Lee, Y.S. Active packaging from chitosan-titanium dioxide nanocomposite film for prolonging storage life of tomato fruit. *Int. J. Biol. Macromol.* **2018**, *112*, 523–529. [CrossRef]

foods

Article

High-Pressure Inactivation of *Bacillus cereus* in Human Breast Milk

Miroslava Jandová [1,2,*], Michaela Fišerová [1], Pavla Paterová [3], Lucie Cacková [3], Pavel Měřička [1], Jan Malý [4], Marian Kacerovský [5], Eliška Kovaříková [6], Jan Strohalm [6], Kateřina Demnerová [7], Jana Kadavá [7], Hana Sýkorová [7], Radomír Hyšpler [8], Dana Čížková [2], Aleš Bezrouk [9] and Milan Houška [6]

1 Tissue Bank, University Hospital Hradec Králové, 500 05 Hradec Králové, Czech Republic; michaela.fiserova@fnhk.cz (M.F.); pavel.mericka@fnhk.cz (P.M.)
2 Department of Histology and Embryology, Faculty of Medicine in Hradec Králové, Charles University, 500 03 Hradec Králové, Czech Republic; cizkovad@lfhk.cuni.cz
3 Department of Clinical Microbiology, University Hospital and Faculty of Medicine in Hradec Králové, Charles University, 500 05 Hradec Králové, Czech Republic; pavla.paterova@fnhk.cz (P.P.); lucie.cackova@fnhk.cz (L.C.)
4 Department of Pediatrics, University Hospital Hradec Králové, 500 05 Hradec Králové, Czech Republic; jan.maly@fnhk.cz
5 Biomedical Research Center, University Hospital Hradec Králové, 500 05 Hradec Králové, Czech Republic; marian.kacerovsky@fnhk.cz
6 Food Research Institute Prague, 102 00 Prague, Czech Republic; eliska.kovarikova@vupp.cz (E.K.); jan.strohalm@vupp.cz (J.S.); milan.houska@vupp.cz (M.H.)
7 Department of Biochemistry and Microbiology, University of Chemistry and Technology Prague, 166 28 Prague, Czech Republic; katerina.demnerova@vscht.cz (K.D.); jana.kadava@vscht.cz (J.K.); hana.sykorova@vscht.cz (H.S.)
8 Department of Clinical Biochemistry and Diagnostics, University Hospital Hradec Králové, 500 05 Hradec Králové, Czech Republic; radomir.hyspler@fnhk.cz
9 Department of Medical Biophysics, Faculty of Medicine in Hradec Králové, Charles University, 500 03 Hradec Králové, Czech Republic; bezrouka@lfhk.cuni.cz
* Correspondence: miroslava.jandova@fnhk.cz; Tel.: +420-739-569-340

Abstract: Although Holder pasteurization is the recommended method for processing breast milk, it does affect some of its nutritional and biological properties and is ineffective at inactivating spores. The aim of this study was to find and validate an alternative methodology for processing breast milk to increase its availability for newborn babies and reduce the financial loss associated with discarding milk that has become microbiologically positive. We prepared two series of breast milk samples inoculated with the *Bacillus cereus* (*B. cereus*) strain to verify the effectiveness of two high-pressure treatments: (1) 350 MPa/5 min/38 °C in four cycles and (2) cumulative pressure of 350 MPa/20 min/38 °C. We found that the use of pressure in cycles was statistically more effective than cumulative pressure. It reduced the number of spores by three to four orders of magnitude. We verified that the method was reproducible. The routine use of this method could lead to an increased availability of milk for newborn babies, and at the same time, reduce the amount of wasted milk. In addition, high-pressure treatment preserves the nutritional quality of milk.

Keywords: pressurization; *Bacillus cereus*; human breast milk; inoculation

1. Introduction

Human breast milk (HBM) is valuable. It is obtained from female donors and is used to feed children in the first days of life who, for various reasons, cannot be fed by their own mother. HBM is subject to strict microbiological monitoring and any unremovable contamination is grounds for the disposal of the milk. Contamination with bacterial spores is a serious problem because the spores cannot be removed using conventional methods [1–4].

The practice of our Human Milk Bank has shown that the most common spore-forming microbe present in breast milk is *B. cereus*, which accounted for up to 64% of all positive microbial findings. There is also the risk of *B. cereus* spores germination during the warming of pasteurized milk [5,6].

Holder pasteurization (62.5 °C/30 min), which is used to treat breast milk in most Human Milk Banks, cannot effectively eliminate all germs, and after pasteurization, 7–14% of milk is still culture-positive. Contaminated milk must then be discarded from further use according to the standard [1]. Additionally, the Holder method also affects the biological quality of human milk. A number of studies dealing with the analysis of immunological components have confirmed that HPT is more suitable for its conservation than conventional Holder pasteurization [1,7–10].

Currently, other methods for the treatment of HBM that reduce microorganisms, such as UV-treatment [11–14], high-temperature treatment [15–21], thermo-ultrasonication [12,22,23], high-intensity pulsed electric field [24], and high-pressure treatment, are being studied [1,2,12]. However, these methods have not yet been introduced into routine practice in Human Milk Banks [1,2].

High-pressure treatment (HPT) represents an option that should preserve the high quality of milk while also removing resistant spores. The high-pressure inactivation of spores is commonly used to eliminate spores from food products or other materials [1,2,25–31]. HPT is an effective method for inactivating *B. cereus* since it penetrates the cell wall and membrane, causing irreversible damage to the cell structure and function. However, it is important to optimize pressure and time parameters to ensure the complete inactivation of microorganisms while minimizing damage to milk quality [32]. Typically, the process involves subjecting spores to high pressure, typically between 100 and 800 MPa, for specific periods of time; this process is effective against a wide range of spores, including those of *B. cereus* and *Clostridioides difficile* [33].

The advantage of high-pressure inactivation is that it does not require the use of chemicals or heat that could impair the quality of the product (including HBM) and thus, can be used to treat products sensitive to heat or chemicals. On the other hand, high-pressure inactivation may not be suitable for all types of products, since some can be damaged by high pressures. It is important to carefully evaluate the product and the specific application before its use [34]. The current goal of HPT is to optimize these methods in order to achieve the best possible results in terms of nutrient retention and microbiological safety [9].

Demazeau et al. (2018), after applying various optimization tests, defined the conditions under which all vegetative forms and bacterial spores (such as *B. cereus*) were inactivated. The optimal parameters were the following: pressure 350 MPa, temperature 38 °C, treatment rate = 1 $MPa{\cdot}s^{-1}$, for four cycles with cycle duration 5 min each, and latency time (with normal pressure) between each cycle—5 min. Additionally, they found that the bioactivity of many main components, including lipase, α-lactalbumin, casein, lysozyme, lactoferrin, and immunoglobulin IgAs, was preserved [2]. Fekraoui et al. investigated the advantages of using cycled HPT compared to continuous HPT on the inactivation of *Bacillus subtilis* (*B. subtilis*) and *B. cereus* spores [35]. In his review work, Billeaud (2021) proposed to use high hydrostatic pressure (HHP) with four pressure cycles in the range of 50–150 MPa to promote the germination of *B. cereus* followed by a pressure of 350 MPa to kill 10^6 CFU/mL of *B. cereus* spores while retaining 80–100% of lipase, lysozyme, and lactoferrin activity, and 64% of immunoglobulin IgA [1]. Furukawa et al. (2021) measured the germinating and inactivating effects of cycled HPT (using six cycles of 5 min compression followed by rapid decompression) compared with continuous pressure using heat sensitivity (i.e., 70 °C/30 min). The results showed that compression could initiate spore germination, and rapid decompression could inactivate germinated spores [25].

Hayakawa et al. measured the effect of high pressure on thermoduric (resistant to high-temperature heat treatment) spores of *Bacillus stearothermophilus*, comparing two modes: Mode I = 800 MPa/60 °C/60 min and Mode II 800 = MPa/room temperature/60 min.

Mode I resulted in a decrease in the spore count (from 10^6 to 10^2 CFU per mL). Cycled pressurization (i.e., six cycles, 5 min each) of 400 MPa/70 °C produced similar results to Mode 1, i.e., spore counts decreased from 10^6 to 10^2, while cycles using 600 MPa produced complete sterilization [26].

Obaidat et al. reported the effect of using moderate hydrostatic pressure (40–140 MPa) at a moderate temperature (37–58 °C) to inactivate the spores of *B. subtilis*. The results showed that spore inactivation was exponentially proportional to the time exposed to pressure; pressures below 100 MPa and temperatures of 60 °C led to spore inactivation [27]. Doona et al. devised a "quasi-chemical" model for bacterial spore germination dynamics using HPT, which helped to promote effective reductions in bacterial spores [33].

The purpose of our research was to compare two methods of high-pressure inactivation-pressurization in cycles vs. continuous pressurization-performed on one device, using identical samples and reducing the number of spore-forming microbes, thus reducing the amount of discarded breast milk associated with it. The novelty of our work consists in verifying both procedures on a large set of data. To our best knowledge, there is not study on the same material with the same microorganisms using a four-peak high-pressure technology and a technology that used only one peak with the same pressure holding time as the individual peaks combined.

2. Materials and Methods

In our work, we used the findings of Demazeau et al. and Fekraoui et al. and verified the effectiveness of these methods on a larger data set [2,35]. In the case of pressurization in cycles, we also analyzed the effectiveness of the method for inactivating spores after each completed cycle. For testing in eight separate experiments, isolates of *B. cereus* came directly from breast milk (28 samples from the University Hospital Hradec Králové—UH HK), and also from the Laboratory of the University of Chemistry and Technology Prague (80 samples, a collection strain of *B. cereus* CCM 869 (WDCM 0001)), Faculty of Food and Biochemical Technology (UCT). Control blanks were prepared for each experiment. Figure 1 shows the algorithm of the entire set of experiments. A total of eight experiments were conducted, whereas experiments No. I–V were performed with milk inoculated at UCT and experiments No. VI–VIII were carried out with milk inoculated at UH HK.

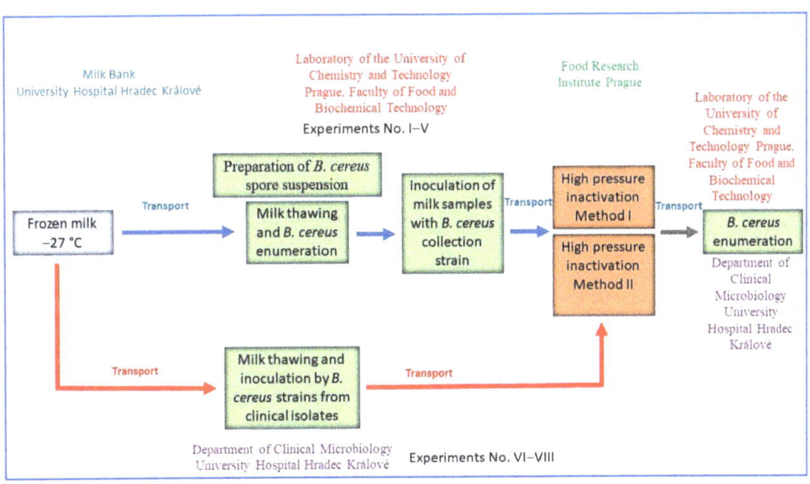

Figure 1. The algorithm of the experiment.

2.1. Sample Preparation—Inoculation with B. Cereus Spores at UCT

2.1.1. Preparation of the *B. cereus* Spore Suspension [36]

For the inoculation of the breast milk samples, the reference strain *B. cereus* CCM 869 (WDCM 0001) was used. The preparation of the spore suspension took place in two phases. First, the strain of *B. cereus* from the original gelatin disk was grown in Brain Heart Infusion broth (BHI; Merck, Darmstadt, Germany) for 48 h at 30 °C. After initial multiplication, cells were centrifuged 6000 g/10 min (Rotanta 460R, Andreas Hettich GmbH & Co. KG, Tuttlingen, Germany) and the pellets were resuspended in a sterile physiological solution. This suspension was then incubated at 30 °C to induce sporulation. The total number of *B. cereus* cells and the concentration of spores were continuously checked.

2.1.2. *B. cereus* Spores Concentration Determination [37]

To determine the concentration of *B. cereus*, a 1 mL aliquot was taken from the suspension. The total number of *B. cereus* was determined by plating on Tryptone Soya Agar (TSA, Oxoid, Hants, UK). To determine the number of spores, cells in the vegetative state were eliminated by a heat treatment of 75 °C for 11 min. This was followed by decimal dilution and plating on TSA medium. After 5 days, the required concentration of 10^5–10^8 spores/mL was reached (specific numbers are presented in the tables for individual experiments).

2.1.3. Preparation of Samples for High-Pressure Inactivation at UCT [37]

Breast milk samples were then inoculated with a suspension of spores of known concentration. First, 1 mL of sample suspension was added to 100 mL of thawed breast milk. After inoculation, the milk samples were divided into 2 bags of 50 mL each (NUK, Dolní Bousov, Czech Republic). A control milk sample (BLANK) was also inoculated in the same way. To verify the initial spore counts in the samples, a 1-mL aliquot was taken from several randomly selected bags, and after heat treatment (75 °C/11 min), was transferred to selective Mannitol Egg Yolk Polymyxin agar (hereafter MYP, Merck, Darmstadt, Germany). The use of selective agar eliminates any accompanying microflora in the milk and allows for the clear differentiation of typical *B. cereus* colonies based on the unique appearance of the colonies on the agar surface (dull pink colonies with a zone of precipitation). Thus, the initial concentration of spores in a 1 mL milk sample was experimentally verified. The number always corresponded to a 100-fold dilution of the suspension used for inoculation (100 mL of milk + 1 mL of initial suspension).

2.2. Sample Preparation—Inoculation with B. cereus Spores at UH HK

2.2.1. Preparation of *B. cereus* Suspension from Collected Clinical Isolates [38]

For this experiment, we used a *B. cereus* strain derived from common clinical samples isolated in our previous research [5,6]. These strains were revitalized and cultivated on Columbia agar (Oxoid, Ltd., Hampshire, UK) for 24 h to achieve typical quantities of *B. cereus*. Using a densitometer, suspensions with a base of 0.5 McF (McFarland) were prepared, followed by a 10-fold dilution.

2.2.2. Preparation of Samples for High-Pressure Inactivation at UH HK [38]

Frozen 100 mL bottles of HBM were thawed in the refrigerator for 24 h, then stabilized in a 22 °C water bath. Before inoculation, the initial concentration of *B. cereus* was determined in all bottles using the cultivation on blood agar (18–24 h at 35 ± 2 °C). The milk samples (in 100 mL bottles) were inoculated with the *B. cereus* suspension prepared as described above, and each sample was divided into two parallel aliquots and placed into bags (NuK, Dolní Bousov, Czech Republic). Samples of 0.5 mL were taken from each inoculated bottle and cultivated on blood agar for 18–24 h at 35 ± 2 °C to determine the exact concentration of bacteria after inoculation. After a 24-h culture, i.e., the following day, we counted the resulting quantity of *B. cereus* on the agars before and after inoculation.

2.3. High-Pressure Treatment Procedure

The samples with a volume of 100 mL were treated in plastic sealed bags (NUK, Dolní Bousov, Czech Republic) (Figure 2) using a high-pressure isostatic press CYX 6/103 (Žd'as join-stock company, Žd'ár nad Sázavou, Czech Republic) (Figure 3), with a chamber volume of 2 L, tempered to 38 °C.

Figure 2. Plastic bags with milk samples intended for pressurization.

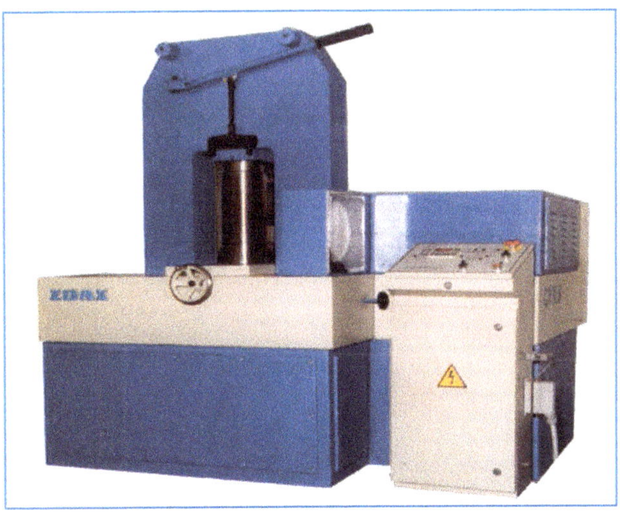

Figure 3. The high-pressure isostatic press CYX 6/103 (Žd'as join-stock company, Žd'ár nad Sázavou, Czech Republic).

2.3.1. Pressurization in Cycles

The samples were subjected to increasing pressure for approximately 5 min during each cycle. After reaching a pressure of 350 MPa, there was a 5-min hold, followed by rapid depressurization. The entire cycle, each lasting 10 min, was repeated four times. The sample was then cooled to 5–8 °C and sent for analysis. Figure 4 shows, in detail, the pressure changes during one cycle, without a pause after decompression. Figure 5 illustrates, in detail, the pressure increase during the first 5 min of the cycle, i.e., pressure was increased in 3-s increments, followed by a 27-s period of equalization. An equalization period was included to achieve an overall average pressurization rate of 1 MPa per second. (Note that Figure 4 only shows the pressure changes during the first 2 min of a single cycle).

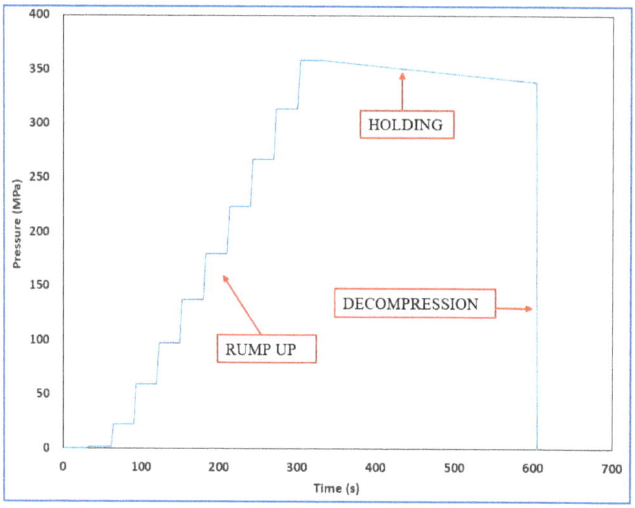

Figure 4. Pressurization cycles—pressure changes during one cycle.

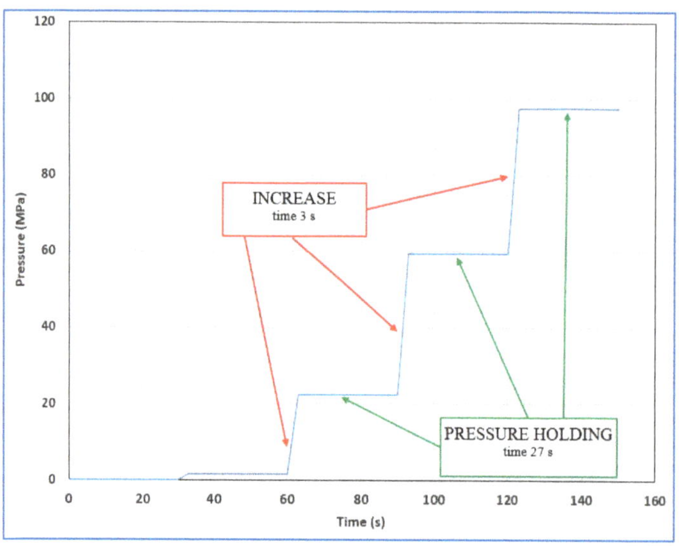

Figure 5. Pressurization cycles—detail changes.

After all four cycles were completed, the samples were subjected to microbiological analysis (Experiments No. I–III). The following experiments (No. IV and No. V) were performed to show how the microbial load decreased after each completed cycle. Individual samples were subjected to a pressure of 350 MPa according to the schedule presented in Table 1. The results of this experiment were also compared with continuous pressurization (i.e., without cycles) at 350 MPa/38 °C/20 min (two samples, No. 9 and 10). Sample BLANK 1 and BLANK 2 underwent the same temperature history but were not pressurized.

Table 1. Chronological course of sample pressurization in four cycles with successive sampling in experiments No. IV and No. V.

	Cycles, Pressure 350 MPa, 38 °C			
Sample No.	**Cycle 1**	**Cycle 2**	**Cycle 3**	**Cycle 4**
1 and 2	5 min			
3 and 4	5 min	5 min		
5 and 6	5 min	5 min	5 min	
7 and 8	5 min	5 min	5 min	5 min

2.3.2. Continuous Pressurization [35]

Samples were subjected to increasing pressure over 5–6 min. After reaching a final pressure of 350 MPa/38 °C, the samples were held at these conditions for 20 min; after depressurization, the samples were cooled to 5–8 °C and sent for analysis.

2.4. *Microbial Analysis of Samples*

2.4.1. Microbial Analysis Performed at UCT [36]

After the HPT of the milk samples, the total number of *B. cereus* in all the samples was determined. A standard methodology was used: a ten-fold dilution of the sample, spreading 200 μL on the surface of selective MYP agar (elimination of accompanying microflora), and the calculation of CFU/mL.

The control samples of milk, i.e., without pressure treatment (BLANK), were processed in the same way. In addition, the number of spores present in the control samples was determined (thermal heating and subsequent spreading on MYP, see previously described procedure). The goal was to verify the effect of time and temperature during sample handling on both the total number of *B. cereus* and the number of spores that can germinate into vegetative forms during the process. The determination of the effectiveness of the pressure treatment could be distorted if there was a significant reduction in the number of spores during the sample handling of the samples.

2.4.2. Microbial Analysis Performed at UH HK [38,39]

It was assumed that only spores survive pressure treatment, so the resulting number corresponds to the number of spores. The number of spores was verified using a quantitative method; the inoculation of the milk samples on blood agar (incubation for 18–24 h/at 35 ± 2 °C) [39] and also by inoculation into thioglycolate broth followed by inoculation on blood agar (incubation for 18–24 h/at 35 ± 2 °C), with results of either "positive" or "negative" [38]. BLANKs underwent the same thermal history but without pressurization.

2.5. *Statistic Methods*

The data were statistically evaluated using MS Excel 2016 (Microsoft Corp., Redmond, WA, USA) and NCSS 10 statistical software (2015, NCSS, LLC., Kaysville, UT, USA, and available online: ncss.com/software/ncss (accessed on 21 April 2023)). The data from normally distributed populations with more than 10 results in a test group were described using the mean and standard deviation of the sample ($\overline{x} \pm$ SD), while the other data were described using the median and the first and third quartiles of $\tilde{x}$ (1st Q, 3rd Q). For the effectiveness of the individual pressurization methods used for the data from the samples inoculated at UCT, the Equal-Variance *T*-Test at $\alpha = 0.05$ was used, and for the data from both pressure methods for the samples inoculated at UH HK, the Wilcoxon Signed-Rank Test at $\alpha = 0.05$ was used.

We also compared residual microbial contamination for both pressurization methods using the Equal-Variance *T*-Test; to adjust for multiple comparisons and keep the α at level 0.05, the Bonferroni correction was used. Thus, the resulting α for a single comparison was 0.017.

3. Results

3.1. Efficiency Comparison of Both Pressure Methods for Samples Inoculated at UCT

Table 2 contains basic descriptive statistics for the data obtained from both methods of pressurization in two experiments (No. I and No. II). We statistically proved that pressurizing in cycles led to a significantly ($\alpha = 0.05$; $p < 0.001$) lower number of spores than continuous pressurization, i.e., pressurizing in cycles appeared to be more effective at inactivating *B. cereus* spores. The results of Experiment I and II are also presented in Figures 6 and 7.

Table 2. Results of the Two-Sample Equal-Variance *T*-Test from data for both methods of pressurization. P1 denotes pressurization in cycles, P2 denotes continuous pressurization, *n* denotes number of evaluated samples. The asterisks denote statistically significant differences.

Experiment No.	I		II	
Initial Count of *B. cereus* Spores CFU/mL	10^6		10^5	
Method	P1	P2	P1	P2
n	10	10	10	10
Minimum CFU/mL	100	1200	53	400
Maximum CFU/mL	290	2900	150	650
(Mean ± SD) CFU/mL	191 ± 82 *	1880 ± 494	110 ± 33 *	535 ± 81

We also selected six representative samples from the entire data set for which a complete set of measurements was performed, and we determined the percentage decrease in *B. cereus* (CFU/mL) compared to the nominal value (Table 3). We demonstrated that both methods led to statistically significant decreases in microbial contaminants ($\alpha = 0.017$, $p = 0.365$). The slightly higher efficiency of pressurization cycles, i.e., method P1, is shown in Figure 8, which, however, was not found to be statistically significant.

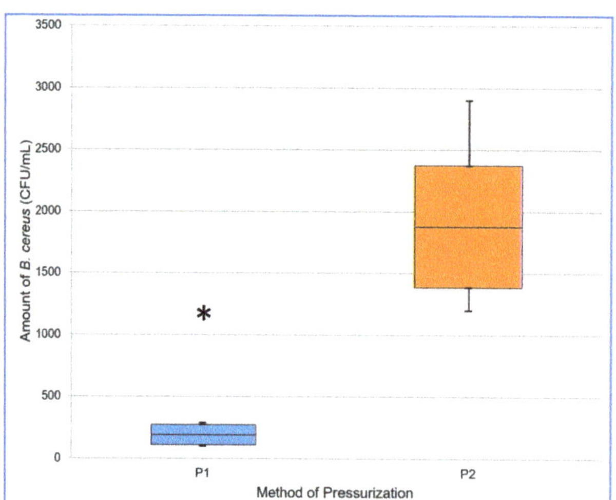

Figure 6. The resulting amount of *B. cereus* (CFU/mL) after applying pressure in cycles (P1) and continuous pressure (P2) in experiment I. The midline of the boxplot denotes the mean, the top line + SD, the bottom line—SD, and the whiskers denote the maximum and minimum values. The asterisk indicates the statistically significant results.

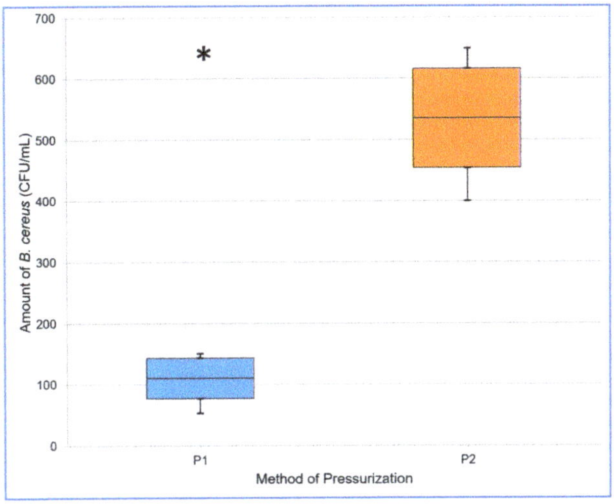

Figure 7. The resulting amount of *B. cereus* (CFU/mL) after applying pressure in cycles (P1) and continuous pressure (P2) in experiment II. The midline of the boxplot denotes the mean, the top line + SD, the bottom line—SD, and the whiskers denote the maximum and minimum values. The asterisk indicates the statistically significant results.

Table 3. Percentage of *B. cereus* CFU/mL after pressurization. P1 denotes pressurization in cycles, P2 denotes continuous pressurization, *n* denotes number of evaluated samples. The replicates are provided in the Supplementary Data (Table S3 in Supplementary Data).

Method	P1	P2
n	6	6
Minimum CFU/mL	0.01	0.04
Maximum CFU/mL	1.21	0.93
(Mean ± SD) CFU/mL	0.24 ± 0.48	0.32 ± 0.34

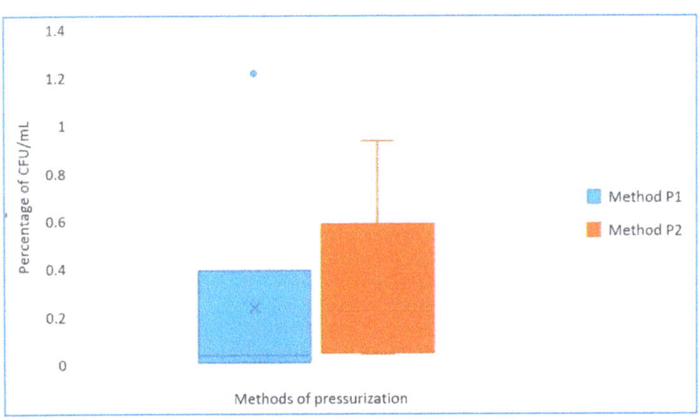

Figure 8. Residual percentage of microbial contamination (CFU/mL) after application of both pressurization methods. P1 denotes pressurization in cycles, P2 denotes continuous pressurization. The blue point in the graph represents an outlier.

3.2. Efficiency Comparison of Both Pressure Methods for Samples Inoculated at UH HK

The breast milk samples were inoculated with a suspension of the *B. cereus* clinical strain isolate in an initial quantity of an average of 92 ± 64 CFU/mL, median 72 (60; 97) CFU/mL. Table 4 summarizes the basic descriptive statistics for data after both methods of pressurization. Because the median and quartile values were zero, we also present the mean values for informative purposes. The method of pressurizing in cycles has been shown to be significantly ($\alpha = 0.05$, $p = 0.024$) more effective at inactivating *B. cereus* spores.

Table 4. Descriptive statistics for samples inoculated at UH HK after both methods of pressurization (Experiments No. VI–VIII). P1 denotes pressurization in cycles, P2 denotes continuous pressurization, *n* denotes number of evaluated samples. The asterisks denote statistically significant differences.

Method	P1	P2
n	28	28
Median (Q1; Q3)	0 (0;0)	0 (0;0)
(Mean ± SD) CFU/mL	0.04 ± 0.19 *	0.36 ± 0.95

3.3. Results of Microbial Analysis of Samples with Successive Sampling: Individual Cycles vs. Continuous Pressurization

Table 5 shows the number of *B. cereus* spores after one, two, three, and four pressurization cycles and after continuous pressurization, including values for BLANK 1 and BLANK 2 (i.e., temperature but no pressure). The initial inoculated *B. cereus* spore count in milk was of the order of 10^4 CFU/mL (Experiment IV) and 10^6 CFU/mL (Experiment V). The presented results showed that after the application of the first and second cycle, the number of spores decreased to 10^3 CFU/mL, and after the third and fourth cycle, to 10^2 CFU/mL. The pressurization applied in cycles thus inactivated spores most effectively after the third cycle. Continuous pressure application was able to only decrease counts to 10^3 CFU/mL.

Table 5. The final amount of *B. cereus* spores after high-pressure treatment in Experiment No. IV and V. P1 denotes cycled pressurization and P2 denotes continuous pressurization.

		The Amount of *B. cereus* Spores in CFU/mL	
Experiment No.	Sample No.	P1	P2
IV	1	1.7×10^2	
	2	1.4×10^2	
	3	25	
	4	10	
	5	10	
	6	5	
	7	5	
	8	<5	
	9		<5
	10		20
	BLANK 1	2.9×10^6	2.4×10^6
	BLANK 2	<5	<5
V	1	4.3×10^3	
	2	5.3×10^3	
	3	1.6×10^3	
	4	1.1×10^3	
	5	5.1×10^2	
	6	4.5×10^2	
	7	2.7×10^2	
	8	1.8×10^2	
	9		1.3×10^3
	10		1.1×10^3
	BLANK 1	5.4×10^2	6.7×10^6
	BLANK 2	1.2×10^2	2.30×10^6

3.4. Pressurization Methods over Time

Figures 9 and 10 show the time courses of pressurization in summary for all runs. Figure 9 is a graph of the pressurization cycles and Figure 10 shows continuous pressurization. The figures show that the time courses of the individual pressures were comparable in repeated experiments. Individual graphs of each experiment are presented separately in the Supplementary Data File. The beginning of each cycle cannot be the same due to the design of the device, which depends on where the multiplier piston stopped during the previous pressurization cycle.

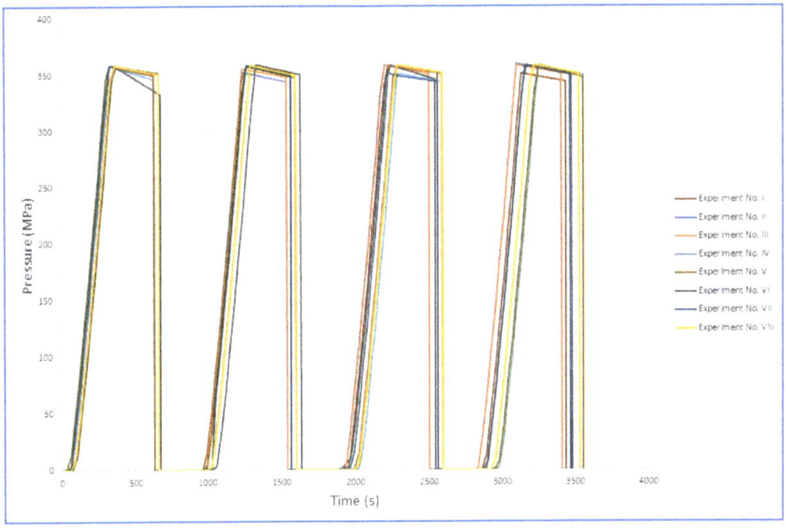

Figure 9. Cumulative visualization of pressurization in cycles.

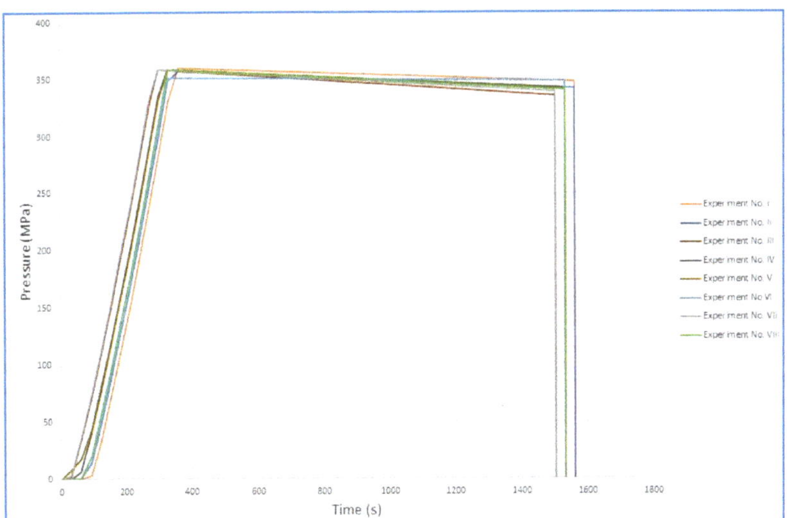

Figure 10. Cumulative visualization of continuous pressurization.

4. Discussion

It is well known that HPT activates the receptors of the inner membrane of spores, which leads to their germination. Subsequently, the high pressure causes the destruction of some vital enzymes and disturbs genetic mechanisms, such as transcription and translation, leading to the inactivation of spores [9,29,40]. The higher efficiency of pressurization in cycles compared to continuous ones aligns with the phenomenon known as sensitization that can occur with certain stress treatments, including pressure cycles in the context of bacterial spores. Sensitization refers to the increased susceptibility of microbial cells to subsequent stress or damage after exposure to an initial sublethal stress. In the case of pressurization in cycles compared to continuous treatment, the hypothesis is that the first pressure cycle may trigger the initiation of spore germination or activate cellular processes that make the spores more responsive to environmental influences. As a result of the initial pressure treatment, the spores may become more sensitive to external stresses, making them more susceptible to subsequent pressure cycles. Subsequent pressure cycles can then have a more significant impact on the already sensitized spores. This cumulative effect enhances the overall efficacy of the pressurization process [40–42]. The presented data show the effectiveness of high-pressure treatment on *B. cereus* spores, i.e., reducing counts to less than 1% of the initial values (Table 3, Figure 8). The use of pressurization in cycles led to lower CFU numbers than continuous pressurization, i.e., pressurization in cycles appeared to be more effective for the inactivation of *B. cereus* spores. Overall, we observed that pressurization led to a reduction in the number of *B. cereus* spores by three to four orders of magnitude, which was similarly observed by other authors, including Hayakawa et al. [26]. Demazeau et al. reported a reduction of up to six orders for both *Staphylococcus aureus* and *B. cereus* [2].

Although the input samples inoculated with *B. cereus* collection strain at UH HK had the quantities commonly found in clinical isolates, the higher efficiency of our method compared to conventional Holder pasteurization was clearly demonstrated. The cumulative visualization of pressurization cycles vs. continuous pressurization presented in Figures 9 and 10 demonstrates the reproducibility of both pressurization methods. As for the samples inoculated at UCT, the actual number of spores after milk inoculation was verified using several randomly selected samples with the assumption that all inoculated milk samples were equivalent if the same procedure was followed. Only samples with completed sets of measurements were used for statistical evaluation. For example, in experiment No. III, there was a loss of spores after milk inoculation, so the data from that experiment were excluded from statistical evaluation. Complete data are available at the Supplementary Data (Tables S1–S3).

The authors present the results of their experimental study aimed exclusively at the evaluation of the high-pressure treatment on *B. cereus*, that can cause severe infection in premature babies [3,43–48]. This technology is an alternative to Holder pasteurization, which has been used for many years by Human Milk Banks as the standard method recommended by EMBA [49]. One of the reasons for its popularity is its effectiveness on the HIV, CMV, and HTLV viruses [50]. Moreover, its effectiveness on the Ebola and Zika viruses was recently demonstrated [51,52].

However, Holder pasteurization is ineffective against *B. cereus* spores [53]. Demazeau et al. has noted the effectiveness of the high pressure of 350 MPa on *B. cereus* spores [2]. The long-term experience of Human Milk Banks [49] as well as the results of our previous studies [5,6] showed that pasteurized milk safety can be achieved if routine post-pasteurization evaluation is performed and if samples (bottles) with positive microbiological findings are discarded. The total discard rate due to microbial positivity in our 5-year follow-up ranged between 8 and 10%, with spore-forming microbes accounting for almost 72% of all positive findings [5]. This was lower than that described by other authors [48,54]. The proportion of *B. cereus* contaminated milk in our Bank was also lower than that described by other authors [46,47,55]. The discard of the pasteurized milk sometimes led to temporarily lowered milk availability. In the case of the Milk Bank UH HK, financial losses can reach up to almost half a million CZK (~20,000 €) per year.

Despite the high discard rate, the amount of milk delivered annually from our bank is sufficient to cover the needs of Pediatric Intensive Care Units [5,6]. However, the discard rate has prevented the use of banked milk for a broader range of newborns, including healthy babies that cannot be breast-fed.

Our previous study showed that the quantity of *B. cereus* in discarded pasteurized milk was low, ranging from 1 to 100 CFU/mL [6]. The values below 10 CFU/mL, which is the limit of initial post-pasteurization *B. cereus* contamination used in Italy, Sweden, and UK [49], were found in 80% of cases [5,6]. Our initial *B. cereus* post-pasteurization limit is below 1 CFU/mL [6], and the same limit is also used in France, Australia, and the USA [49]. This same limit is also planned for use in clinical applications using the high-pressure treatment of human milk in the future.

If the quantitative data from our previous studies mentioned above are taken into account, high-pressure treatment would lead to a reduction in the number *B. cereus* spores below 1 CFU/mL limit and/or to negativity in the majority of post-high-pressure treatment bacteriological assessments. By accepting samples with a final "negative" result, we cannot know with certainty that the samples are free of *B. cereus* spores [49]. For this reason, the manipulation of high-pressure treated milk should be similar to the manipulation of pasteurized milk [6], preferably stored frozen and used within 1 h after thawing and warming to 37 °C.

High-pressure treatment has some disadvantages when compared to Holder pasteurization. While it has been shown to be effective on HIV and CMV viruses [10,56], its effect on other viruses such as Zika or Ebola has not yet been studied [17].

Table 6 lists the advantages and disadvantages of both methods, i.e., Holder pasteurization and high pressurization. The Table shows that the high-pressure method is more effective on spore-forming microbes, and its use could lead to financial savings. At the same time, it retains more bioactive substances [1]. A detailed review comparing both methods was published by Weselowska et al., 2019 [17], and the results of our study will be published later.

Table 6. A comparison of Holder pasteurization and high pressurization inactivation [5,27].

Parameter	Holder Pasteurization	High-Pressure Inactivation
Duration	30 min	15–20 min
Temperature	62 °C	38 °C
Spore inactivation	No	Yes
Gentleness to bioactive substances	No	Yes
Device operation	less demanding	more demanding

The use of Holder pasteurization is prescribed in the Czech Republic by a Decree of the Ministry of Health; however, this does not prevent the introduction of innovative methods if their efficiency is demonstrated relative to EU rules.

The processing of breast milk is currently regulated at the European Union level by European Parliament and Council Regulations No. 852/2004 and No. 178/2002, and by European Commission Regulation No. 2073/2005 [57–59], which mandate the use of procedures based on Hazard Analysis and Critical Control Points (HACCP). Microbiological criteria can be used in the validation and verification of procedures based on HACCP principles and other hygiene control measures. However, a legislative change is now being prepared at the European Union level, based on a Recommendation by the European Directorate for the Quality of Medicines & HealthCare, where breast milk should be classified as a substance of human origin (SOHO) [38]. From this point of view, high-pressure treatment could be introduced into common practice, provided that the method is directly validated for milk banking conditions. Therefore, the high-pressure method should be compared along with Holder pasteurization. We fully agree with the statement by Wesolowska that human participant studies are needed to assess promising new human milk processing techniques [17].

5. Conclusions

High pressurization methods are more effective against *B. cereus* spores compared to classic Holder pasteurization, and the application of pressure over several cycles is even more effective. On a set of 108 breast milk samples, we verified the effectiveness of two HPTs: (1) 350 MPa/5 min/38 °C in four cycles and (2) cumulative pressure of 350 MPa/20 min/38 °C. These methods reduced the number of spores by 3–4 orders of magnitude; HPT is more effective in cycles. Additionally, we verified that the method was reproducible. The implementation of this technique into routine practice could lead to an increased availability of breast milk for newborn babies, and at the same time, reduce the costs associated with discarded contaminated milk.

For common practice of the Human Milk Bank, we recommend using HPT of 350 MPa/5 min/38 °C in three cycles.

Supplementary Materials: The following supporting information can be downloaded at: https://www.mdpi.com/article/10.3390/foods12234245/s1, Figure S1:_UH HK_Experiment VI_Pressurization in cycles; Figure S2:_UHHK_Experiment VI_Continuous pressurization; Figure S3: UHHK_Experiment VII_Pressurization in cycles; Figure S4:_UH HK_Experiment VII_Continuous pressurization; Figure S5: _UH HK_Experiment VIII_Pressurization in cycles; Figure S6:_UH HK_Experiment VIII_Continuous pressurization; Figure S7:_UCT_Experiment I_Pressurization in cycles; Figure S8:_UCT_Experiment I_Continuous pressurization; Figure S9:_UCT_Experiment II_Pressurization in cycles; Figure S10: _UCT_Experiment II_Continuous pressurization; Figure S11:_UCT_Experiment III_Pressurization in cycles; Figure S12:_UCT_Experiment III_Continuous pressurization; Figure S13:_UCT_Experiment_IV _Pressurization in cycles; Figure S14:_UCT_Experiment IV_Continuous pressurization; Figure S15: _UCT_Experiment V_Continuous pressurization; Figure S16:_UCT_Experiment V_Chronological course of sample pressurization in 4 cycles with successive sampling; Table S1: Results of microbiological analysis performed in samples inoculated at Department of Biochemistry and Microbiology, University of Chemistry and Technology Prague; Table S2: Results of microbiological analysis performed in samples inoculated at Department of Clinical Microbiology, University Hospital Hradec Králové; Table S3: Results of residual of *B. cerus* (in % CFU/mL).

Author Contributions: Investigation, project administration, and writing—original draft, M.J.; resources, M.F.; formal analysis, A.B., P.P., L.C., K.D., J.K. and H.S.; methodology, E.K. and J.S.; writing—review and editing, P.M., J.M., M.K., R.H. and D.Č.; supervision, M.H. All authors have read and agreed to the published version of the manuscript.

Funding: This project was supported by MH CZ-DRO (UHHK, 00179906) and by the Ministry of Agriculture of the Czech Republic, institutional support MZE-RO0323.

Data Availability Statement: The data presented in this study are available on request from the corresponding author.

Conflicts of Interest: We confirm that this work is original and has not been published elsewhere, nor is it currently under consideration for publication elsewhere. We have no conflict of interest to disclose.

References

1. Billeaud, C. High Hydrostatic Pressure Treatment Ensures the Microbiological Safety of Human Milk Including Bacillus Cereus and Preservation of Bioactive Proteins Including Lipase and Immuno-Proteins: A Narrative Review. *Foods* **2021**, *10*, 1327. [CrossRef] [PubMed]
2. Demazeau, G.; Plumecocq, A.; Lehours, P.; Martin, P.; Couëdelo, L.; Billeaud, C. A New High Hydrostatic Pressure Process to Assure the Microbial Safety of Human Milk While Preserving the Biological Activity of Its Main Components. *Front. Public Health* **2018**, *6*, 306. [CrossRef] [PubMed]

3. Decousser, J.-W.; Ramarao, N.; Duport, C.; Dorval, M.; Bourgeois-Nicolaos, N.; Guinebretière, M.-H.; Razafimahefa, H.; Doucet-Populaire, F. Bacillus Cereus and Severe Intestinal Infections in Preterm Neonates: Putative Role of Pooled Breast Milk. *Am. J. Infect. Control.* **2013**, *41*, 918–921. [CrossRef] [PubMed]
4. Mullié, C.; Obin, O.; Outurquin, G.; Grognet, S.; Léké, A.; Adjidé, C. Breastmilk Donations: Bacteriological Assessment, Analysis of Causes of Non-Compliance and Suggestions for Improvement. *Arch. Pédiatrie* **2018**, *25*, 263–268. [CrossRef]
5. Jandová, M.; Měřička, P.; Fišerová, M.; Landfeld, A.; Paterová, P.; Hobzová, L.; Jarkovská, E.; Kacerovský, M.; Houška, M. Bacillus Cereus as a Major Cause of Discarded Pasteurized Human Banked Milk: A Single Human Milk Bank Experience. *Foods* **2021**, *10*, 2955. [CrossRef] [PubMed]
6. Jandová, M.; Měřička, P.; Fišerová, M.; Landfeld, A.; Paterová, P.; Hobzová, L.; Jarkovská, E.; Kacerovský, M.; Houška, M. Quantitative Risk Assessment of Bacillus Cereus Growth during the Warming of Thawed Pasteurized Human Banked Milk Using a Predictive Mathematical Model. *Foods* **2022**, *11*, 1037. [CrossRef] [PubMed]
7. Permanyer, M.; Castellote, C.; Ramírez-Santana, C.; Audí, C.; Pérez-Cano, F.J.; Castell, M.; López-Sabater, M.C.; Franch, À. Maintenance of Breast Milk Immunoglobulin A after High-Pressure Processing. *J. Dairy Sci.* **2010**, *93*, 877–883. [CrossRef]
8. Delgado, F.J.; Contador, R.; Álvarez-Barrientos, A.; Cava, R.; Delgado-Adámez, J.; Ramírez, R. Effect of High Pressure Thermal Processing on Some Essential Nutrients and Immunological Components Present in Breast Milk. *Innov. Food Sci. Emerg. Technol.* **2013**, *19*, 50–56. [CrossRef]
9. Peila, C.; Emmerik, N.E.; Giribaldi, M.; Stahl, B.; Ruitenberg, J.E.; Van Elburg, R.M.; Moro, G.E.; Bertino, E.; Coscia, A.; Cavallarin, L. Human Milk Processing: A Systematic Review of Innovative Techniques to Ensure the Safety and Quality of Donor Milk. *J. Pediatr. Gastroenterol. Nutr.* **2017**, *64*, 353–361. [CrossRef]
10. Viazis, S.; Farkas, B.E.; Jaykus, L.A. Inactivation of Bacterial Pathogens in Human Milk by High-Pressure Processing. *J. Food Prot.* **2008**, *71*, 109–118. [CrossRef]
11. Christen, L.; Lai, C.T.; Hartmann, B.; Hartmann, P.E.; Geddes, D.T. Ultraviolet-C Irradiation: A Novel Pasteurization Method for Donor Human Milk. *PLoS ONE* **2013**, *8*, e68120. [CrossRef] [PubMed]
12. Kontopodi, E.; Stahl, B.; Van Goudoever, J.B.; Boeren, S.; Timmermans, R.A.H.; Den Besten, H.M.W.; Van Elburg, R.M.; Hettinga, K. Effects of High-Pressure Processing, UV-C Irradiation and Thermoultrasonication on Donor Human Milk Safety and Quality. *Front. Pediatr.* **2022**, *10*, 828448. [CrossRef] [PubMed]
13. Stinson, L.F.; Geddes, D.T.; Furfaro, L.L. Effect of Holder Pasteurization and UV-C Irradiation on Bacteriophage Titres in Human Milk. *FEMS Microbiol. Lett.* **2023**, *370*, fnad057. [CrossRef] [PubMed]
14. Barbarska, O.; Strom, K.; Oledzka, G.; Calvo, J.; Gayà, A.; López-Mendoza, M.C.; Rutkowska, M.; Rosiak, E.; Wesolowska, A.M. Effect of Nonthermal Processing on Human Milk Bactericidal Activity Against Escherichia Coli. *J. Pediatr. Gastroenterol. Nutr.* **2020**, *70*, 864–867. [CrossRef] [PubMed]
15. Manzardo, O.A.; Toll, L.J.; Müller, K.; Nickel, E.; Jonas, D.; Baumgartner, J.; Wenzel, F.; Klotz, D. A Novel Heat Treatment Protocol for Human Milk. *Front. Pediatr.* **2022**, *10*, 990871. [CrossRef] [PubMed]
16. Moro, G.E.; Arslanoglu, S. Heat Treatment of Human Milk. *J. Pediatr. Gastroenterol. Nutr.* **2012**, *54*, 165–166. [CrossRef]
17. Wesolowska, A.; Sinkiewicz-Darol, E.; Barbarska, O.; Bernatowicz-Lojko, U.; Borszewska-Kornacka, M.K.; van Goudoever, J.B. Innovative Techniques of Processing Human Milk to Preserve Key Components. *Nutrients* **2019**, *11*, 1169. [CrossRef]
18. Goldblum, R.M.; Dill, C.W.; Albrecht, T.B.; Alford, E.S.; Garza, C.; Goldman, A.S. Rapid High-Temperature Treatment of Human Milk. *J. Pediatr.* **1984**, *104*, 380–385. [CrossRef]
19. Giribaldi, M.; Coscia, A.; Peila, C.; Antoniazzi, S.; Lamberti, C.; Ortoffi, M.; Moro, G.E.; Bertino, E.; Civera, T.; Cavallarin, L. Pasteurization of Human Milk by a Benchtop High-Temperature Short-Time Device. *Innov. Food Sci. Emerg. Technol.* **2016**, *36*, 228–233. [CrossRef]
20. Escuder-Vieco, D.; Espinosa-Martos, I.; Rodríguez, J.M.; Fernández, L.; Pallás-Alonso, C.R. Effect of HTST and Holder Pasteurization on the Concentration of Immunoglobulins, Growth Factors, and Hormones in Donor Human Milk. *Front. Immunol.* **2018**, *9*, 2222. [CrossRef]
21. Lv, R.; Zou, M.; Chantapakul, T.; Chen, W.; Muhammad, A.I.; Zhou, J.; Ding, T.; Ye, X.; Liu, D. Effect of Ultrasonication and Thermal and Pressure Treatments, Individually and Combined, on Inactivation of Bacillus Cereus Spores. *Appl. Microbiol. Biotechnol.* **2019**, *103*, 2329–2338. [CrossRef] [PubMed]
22. Mank, E.; Kontopodi, E.; Heijboer, A.C.; Van Elburg, R.M.; Hettinga, K.; Van Goudoever, J.B.; Van Toledo, L. Thermoultrasonication, Ultraviolet-C Irradiation, and High-Pressure Processing: Novel Techniques to Preserve Insulin in Donor Human Milk. *Clin. Nutr.* **2021**, *40*, 5655–5658. [CrossRef] [PubMed]
23. Czank, C.; Simmer, K.; Hartmann, P.E. Simultaneous Pasteurization and Homogenization of Human Milk by Combining Heat and Ultrasound: Effect on Milk Quality. *J. Dairy Res.* **2010**, *77*, 183–189. [CrossRef]
24. Xu, S.; Walkling-Ribeiro, M.; Griffiths, M.W.; Corredig, M. Pulsed Electric Field Processing Preserves the Antiproliferative Activity of the Milk Fat Globule Membrane on Colon Carcinoma Cells. *J. Dairy Sci.* **2015**, *98*, 2867–2874. [CrossRef] [PubMed]
25. Furukawa, S.; Nakahara, A.; Hayakawa, I. Effect of Reciprocal Pressurization on Germination and Killing of Bacterial Spores. *Int. J. Food Sci. Tech.* **2000**, *35*, 529–532. [CrossRef]
26. Hayakawa, I.; Kanno, T.; Yoshiyama, K.; Fujio, Y. Oscillatory Compared with Continuous High Pressure Sterilization on Bacillus Stearothermophilus Spores. *J. Food Sci.* **1994**, *59*, 164–167. [CrossRef]

27. Obaidat, R.; Yu, D.; Aljawhiri, S.; Macgregor, R. Moderate Hydrostatic Pressure–Temperature Combinations for Inactivation of *Bacillus Subtilis* Spores. *High Press. Res.* **2015**, *35*, 317–329. [CrossRef]
28. Doona, C.J.; Feeherry, F.E.; Ross, E.W.; Kustin, K. Chemical Kinetics for the Microbial Safety of Foods Treated with High Pressure Processing or Hurdles. *Food Eng. Rev.* **2016**, *8*, 272–291. [CrossRef]
29. Black, E.P.; Koziol-Dube, K.; Guan, D.; Wei, J.; Setlow, B.; Cortezzo, D.E.; Hoover, D.G.; Setlow, P. Factors Influencing Germination of *Bacillus subtilis* Spores via Activation of Nutrient Receptors by High Pressure. *Appl. Environ. Microbiol.* **2005**, *71*, 5879–5887. [CrossRef]
30. Heinz, V.; Knorr, D. Effects of High Pressure on Spores. In *High Pressure Processing in Foods*, 1st ed.; Nottingham University Press: Nottingham, UK, 2002; pp. 77–113. ISBN 1-897676-50-6.
31. Roobab, U.; Inam-Ur-Raheem, M.; Khan, A.W.; Arshad, R.N.; Zeng, X.; Aadil, R.M. Innovations in High-Pressure Technologies for the Development of Clean Label Dairy Products: A Review. *Food Rev. Int.* **2023**, *39*, 970–991. [CrossRef]
32. Modugno, C.; Peltier, C.; Simonin, H.; Dujourdy, L.; Capitani, F.; Sandt, C.; Perrier-Cornet, J.-M. Understanding the Effects of High Pressure on Bacterial Spores Using Synchrotron Infrared Spectroscopy. *Front. Microbiol.* **2020**, *10*, 3122. [CrossRef] [PubMed]
33. Doona, C.J.; Feeherry, F.E.; Kustin, K.; Chen, H.; Huang, R.; Philip Ye, X.; Setlow, P. A Quasi-Chemical Model for Bacterial Spore Germination Kinetics by High Pressure. *Food Eng. Rev.* **2017**, *9*, 122–142. [CrossRef]
34. European Directorate for the Quality of Medicines; Consell d'Europa. *Guide to the Quality and Safety of Tissues and Cells for Human Application*, 5th ed.; Council of Europe: Strasbourg, France, 2022; p. 701. ISBN 978-92-871-9303-2.
35. Fekraoui, F.; Ferret, É.; Paniel, N.; Auvy, O.; Chamontin, C.; André, S.; Simonin, H.; Perrier-Cornet, J.-M. Cycling versus Continuous High Pressure Treatments at Moderate Temperatures: Effect on Bacterial Spores? *Innov. Food Sci. Emerg. Technol.* **2021**, *74*, 102828. [CrossRef]
36. *ISO 7932:2004*; Microbiology of Food and Animal Feeding Stuffs—Horizontal Method for the Enumeration of Presumptive Bacillus Cereus—Colony-Count Technique at 30 °C. International Organization for Standardization: Geneva, Switzerland, 2004.
37. Baron, F.; Cochet, M.-F.; Ablain, W.; Grosset, N.; Madec, M.-N.; Gonnet, F.; Jan, S.; Gautier, M. Rapid and Cost-Effective Method for Micro-Organism Enumeration Based on Miniaturization of the Conventional Plate-Counting Technique. *Lait* **2006**, *86*, 251–257. [CrossRef]
38. Ministry of Health of the Czech Republic. 2.6.1 Sterility Testing. In *Czech Pharmacopoeia*, 1st ed.; Grada Publishing, A.S.: Prague, Czech Republic, 2017; p. 1000. ISBN 978-80-271-0500-7.
39. Ministry of Health of the Czech Republic. 2.6.12 Microbiological Testing of Non-Sterile Products (Total Number of Living Microorganisms). In *Czech Pharmacopoeia*, 1st ed.; Grada Publishing, A.S.: Prague, Czech Republic, 2017; p. 1000. ISBN 978-80-271-0500-7.
40. Paidhungat, M.; Setlow, B.; Daniels, W.B.; Hoover, D.; Papafragkou, E.; Setlow, P. Mechanisms of Induction of Germination of *Bacillus Subtilis* Spores by High Pressure. *Appl. Environ. Microbiol.* **2002**, *68*, 3172–3175. [CrossRef]
41. Nguyen Thi Minh, H.; Dantigny, P.; Perrier-Cornet, J.; Gervais, P. Germination and Inactivation of *Bacillus Subtilis* Spores Induced by Moderate Hydrostatic Pressure. *Biotech Bioeng.* **2010**, *107*, 876–883. [CrossRef] [PubMed]
42. Setlow, P. Germination of Spores of Bacillus Species: What We Know and Do Not Know. *J. Bacteriol.* **2014**, *196*, 1297–1305. [CrossRef] [PubMed]
43. Ramarao, N.; Belotti, L.; Deboscker, S.; Ennahar-Vuillemin, M.; De Launay, J.; Lavigne, T.; Koebel, C.; Escande, B.; Guinebretière, M.H. Two Unrelated Episodes of Bacillus Cereus Bacteremia in a Neonatal Intensive Care Unit. *Am. J. Infect. Control.* **2014**, *42*, 694–695. [CrossRef]
44. Fournier, S.; Faraut-Derouin, V.; Casetta, A.; Frange, P.; Doit, C.; Fortineau, N.; Romain, O.; Patkai, J.; de Cillaz, C.; Rigourd, V.; et al. Bactériémies à Bacillus Cereus En Réanimation Néonatale à l'AP-HP En 2016. *Bull. Epidemiol. Hebd* **2018**, *25–26*, 536–540.
45. Glasset, B.; Herbin, S.; Granier, S.A.; Cavalié, L.; Lafeuille, E.; Guérin, C.; Ruimy, R.; Casagrande-Magne, F.; Levast, M.; Chautemps, N.; et al. Bacillus Cereus, a Serious Cause of Nosocomial Infections: Epidemiologic and Genetic Survey. *PLoS ONE* **2018**, *13*, e0194346. [CrossRef]
46. Lewin, A.; Delage, G.; Bernier, F.; Germain, M. Banked Human Milk and Quantitative Risk Assessment of *Bacillus Cereus* Infection in Premature Infants: A Simulation Study. *Can. J. Infect. Dis. Med. Microbiol.* **2019**, *2019*, 6348281. [CrossRef]
47. Cormontagne, D.; Rigourd, V.; Vidic, J.; Rizzotto, F.; Bille, E.; Ramarao, N. Bacillus Cereus Induces Severe Infections in Preterm Neonates: Implication at the Hospital and Human Milk Bank Level. *Toxins* **2021**, *13*, 123. [CrossRef] [PubMed]
48. Mallardi, D.; Piemontese, P.; Liotto, N.; Colombo, R.M.; Dodaro, A.; Schiavello, A.; Tabasso, C.; Plevani, L.; Bezze, E.; Menis, C.; et al. New Operating Approach to Limit Bacillus Cereus Contamination of Donor Human Milk. *J. Hum. Lact.* **2022**, *38*, 102–107. [CrossRef]
49. Weaver, G.; Bertino, E.; Gebauer, C.; Grovslien, A.; Mileusnic-Milenovic, R.; Arslanoglu, S.; Barnett, D.; Boquien, C.-Y.; Buffin, R.; Gaya, A.; et al. Recommendations for the Establishment and Operation of Human Milk Banks in Europe: A Consensus Statement From the European Milk Bank Association (EMBA). *Front. Pediatr.* **2019**, *7*, 53. [CrossRef] [PubMed]
50. Lanari, M.; Sogno Valin, P.; Natale, F.; Capretti, M.G.; Serra, L. Human Milk, a Concrete Risk for Infection? *J. Matern.-Fetal Neonatal Med.* **2012**, *25*, 67–69. [CrossRef] [PubMed]
51. Hamilton Spence, E.; Huff, M.; Shattuck, K.; Vickers, A.; Yun, N.; Paessler, S. Ebola Virus and Marburg Virus in Human Milk Are Inactivated by Holder Pasteurization. *J. Hum. Lact.* **2017**, *33*, 351–354. [CrossRef] [PubMed]
52. Pfaender, S.; Vielle, N.J.; Ebert, N.; Steinmann, E.; Alves, M.P.; Thiel, V. Inactivation of Zika Virus in Human Breast Milk by Prolonged Storage or Pasteurization. *Virus Res.* **2017**, *228*, 58–60. [CrossRef]

53. Vidic, J.; Chaix, C.; Manzano, M.; Heyndrickx, M. Food Sensing: Detection of Bacillus Cereus Spores in Dairy Products. *Biosensors* **2020**, *10*, 15. [CrossRef]
54. Dewitte, C.; Courdent, P.; Charlet, C.; Dumoulin, D.; Courcol, R.; Pierrat, V. Contamination du lait maternel par une flore aérobie: Évaluation des pertes pour un lactarium. *Arch. Pédiatrie* **2015**, *22*, 461–467. [CrossRef]
55. Adjidé, C.C.; Léké, A.; Mullié, C. *Bacillus Cereus* Contamination of Pasteurized Human Milk Donations: Frequency, Origin, Seasonal Distribution, Molecular Typing of Strains and Proposed Corrective/Preventive Actions. *J. Matern. Fetal Neonatal Med.* **2022**, *35*, 1554–1561. [CrossRef]
56. Terpstra, F.G.; Rechtman, D.J.; Lee, M.L.; Hoeij, K.V.; Berg, H.; Engelenberg, F.A.C.V.; Wout, A.B.V. Antimicrobial and Antiviral Effect of High-Temperature Short-Time (HTST) Pasteurization Applied to Human Milk. *Breastfeed. Med.* **2007**, *2*, 27–33. [CrossRef] [PubMed]
57. European Commission. *Commission Regulation (EC) No 2073/2005 of 15 November 2005 on Microbiological Criteria for Foodstuffs*; European Commission: Luxembourg, 2006; Volume 2073/2005.
58. European Parliament and Council. *Regulation (EC) No. 852/2004 of the European Parliament and of the Council on the Hygiene of Foodstuffs*; European Parliament and Council: Luxembourg, 2004; Volume 852/2004.
59. European Parliament and Council. *Regulation (EC) No 178/2002 of the European Parliament and of the Council of 28 January 2002 Laying down the General Principles and Requirements of Food Law, Establishing the European Food Safety Authority and Laying down Procedures in Matters of Food Safety*; European Parliament and Council: Luxembourg, 2002; Volume 178/2002.

foods

Review

Progress in Low-Impact Processing Technologies to Deliver More Sustainable and Healthy Food Tomorrow

Marco Dalla Rosa [1,2,*], Santina Romani [1,2], Pietro Rocculi [1,2], Urszula Tylewicz [1,2] and Silvia Tappi [1,2]

[1] Department of Agriculture and Food Sciences, Alma Mater Studiorum University of Bologna, Piazza Goidanich 60, 47521 Cesena, Italy; santina.romani2@unibo.it (S.R.); pietro.rocculi3@unibo.it (P.R.); urszula.tylewicz@unibo.it (U.T.); silvia.tappi2@unibo.it (S.T.)

[2] Interdepartmental Centre for Agrifood Industrial Research, Alma Mater Studiorum University of Bologna, Via Quinto Bucci 336, 47521 Cesena, Italy

* Correspondence: marco.dallarosa@unibo.it

Abstract: Following the debate on food processing, resulting in a negative definition of ultra-processed products, the improvement of the food system could be pursued through the co-creation of new food solutions aimed at enhancing human health and increasing safety and sustainability, in particular by using neglected foodstuff, crops or by-products, and applying mild processing technologies. The proper management of mild/non-thermal processing technologies, such as dynamic and hydrostatic high-pressure, vacuum impregnation, ultrasound, pulsed electric field and cold plasma applications, can result in a less negative effect with respect to the traditional thermal treatments, and, in some cases, the overall functionality can be improved. In many cases, these treatments can induce structural changes that improve the bioaccessibility and/or the bioavailability of bioactive compounds such as probiotic microorganisms. Moreover, non-thermal pretreatments, also combined with mild thermal drying technology, could lead to a significant reduction in the total request of energy, even when considering the energy input for their application. A selected review of results published in the last few years on those strategies is presented, considering studies carried out within the frame of different national and EU projects.

Keywords: non-thermal technologies; sustainability; healthy food; ultra-processed food

1. Introduction

1.1. Challenges

As is well known, the agri-food production system is called to tackle different challenges. UN recommendations have also been developed following Food and Agriculture Organization of the United Nations (FAO) alerts on food production sustainability, which have been published in some FAO reports, in particular that published in 2011 on Livestock in Food security [1]. Since the publication of this report, there has been concern about the environmental sustainability of actual food habits in so-called developed countries, considering that 75% of available lands are already used for animal production, and 18% of global gas emissions (14.5% of which result from human activities) are related to animal (cattle and ruminants in general) production.

Moreover, these concerns might even increase, as the global population is projected to reach 9.6 billion by 2050, about 1.4 billion more than today, together with increasing urbanization, thus making it more and more difficult to access locally produced food, mainly in developing countries [2]. Furthermore, the increase in the world's population will be accompanied by an increase in the middle class in those countries where economic

https://doi.org/10.3390/foods14132332

development is higher; purchasing power goes hand-in-hand with an increasing demand for richer food, both from a quality and nutritional point of view.

Therefore, it is necessary to provide evidence of the need for change. Firstly, to guarantee the availability of sufficient and secure food for all in the future; to accomplish this, scientists, geneticists, agronomists and engineers shall be involved in actions to tackle climate change and resource reduction in order to pursue a "sustainable intensification" of food production. Furthermore, the relationship between food/nutrition and sustainability in a rational approach is related to food losses and wastes. Nowadays, it is well known that around 30% of global food production is lost or wasted along the food chain, from production to the consumer's dish [3]. Farmers, growers, food producers and processors are mainly involved in introducing technological innovations to reduce food losses from the field, where raw materials are produced, along with the food chain and distribution [4].

1.2. Nutrition Needs

Recently, the European Food Safety Authority (EFSA) published an update of the Dietary Reference Values for the EU, which includes recommended values for microelements like sodium and chloride, which are examples of the challenges posed by the Reference Intake Levels of Nutrients and Energy (LARN) recommendations when there is a complex physiological interaction between several nutrients. The updated LARNs have been differently designed; other than different recommendations for the general population, it includes those for infants (7–11 months), children and adolescents (1–17 years), adults (≥18 years), pregnant women and lactating women [5]. Following these recommendations, the Healthy Eating Pyramid was created, addressing other aspects of a healthy lifestyle (weight control and regular exercise, vitamin D and multivitamin supplements, moderate alcohol consumption, etc.), and is therefore a useful tool for health professionals and health educators. Another tool is the Healthy Eating Plate, which can be used to create a balanced meal by following a few simple recommendations. The meal should consist of vegetables and fruits ($^{1}/_{2}$ of your plate), whole grains, which should be preferred ($^{1}/_{4}$ of your plate), protein sources ($^{1}/_{4}$ of your plate) and healthy plant oils, such as olive oil, in moderation [6].

1.3. Food Production and Environment

Owing to the aforementioned advice, the recommendation to drastically reduce meat consumption emphasizes the need to explore new sources of protein that are suitable for healthily supplying the human body with the required proteins and nutrients. Furthermore, a swift global food transformation towards healthy diets from sustainable food systems is even more necessary, and without such a food transformation, the world will not meet the targets set in the United Nations SDGs and the Paris Climate Agreement. In fact, the agri-food system is among the major drivers of climate change and other critical climate factors, such as changes in land use, the depletion of freshwater resources and the pollution of both terrestrial and aquatic ecosystems; Ref. [7] analyzed several options in order to reduce the environmental effects of the food system, including dietary changes towards healthier diets, with high quantities of plant-based products, improvements in technologies and management, and reductions in food loss and waste. Their analysis showed that it may be possible to stay within planetary boundaries, considering the environmental pressure expected by 2050, with a combination of high-ambition measures for greenhouse gas emissions and nitrogen and phosphorus applications, and with a combination of medium-ambition measures for cropland and blue water use. Analyzing the options at the planetary level shows in detail the possible combinations of the different measures. The analysis showed that an ambitious dietary change toward a more plant-based and flexitarian diet is needed to stay within the midpoint of the greenhouse gas (GHG) limit, in combination with other factors like food loss and waste reduction, and technological improvements, especially to stay within the mean values of cropland and blue water boundaries [7,8]. On the other hand, there is a need for technological improvement due to the necessity to achieve a reduction in energy and production factors so to increase food processing sustainability. Some innovative processing technologies, based on non-thermal physical principles, promise a lower energy input in food processing, thus combining the "mild" treatment of food with a better utilization of energy and water resources, which should lead to more sustainable processes in terms of the environmental impact. The proper validation of the range of these emerging and novel technologies is, however, needed through accurate and complete robust data collection so to ensure the full reliability of the introduction of these new technologies.

1.4. Food Classification

Consumers demand foods that are healthy and palatable, but which are also able to satisfy the needs linked to changing lifestyles in an increasingly complex relationship between work, interests, different activities and the time to prepare food.

In this context, the food industry is increasingly moving towards offering processed foods to provide safe, healthy and palatable food with longer service and shelf-life characteristics to be consumed outside of the home or at home with reduced preparation time. Nevertheless, in recent years numerous epidemiological studies have highlighted the relationship between industrial products, generically considered as "processed" or "ultra-processed" (UPF), and health risks. The increase in obesity and other food-related problems (e.g., cardiovascular disease and diabetes) has been related to the intensity of processing according to the NOVA classification [9,10]. A great debate has started around the topic, with different viewpoints held between nutritionists, food scientists and food engineers; however, it appears quite clear to the majority of the food science community that only classifying foods according to the level of processing, considering the number of ingredients, is not only wrong but also misleading, and could be an obstacle for the food industry to move forward with innovations [11]. To this end, all functional foods will be classified as UPFs [12]. In fact, from one side, consumers view UPFs negatively, while new food trends demand fewer ingredients, "clean labels" and "clean eating" diets [13]. However, the food processing industry could use the occasion of the "processing debate" to change the narrative in those consumers that view these products negatively, even while new food trends demand fewer ingredients, "clean labels" and "clean eating" diets. Many food companies can already point to ongoing initiatives around product reformulation. A greater attention to food processing could spur innovation in food processing technology, as well as the introduction of novel, low-impact technologies, even if adding one or more processing steps [14].

A conceptualization of processed foods was drawn by [15], where processed foods were divided as a function of the following different aspects: the extent and nature of change, as well as the place and purpose of processing. From the analysis of the literature on UPFs and NOVA classifications, and the assumption of the effect on consumers' health, and considering the viewpoint of food science and technology, processing and nutritional value do not have a linear relationship, and these concepts need to be dissociated [15]. Furthermore, the impact of UPFs on greenhouse gas emissions seems to be similar to that produced with minimal processing. The advancements in food processing technologies can affect this impact on the entire supply chain, thus reducing the potential threat to sustainability and biodiversity [12].

2. Role of Innovation to Introduce Low-Impact Technologies

Following the increasing demand for healthy and functional food, in combination with enhancing food system sustainability, intensive research efforts have been made to develop new processing technologies as an alternative to conventional thermal processing in order to obtain safe products, with respect to conventionally processed products, with sensory and nutritional properties much more like those of fresh products [16]. The validation and potential implementation of non-thermal technologies in the food production chain, including agri-food waste, to enhance the processing efficiency at lower treatment intensities were reviewed by [17].

In a basic semantic structure, we could imagine a relationship between minimal processing throughout mild and non-thermal (bio)technologies —> the co-creation of new food solutions —> the enhancement gut health and the increase in safety and sustainability.

2.1. Cold Atmospheric Plasma

Cold atmospheric plasma (CAP) technology is based on the ionization of a gas mixture that leads to the formation of excited molecules, ions, electrons and radical species that co-exist with electromagnetic radiation (UV and visible light). It is considered an innovative technology that can be applied in food processing for decontamination and stabilization in the food and packaging industry [18,19]. Recently, research on the application of cold plasma has been focused much more on different aspects than decontamination, including toxicant degradation, enzymatic inactivation, functionalization, quality improvement and nutrient extraction [20,21].

In the frame of a collaborative research project (Italian National Project PRIN PLASMAFOOD), a system composed by a high-voltage generator, a treatment chamber where you can set the temperature and the operating gas, was realized. Finally, there is the plasma source itself, as Surface Dielectric Barrier Discharge (SDBD), with optical absorption spectroscopy to monitor the process, in order to find a connection between the efficacy of our treatment and the concentration of different reactive species. The equipment was developed in collaboration with Alma Plasma, a spin-off of the University of Bologna (https://site.unibo.it/idea/en/our-innovative-businesses-start-ups-and-spin-offs/almaplasma-srl (accessed on 26 June 2025)) [22].

Cold plasma treatment was used first as a decontamination treatment to extend the food shelf-life, like in the case of fresh-cut fruit [16,23], where the microbial growth and enzymatic activities of fresh-cut melon, apples and kiwifruit were studied. The results of yeasts, lactobacilli and lactococci on melon samples demonstrated that the tested cold plasma treatment was very promising in order to stabilize fresh-cut melon samples, allowing for efficient decontamination. Furthermore, a significant increase in microbial shelf-life was observed by optimizing the treatment time, which was due to delayed growth during the storage of the surviving spoilage microflora. Another very interesting approach to assess the cold plasma effect on product quality was the use of isothermal calorimetry (TAM) in order to evaluate the metabolic response to stress in fresh-cut tissue through the determination of metabolic heat production. The results regarding this approach showed that the heat production of the treated samples was significantly lower compared to the controls for all 24 h of the analysis, proportionally to the treatment time. The calculation of the total metabolic heat produced by the fruit tissues during storage at 10 °C and the differences among the samples were more pronounced after 24 h of analysis compared to after 12 h. Considering the enzymatic activity, a slight reduction was observed, but this effect was dependent on the type of enzyme considered, since peroxidase (POD) and pectin methylesterase (PME) activities were assessed. Cold plasma was also used for virus decontamination, both as gas and as plasma-activated water (PAW). Hepatitis A virus (HAV) and Noroviruses, using murine Norovirus (MNV) as a surrogate for human Noroviruses, were tested. The viral stocks used were HAV strain HM175 and MNV strain MNV-1. Reduced infectivity and viral integrity PCR (EMA-rt-q-PCR) were evaluated [24] according to the literature [25]. As a consequence of the experience carried out on foodborne viruses, during the COVID-19 pandemic, the application of CAP was forwarded on the decontamination of SARS-CoV-2 from food and food packaging surfaces. CAP treatment for 10 min completely degraded the inoculated SARS-CoV-2 RNA molecules on packaging materials and packaged products, whose concentration resulted below the detection limits for each target sequence as detected by the RT-PCR reaction performed using the n-COVID Allplex SARS-CoV-2 assay mix [22].

The application of CAP to reduce enzymatic activity was also studied; Ref. [26] carried out research on the inactivation of polyphenoloxidase (PPO) in sugar model systems to further elucidate the results already shown in the literature [27,28]. PPO inactivation data after different exposure times to CAP as a function of the reactive species (O_3 concentration) owing to the CAP treatments, fitted by a first-order model using a Peleg model, showed that the CAP treatments significantly ($p < 0.05$) reduced the PPO activity, with inactivation depending on the processing time ($p < 0.05$) and ozone concentration in the treatment chamber. Spectroscopic analyses revealed that the loss of the PPO activity was due to a change in the protein tertiary structure and the loss of alpha-helices structures [26].

Cold plasma has been applied to decontaminate fish and seafood products, generally considered as the most perishable foodstuffs, but which is also particularly sensitive to quality change during plasma treatments [29–33]. In general, the application of cold plasma caused a reduction from 1 to 3 log CFU/g of naturally present or inoculated microorganisms on fishery products. The effect depended on the process parameters and the type of the microorganism to be inactivated. However, the negative effect of cold plasma on these products, rich in highly unsaturated fatty acids, is lipid oxidation, which could increase in the treated products due to the highly oxidative power of plasma reactive species [31]. In the framework of the FutureEUAqua EU project, Ref. [34] investigated the effect of cold plasma generated (SDBD system) with different gas mixtures on the safety, quality and nutritional aspects of fresh sea bream fillets. The gas mixtures used were air (80% N_2 and 20% O_2) and Argon (80% Argon and 20% O_2), and *E. coli* and *L. innocua* were investigated as inoculated microorganisms. The most effective plasma treatment appeared to be Argon-20, which had the highest mortality rate for all microorganisms, with a reduction slightly higher than 1 log CFU/g, with *E. coli* being the most sensitive microorganism, although the differences were not statistically significant with the treatment with Air-20. Fatty acids (FAs) and protein quality were tested in both not-digested and digested samples (according to the INFOGEST® protocol) to assess the bioaccessibility of FAs. Regardless of the treatment, no differences in the fatty acid composition were detected between the control and treated samples, and the only difference emerging from the NMR spectral data was an overall 10% reduction in protein hydrolysis in Air-20 fillets compared to Argon-20 and controls. With the exposure of seafood to CAP, TBARS concentrations significantly increased up to about 3.5–4.0 mg MDA/kg, without significant differences among the samples, and this result agrees with most of the literature. Since lipid oxidation might affect the sensorial properties of the product, a useful strategy to protect CAP-treated seafood from lipid oxidation, other than the optimization of processing conditions, could be the application of natural antioxidant compounds before treatment.

A further enforcement of the use of cold plasma has been the research on the reduction in mycotoxins in intermediate moisture and dried food so to enhance their healthiness; Ref. [35] investigated the efficacy of cold atmospheric plasma (CAP) treatment in degrading the emerging *Alternaria* toxins as pure molecules and as natural contaminants in dried tomatoes. The effects of CAP treatments under O_3 and NO_x regimes on naturally contaminated dried tomatoes with low and high tenuazonic acid contaminations (TeA, the most abundant *Alternaria* mycotoxin found in naturally contaminated products) were the reduction in toxins by about 40% after 15 min and up to 55% after 60 min of treatment. The 15 min treatment on dried tomatoes artificially contaminated at two concentration levels of TaA led to a more than 20% and more than 35% toxin reduction according to the initial toxin concentration, confirming that a higher substrate concentration resulted in a greater mycotoxin degradation. Lower effects have been found on other *Alternaria* toxins, such as alternariol (AOH), alternariol monomethyl ether (AME), altenuene (ALT) and tentoxin (TEN). Thus, the results showed that CAP treatment can be an effective method to reduce different *Alternaria* mycotoxins in sun-dried tomatoes, ensuring food safety and reducing mycotoxin contamination in agricultural products.

As previously described, recently, cold plasma treatments were studied to obtain further objectives than decontamination. Following this approach, Refs. [36,37] studied the modification of the functional properties of starches of different origins affected by plasma-activated water (PAW) treatments in order to enhance their technological suitability. Using PAW treatments of potato starch in combination with annealing (ANN) showed a synergistic effect in terms of the thermal stability and shear resistance, leading to a lower breakdown viscosity, higher elasticity and gelatinization enthalpy, probably due to crosslinking. Modified potato starch obtained using those treatments could be used as a more sustainable material in food applications for baked goods, as well as a thickening agent for candies and ice cream, and which is generally used in many processed foods, as they are better able to withstand the high thermal and mechanical processing conditions in the food industry. In addition, the PAW and ANN treatments can significantly enhance the behavior of potato starch in terms of water interaction and the resulting functional properties. Finally, studies on the functional properties of the pasting properties of normal maize, waxy maize and potato starches subjected to PAW treatments throughout the deep rheological analysis of three starches, characterized by different botanical origins and different compositional and structural properties, led to very interesting results. Cereal and tuber/roots starches can be successfully modified by PAW, resulting in a promising strategy for starch modification as a "green" alternative for existing methods of starch modification in the food industry. The results obtained in those studies, however, show that the degree of modification is strongly dependent on the starch type (the botanical, amylose and amylopectin content). Therefore, process optimization based on the specific substrate appears necessary for obtaining tailored functionality.

2.2. Pulsed Electric Field (PEF)

The exposure of biological cells to an external electric field affects the membrane permeability; the application of short electric pulses (pulsed electric field (PEF) treatment) causes a phenomenon called electroporation, when the formation of transient or permanent pores occurs, leading to an impairment of the membrane semipermeability [38–40]. Depending on the level of the applied electric field with an intensity above a threshold value, the electroporation can be reversible (transient pores) or irreversible (permanent pores), so that permanent pores are formed and the membrane is damaged, leading to cell death [41,42]. The electric field strength is typically in the range from 0.5 to 1.5 kV/cm for reversible electroporation, and from 1.0 to 5.0 kV/cm for irreversible permeabilization, in plant or animal tissues. The irreversible electroporation (with an electric field strength of 20–50 kV/cm) is also widely applied for enzyme and microorganism inactivation [43]. Nevertheless, in the last years more work has been performed on the benefit that PEF technology as reversible electroporation can give to mass and heat transfer, and on the recovery on bioactive compounds from by-products or neglected products, both crops and animal products, including seafoods.

An example of the use of low-intensity PEF to improve the mass and heat transfer is reported by [16], as research led in the context of the FP7 ERA-Net CORE Organic Plus project.

Reversible PEF was applied in combination with a mild process like the osmotic dehydration (OD, where water removal occurs without any state change at ambient temperature) of organic strawberry, considering mass process transfers and water distribution, as assessed by means of low-frequency Time-Domain Nuclear Magnetic Resonance (TD-NMR). Regarding the mass transfer improvement, the results showed an increase in water and solid kinetics due to the permeabilization of the cell membranes induced by the PEF treatment. The application of an electric field intensity of 100 V cm^{-1} was already sufficient to increase the water loss by 12% after 60 min of osmotic dehydration. PEF treatment effects should be considered time-dependent; the formation and growth of pores in the membrane are not immediate but continue for several minutes after the treatment. With regard to the water distribution owing to osmotic dehydration, TD-NMR showed a decrease in the mean transverse relaxation time (T2) values of the water populations during the osmotic treatment due to the water removal and the different water–solute–biopolymer interactions. In the PEF-treated samples, a fast T2 decrease was observed immediately after the treatment due to the different water–solute–biopolymer interactions induced by the loss of compartmentalization within the strawberry tissue. Thus, the results showed that the OD efficiency could be highly improved by the PEF pretreatments by facilitating the diffusion of the inner water by affecting the membrane permeability. Finally, the highest quality properties, as determined by the firmness and color, were obtained using a trehalose and $CaCl_2$ solution as the osmotic agent. Cell viability was partially maintained in the strawberry samples treated with an electric field strength intensity of 100 V cm^{-1}.

In [17], the combination of PEF and osmotic dehydration treatments prior to the final air-drying process (led at low–medium temperatures) to obtain shelf-stable products using kiwifruit waste (undersized fruits) has been reported. Different mathematical models were used to fit the drying kinetics, and the results showed that the PEF/OD-treated sample had higher drying kinetic rates, presenting better quality and acceptability as measured from both instrumental and sensorial analyses.

The combination of PEF with mild air drying was performed on vegetables by [44], applying PEF pretreatments prior to low–medium-temperature air drying (40°, 55° and 70 °C). The results showed an acceleration of the drying kinetics from 32% to 48%, according to the drying temperature, to reach the final target water activity (a_w = 0.30) in the dried samples of black cabbage. In carrots, the application of PEF before convective drying at 70 °C reduced the drying time by 6.9–8.2% in comparison with the untreated material [45]. In addition, better color retention was recorded in the PEF-treated samples in comparison with the untreated dried samples.

As mentioned above, the application of low-intensity PEF to achieve reversible electroporation has been widely used to recover bioactive compounds from by-products or neglected products. Following this strategy, the application of PEF was studied to improve bioactive compound recovery from brewery by-products [17,46]. Response surface plots were used to show the combined effects of process variables for free phenolic compound and flavan-3-ol extraction from brewers' spent grains (BSGs). The obtained optimal PEF conditions (2.5 kV/cm, 50 Hz and 14.5 s) improved the yield of free and bound phenolics by 2.7 and 1.7 times, respectively, compared to the control samples without PEF treatment. The total phenolic content (free and bound) obtained in brewers' spent grains after the PEF treatment at optimal conditions (640.46 μg g^{-1} d.w.) was 43.23% higher than in the control sample (363.58 μg g^{-1} d.w.). These phenolic extracts could be useful as ingredients in the food industry because of the low cost and high nutritional value of BSGs [47]. These extracts could be used to enrich bakery products such as bread, biscuits, cookies and pasta products.

Another interesting application of PEF to recover functional compounds from neglected sources is to improve the extraction of chitin, chitosan and carotenoid from seafood by-products [48]. In fact, chitosan has been studied extensively in the last few years due to its interesting properties as an antioxidant, emulsifying agent, ingredient in edible and active films/coatings, flocculating and clarifying agent, and food preservative, having antimicrobial and antifungal properties. In food formulation, chitosan can be used as food fiber, in the immobilization of enzymes and as a stabilizer of color, texture and odor [49,50]. PEF has been used to improve the recovery of chitin and chitosan, which is its deacetylated form, from crustacean by-products [51], following the methodological approach to apply non-thermal technology—including ultrasound, high-pressure processing, pulsed electric fields, cold plasma and supercritical fluid technology—in the extraction of valuable components from wastes and by-products. In fact, crustacean by-products obtained from the processing plant, namely, heads, shells, pleopods and tails, contain different valuable compounds, such as chitin and chitosan, carotenoids, lipids and proteins.

Bioactive compounds or products, obtained from crustacean by-products to be used in seafood products, have been listed by [49]. Extracted compounds, mainly chitosan, from the shells of shrimp, crab and prawn, as well as the shrimp cephalothorax, head and tail, are used in fish sticks, while croaker fish and surimi have different functions, from antioxidants to cryoprotectants and gelling agents. Carotenoid astaxanthin recovery from Red (*Aristeus antennatus*) and Camarote (*Melicertus kerathurus*) shrimp side streams were studied using PEF in combination with accelerated solvent extraction (ASE) in order to valorize a highly valuable by-product but with a general reduction in processing waste [50]. The PEF treatment led to a significant increase in the extraction yield when followed by solvent extraction. The recovery of astaxanthin in *M. kerathurus* and *A. antennatus* increased by 46% and 48%, respectively, compared to the control. Better results were obtained using dimethyl sulfoxide (100% DMSO) with respect to absolute ethanol as a solvent. Similarly, the Trolox equivalent antioxidant capacity (TEAC) and oxygen radical absorbance capacity (ORAC) of shrimp side stream extracts also showed higher activities with the PEF–ASE combination. Nevertheless, these technologies are not yet adapted to the reality of the shellfish processing industry, and extensive research is still needed to find the technological and economic conditions for their implementation. The advantages and disadvantages of PEF technology in extraction processes are reported in Table 1 [49].

Table 1. Advantages and disadvantages of PEF technology in comparison with traditional extraction technologies [49,52].

Technology	Advantages	Disadvantages
PEF	Non-thermal behavior High selectivity Less time and energy consumption High yields for carotenoid extraction Does not require any additional chemicals Can be used in continuous mode	Poorly studied for the compound's extraction from crustacean by-products Possible limited utilization due to the conductivity of matrix High initial investment of PEF equipment Limited extraction of lipophilic compounds

Reversible PEF can promote metabolic stress responses, and therefore influence the production of secondary metabolites in plant products and cell cultures. Generally, higher polyphenol, carotenoid and anthocyanin contents during the storage of pumpkin, apples, berries, tomato and other vegetables as stimulation due to the reversible PEF treatment, as well as an increase in glucosinolates in *Brassicaceae*, was found, probably also due to the increase in the extractability [53–55].

Stimulating the biosynthesis of plant secondary metabolites with antioxidant and other biological properties associated with important health benefits through PEF could be an interesting opportunity for the food industry to meet the increasing demand for foods with a high functional value.

In the case of olives, corn seeds and soybeans, an increase in the yield of oil in olives, and an increased recovery of isoflavonoids in soybean oil and phytosterols in maize germ oil were found, as reviewed by [56].

The application of PEF on seed germination has been reviewed by [57]; reversible PEF treatments were applied on the seeds of barley, wheat, wheatgrass and others, with a very wide range of operating conditions. Among the specific effect on different species, considering wheat seeds, monopolar rectangular pulses with different frequencies, treatment times and total energy were found to be an alternative to chemical treatments for surface disinfection, significantly increasing the germination and seedling rates by 10 and 28%, respectively, compared to the untreated sample, and the PEF treatments allowed for tolerance to cold and salt stress, together with an improved vigor. All of the reviewed literature hypothesizes that, as the main mechanism, PEF can improve seed germination and the growth rate by stimulating a physiological response throughout ROS production and oxidative stress. Nevertheless, the proper mechanism is still far from being fully understood.

A further successful application of PEF was tested by [58,59], with the application of PEF treatment to reduce the formation of acrylamide (AA) in potato during frying (Figure 1).

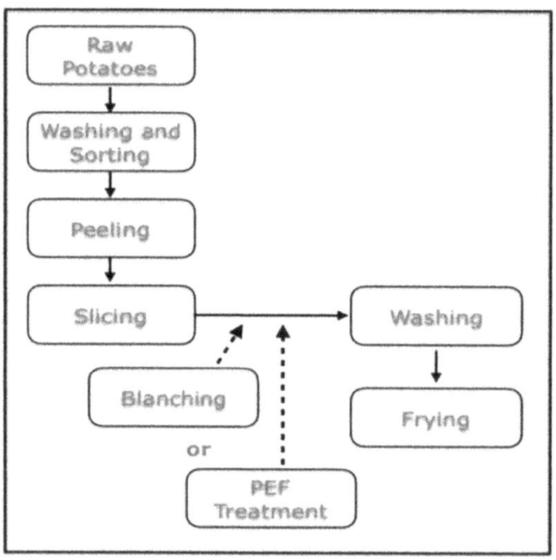

Figure 1. PEF treatment in the pre-frying of potato stick processing (from [58]).

Electroporation was better able to reduce the reagents in the raw material, as sliced potato, in comparison with the blanching treatment (48% vs. 40% reduction, respectively), having a higher effect on the reduction in free asparagine than reducing sugars. On the other hand, blanching—as a thermal treatment—would change other quality properties, such as the texture behavior, during frying, so that would possibly be avoided. Acrylamide reductions were found to be around 30% and 17% in the PEF and blanched potato slices after frying. In another study, Ref. [59] performed a combination of PEF and yeast pre-treatments for sliced potato prior to frying, as it was shown that the yeast *Aureobasidium pullulans* L1 strain can successfully reduce AA formation due to the enzymatic and metabolic activity of the yeast, which is able to reduce the levels of glucose, fructose and asparagine [60]. Nevertheless, the yeast pretreatment alone led just to a slight AA reduction (around 5%), while the yeast treatment combined with PEF, followed by water dipping, was the most effective in reducing AA, leading to a reduction of around 36% and 59%, respectively, for a 5 and 15 min dipping time. Thus, the combination of these non-thermal (bio)technologies significantly increased the safety of a popular processed food like potato crisps.

2.3. Ultrasound Treatment (US)

Ultrasound (US) is a form of energy generated by sound waves with a frequency above the hearing threshold of the human ear (18–100 MHz). In particular, high-intensity (power) ultrasound operates at the frequency range of kHz (18–100 kHz) and waves of high intensity (>1 W/cm^2) [61,62]. The principles of the application of US are based on the fast compression and expansion of the treated material (the sponge effect) and cavitation phenomena, which consist of the formation of large bubbles moving from growth to collapse, causing the liberation of a high quantity of energy. Both phenomena, the compression/expansion of the tissue and cavitation, promote the disruption of cell walls and membranes, resulting in the formation of microscopic channels in the biological material, thereby promoting the alteration of the physical and chemical properties of food products [63]. Thus, ultrasound-assisted extraction (UAE) is commonly used nowadays in food technology as a substitute for conventional extraction techniques so to improve the recovery of bioactive compounds from plant materials, but also to increase food safety, since US can be successfully used in food enzymes and microbial inactivation [16,64]. A list of US applications in fruit and vegetable products subjected to different processing parameters is reported in Table 2. These data are a synthesis of selected results reported in the cited publications highlighting the US treatments of various fruits and vegetables in terms of improving the bioactive compound content, and thus the potential impact on consumer health, which have been widely reviewed in the literature [65–67]. In particular, flavonoids undoubtedly have great potential to prevent/treat many chronic diseases associated with oxidative stress and inflammatory components.

Table 2. Ultrasound (US) application in fruit and vegetable products subjected to different processing parameters (modified from [16]). TPC = total phenolic content.

Product	US Parameters	Results Obtained by US Application	References
Acerola juice	18 kHz, 2000 to 3000 W/L	Increase in the availability of pro-vitamin A and vitamins B3, B5, C and E	[68]
Carrots	20 °C, 120 min	Increase in the extraction rate of TPC (2186 mg of gallic acid/100 g_{dw}), vitamin C (148 mg/100 g_{dw}) and β-carotene (12 mg/100 g)	[69]
Carrots	21 and 35 kHz frequency for 10, 20 and 30 min	Increase in the carotenoid content at 35 kHz	[70]
Cashew apple bagasse	226 W/cm^2, bagasse/water ratio of 1:4 (*w*/*w*), 6 min	Increase in the extraction of vitamin C	[71]

Table 2. *Cont.*

Product	US Parameters	Results Obtained by US Application	References
Cranberry	21 kHz, 180 W, 30–60 min	Retention of Vit. C with 30 min, while loss with longer treatment Good retention of polyphenols, anthocyanins and flavonoids for both durations	[72]
Cranberry	21 kHz, 180 W+ blanching	Increase in bioactive compounds such as Vit. C, polyphenols, anthocyanins and flavonoids	[72]
Green tea	22–83 °C, tea-to-water ratio (12–73 g L^{-1}), amplitude (23–77%)	Max. polyphenol content (12,318 mg L^{-1}) and max. flavonoids (3774 mg L^{-1}) at 77 °C, tea-to-water ratio of 73 g L^{-1} and amplitude of 77%	[73]
Guava leaves	24 kHz, 200 W, 40 min, ethanol/water (*v*/*v*) ratio of 60%	Increase in the extraction of flavonols and flavan-3ols	[74]
Lemon by-products	150–250 W, 45–55 °C, 35–45 min	Optimal TPC extraction of 18.10 ± 0.24 mg GAE/g_{dw} at US power of 250 W, 50 °C and 45 min Optimal rutin extraction of 3.20 ± 0.12 mg/g_{dw} at US power of 150 W, 48 °C and 35 min	[75]
Peaches	37 kHz, 10–30 min, 30–50 °C, ultrasonic power of 30–70%	The optimal conditions for extractions of TPC were 41.53 °C, 43.99% and 27.86 min	[76]
Pumpkins	37 kHz, 10–30 min, 30–50 °C, ultrasonic power of 30–70%	The optimal conditions for extractions of TPC were 41.45 °C, 44.60% and 25.67 min	[76]
Sour cherry	25 kHz (0.4 W/cm^2), 30–120 min	At short times, no effect on bioactive compounds; prolonged application provoked loss of about 10% of total polyphenols	[77]
Spinach	37 and 80 kHz, 5–30 min, 30–50 °C, ultrasonic power of 30–70%	Max. total phenol (33.96 ± 11.30 mg gallic acid/g_{dw}) and flavonoids (27.37 ± 11.85 mg/g_{dw}) at 37 kHz, 30 min, 40 °C and 50%	[78]
Strawberry	40–70 °C and 30 and 60 W	>65% of Vit. C retention in dried product	[79]
Sweet potato	28 kHz, 300 W, 20–60 min	>70% of Vit. C retention in osmodehydrated product; however, higher loss of carotenoids	[80]

It is well established that ultrasound treatment can increase the mass transfer during the osmotic dehydration of different plant tissues, but the complete overview of the effects that this technology can have on the quality of a specific fruit, like kiwifruit, was largely missed. The study of the selected chemical and physical properties of differently treated kiwifruit is described by [81]. The water activity, freezable water content, texture, color and chlorophyll content of ultrasound-treated kiwifruit were investigated. In addition, the effect of combined treatments (US pretreatment and OD process) on the above-mentioned physico-chemical kiwifruit characteristics was also studied. Generally, the obtained results allow us to claim that US reduced the dehydration time and reduced the oxidation of bioactive components exceptionally at lower temperatures due to the thermolabile character of most substances. Some nutrients can be lost in the liquid phase during the preliminary US treatment. Nevertheless, in most cases, the influence of US pretreatment has been shown in dried products, where the untreated samples exhibited lower quality attributes than the US-treated samples [81]. Ultrasound-assisted osmotic dehydration was also applied to cranberries, where a 30 min sonication at 21 kHz and a total power generated by sonotrodes of 180 W, corresponding to an intensity of 3.6 W/g, was used before the OD treatment was carried out, using sucrose and trehalose as an osmotic solution [81]. The results showed that the best mass transfer values in terms of water removal were obtained in cranberry samples subjected to the combination of OD with sucrose and US treatment. Furthermore,

US pretreatment led to a lower weight reduction in samples during storage if treated with any osmotic solution in comparison to those without US pretreatment. The US pretreatment also promoted benefits when considering qualitative characteristics, for example, in the color lightness maintenance, compared to the untreated fruits. Going more deeply into the mechanism of the influence of US treatment on the water state and quality properties, US- and OD-processed cranberry fruits (whole and cut) were studied by differential scanning calorimetry (DSC), SEM microscopy and TD-NMR measurements. While whole fruits demonstrated only a slight effect of the ultrasound treatment, owing to the very tough skin of cranberry, the OD process of cut fruits showed a larger action of the US treatment on the mass transfer. US-assisted OD samples were shown to release water from the vacuoles into extracellular spaces through TD-NMR analysis, and the uptake of sugar as mass transfer from the solution to the fruits showed a significant increase by sonication.

As previously cited, ultrasound-assisted extraction (UAE) has also largely been studied to increase the yields of functional compounds extracted from by-products, underestimated raw material or neglected products [82]. In [74], as an example, the optimal conditions for the application of UAE to the extraction of phenolic compounds from *Psidium guajava* L. leaves were studied. Using Response Surface Methodology (RSM), it was possible to design optimized extraction processing conditions by combining the ethanol/water ratio and US power as processing factors, with flavonols, flavan-3-ols, DPPH, TEAC and the sum of phenolic compounds (SPCs) as dependent factors. This is very interesting to highlight, as the optimal conditions found by RSM could be different, even significantly different, when taking into account the different functional compounds or properties, as reported in Table 3. These results are very promising because the increase in flavonoids has the great potential to prevent/treat many chronic diseases associated with oxidative stress and inflammatory components, such as cardiovascular diseases, cancer, mental health, gastrointestinal diseases and others [66].

Table 3. Optimal conditions obtained by RSM for ultrasound-assisted extraction of different functional compounds/properties from guava leaves (adapted from [74]).

RSM Optimal Conditions	DPPH (μmol Trolox/g Leaf d.w.)	TEAC (μmol Trolox/g Leaf d.w.)	SPC (mg/g Leaf d.w.)	Flavonols (mg/g Leaf d.w.)	Flavan-3-ols (mg/g Leaf d.w.)
Time (min)	22	45	41	38	37
EtOH/water ratio (% (v/v))	54	58	62	62	63
US power (W)	80	180	230	235	228

2.4. *Dynamic/Hydrostatic High-Pressure Processing*

High-pressure (HP) advanced technologies are promising processing technologies that could by widely used in the food industry due to their valuable features.

High hydrostatic pressure (HHP) consists of the treatment of food products, which are packed and placed in a liquid-filled pressure chamber at a high pressure of 100–1000 MPa at chilled or mild process temperatures (<45 °C), in order to achieve microbial inactivation, modification and extraction. During the treatment, the high pressure compresses the food, causing a reduction in the volume, which counteracts the external forces through the food's chemical structure [83,84]. Conversely, high-pressure homogenization (HPH) is a food processing technique that uses high pressure to reduce particle sizes and inactivate microorganisms, thus improving product stability and safety. HPH involves forcing a liquid food product through a small valve at high pressures (typically 100–300 MPa). The intense shear forces, turbulence and cavitation are then created in the liquid food. These forces break down large particles and droplets into smaller, more uniform sizes, resulting in a more stable emulsion or suspension [85–87].

Recently, significant research efforts have been devoted to studying the development and the optimization of HP homogenization and HP hydrostatic processes [88,89]. Nevertheless, the impact of the high-pressure effect on the nutritional properties for poorly investigated raw material is still unknown, even though it is generally considered that, when compared to a corresponding thermal treatment, HHP seems to reduce vitamin loss and improve its maintenance during storage. In orange juice pressed at 100–400 MPa from 30 to 60 °C up to a 5 min treatment time, an increase in the hesperidin concentration as a function of the applied pressure was observed, but not for flavanones. After 10 min of treatment of HHP at 400 MPa on apple juice, an increase in hydroxycinnamic, procyanidin acid and catechin contents was reported, as reviewed by [16].

A metabolomic approach to evaluate HPP/HHP treatments has been performed by [90] on minced gray mullet (*Mugil cephalus*) during chilled storage. Samples pasteurized by high hydrostatic pressure (HHP) at 0, 400, 500 and 600 MPa were analyzed using proton NMR-based metabolomic techniques, alongside microbial analysis. HHP showed bacteriostatic properties reducing the bacterial counts and a positive influence on the formation of volatile compounds, thus demonstrating the efficiency of HHP as a non-thermal technology to preserve fish freshness. These findings highlight HHP as an efficient approach for preserving the freshness of fish, which is also related with the elimination of common seafood pathogens, such as Vibrio and Listeria spp., and slowing the growth of spoilage microorganisms [90]. Concerning the effects on the bio-metabolisms, HHP was able to significantly reduce ($p < 0.05$) the production of spoilage-related molecules such as TMA, tyramine, hypoxanthine and ethanol. In fact, the key metabolic pathways, a part of the lysine metabolism, were influenced by the HHP treatment. The results showed the ability of the NMR-based metabolomics strategy to effectively identify processing effects on food matrices. Very interestingly, there was a schematic presentation of the effect of the HHP treatment on the different metabolism pathways, such as the amino acid, carbohydrate, pyruvate, nitrogen and nucleotide metabolisms, and on the level of efficacy in stimulating the increase in metabolite concentrations extracted from different treatments. Nevertheless, further research and development are needed to find a preservation method able to inhibit the lysine metabolism, as pointed out by [91].

High-pressure technologies are also used in dynamic processes for fluid food in the form of high-pressure homogenization (HPH). The application of this technology can impact both the stability, with partial sanitation prolonging the product shelf-life, and the improvement of the nutritional, sensorial and textural properties of foods, like juices, beverages, emulsions and purees. On the other hand, although phenolic compounds have a key role in the health benefits of fruit juice consumption, little is known about the effect of processing on their bioaccessibility; Ref. [92] reported on the impact of high-pressure homogenization processing at 20 MPa on the nutritional and functional value of mandarin juice with and without the addition of trehalose and probiotic *Lactobacillus salivarius*. During digestion, there is a release of phenolic compounds from the food matrix, which is an important prerequisite for their effectiveness in the human body, and so it is essential to find technological treatments that not only maintain the concentration of phytochemicals, but also their bioaccessibility. HPH treatments were shown to reduce the concentration of total phenolics and the main flavonoids, but also to significantly increase their bioaccessibility after in vitro digestion, probably due to the disruption of cells and membranes, which can increase the availability of functional compounds. Further positive results were related to the total antioxidative capacity (TAC) not reduced by the HPH treatment, and the effect of the treated juices against ROS generation and lipid oxidation in cultured liver cells (HepG2 cells) in a basal condition. HPH-treated citrus juices with the addition of a probiotic were the most protective samples both in terms of the accessibility of bioactive molecules and the concentration of reactive oxygen species (ROS) among stressed cells [40]. Research regarding the addition of probiotics was also carried out to study the effect of the HPH treatment as an encapsulation process that is able to give further functionality to citrus juice with microencapsulated probiotic microorganisms to be used in the vacuum impregnation of the fruit matrix; Ref. [93] investigated the probiotic survival and in vitro digestion of encapsulated *Lactobacillus salivarius* spp. *salivarius* included into an apple sample by vacuum impregnation, dried by low-temperature air drying and stored for 30 days (Figure 2).

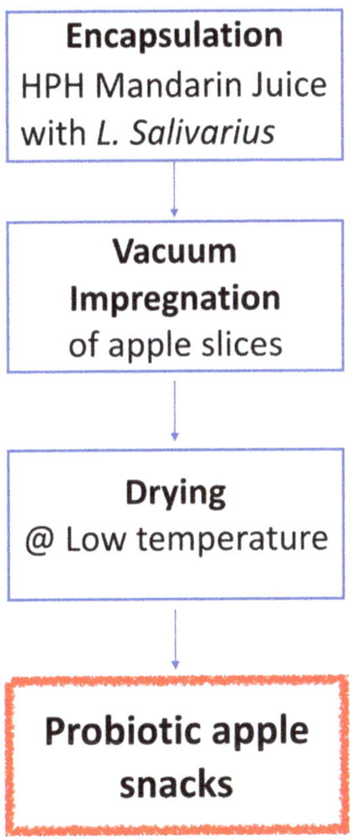

Figure 2. Flow diagram of processing to obtain a snack with impregnation of citrus juice with HPH-encapsulated probiotic *Lactobacillus salivarius* (from [93]).

The encapsulation of the probiotic was performed by HPH at 70 MPa on an emulsion to obtain microcapsules, and added to mandarin juice, with vacuum impregnation at 50 mbar for 10 min. Drying was carried out at 40 °C to reach a water activity value of around $a_w = 0.50$. The experimental design resulted from the conceptualization of the strategy of research previously defined as improving the food functionality by studying the structure–property relationships [93]. The survival of inoculated microorganisms throughout the process to obtain the dried impregnated apple samples was assessed on *L. salivarius* with and without microencapsulation, resulting in a significantly better survival in the samples with the microencapsulated probiotic; however, in both cases, an evident reduction (from around 8 Log CFU/g to around 7 Log CFU/g of dried fruit) was due to the drying process, even though it was carried out at a low temperature. During storage, after 30 days, the microencapsulation was able also to protect the probiotic from degradation phenomena in the dried apple samples (with a percentage of survival of 39% in comparison to 19% for the non-encapsulated probiotic). The simulated gastrointestinal digestion of dried apple samples encapsulated and non-encapsulated with *L. salivarius* showed that the encapsulated probiotic had a higher resistance, and the number of *L. salivarius* spp. *salivarius* in the impregnated and dried apple was enough, thus maintaining that the specific functionality of *L. salivarius* is mainly a potential effect against *Helicobacter pylori* infection.

Furthermore, Ref. [94] demonstrated that the application of 200 MPa (ultra-high-pressure homogenization, UHPH) for three cycles made it possible to obtain a stable kiwifruit juice in refrigerated storage for more than 40 days, and increased the shelf-life by 1 week at room temperature compared to the control, while at the same time increasing the availability of polyphenols and the antioxidant activity, and better preserving the color. In fact, HPH promoted the disarrangement of the cell clusters into single cells and/or cell fragments. Storage at three temperatures from 5 to 25 °C also showed microbial load under limits at non-refrigeration temperatures (15 and 25 °C), with a shelf-life of 9 and 5 days, respectively. From a rheological viewpoint, the release and solubilization of cell wall constituents, such as pectin and proteins, caused the increase in the volume fraction of particles and led to the improvement of particle interactions, thus increasing the viscosity. The obtained results on the antioxidant capacity (TEAC) agree with the previous literature, where the UHPH process can improve the extractability of antioxidant components through the disruption of the cell wall components. A UHPH treatment of 200 MPa for three cycles showed a slight decrease in the TPC in kiwifruit juice. On the other hand, an increase in the initial antioxidant activity was observed in the samples treated by HPH, and this might be explained by a partial inactivation of polyphenoloxidase and peroxidase enzymes, thereby causing the degradation of phenolic compounds in the vegetable matrix.

3. Conclusions

Non-thermal pretreatments, such as cold plasma, pulsed electric fields, high-pressure processing and ultrasound processing, also combined with mild thermal drying technology, would bring a significant reduction in the total request of energy, even when considering the energy inputs for their application, as reported by several authors in the literature and as reviewed by [95]; the advantages in terms of the quality and carbon footprint reduction are summarized in Table 4.

Table 4. Effects on food quality and carbon footprint elements of non-thermal processing (adapted from [95]).

Effects on Food Quality	Carbon Footprint Reduction
Minimal quality loss	Less wastewater
Increasing bioavailability	Increasing energy and water savings
Reduction in processing contaminants	Lower environmental impact
Maintenance of nutritional values	Decreased operational costs
Maintenance of sensorial properties	Decreased electricity
Inactivation of microorganisms	Less time-consuming
Improvement of heat and mass transfer	Inexpensive
Improvement of firmness and texture	Non-hazardous
Decreased color change	Minimal source demands
Increased shelf-life	Simple processing design

Dynamic and hydrostatic high-pressure, vacuum impregnation, ultrasound, pulsed electric field and cold plasma applications can result in a less negative effect with respect to traditional thermal treatments, and the overall functionality can also be improved by recovering functional components from by-products or wastes, following the circular bioeconomy approach.

In Figure 3, a comparison of the positive and negative properties of the technologies considered is reported.

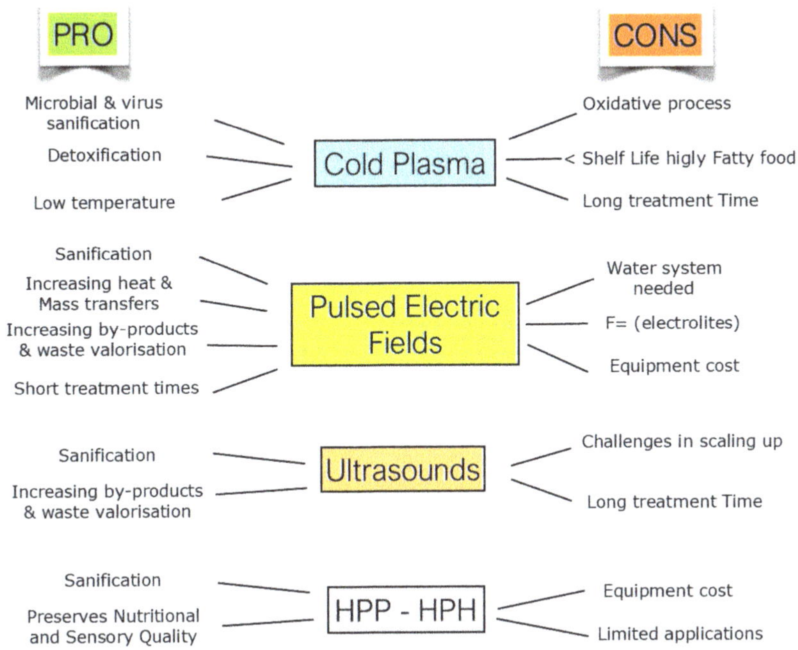

Figure 3. Comparison of pros and cons of investigated technologies.

These treatments can produce structural changes that improve the bioaccessibility and/or the bioavailability of bioactive compounds, such as probiotic microorganisms, to improve food healthiness and gut microbioma.

Non-thermal low-impact processing technologies could help the food industry to pursue a better recognition in sustainability, and to offer better food functionality, thereby recovering favorable consideration in the framework of the ultra-processing debate.

However, an interdisciplinary approach among food engineers, microbiologists, food chemists, bio-NMR specialists and nutritionists is necessary to research the best processing conditions, and to increase and assess the sustainability of processing in the food industry.

Author Contributions: Conceptualization, M.D.R., P.R., S.R., U.T. and S.T.; writing—original draft preparation, M.D.R.; writing—review and editing, M.D.R., S.T. and U.T.; funding acquisition, M.D.R., P.R., S.R. and S.T. All authors have read and agreed to the published version of the manuscript.

Funding: This research was funded by the National Recovery and Resilience Plan (NRRP), Mission 4 Component 2 Investment 1.3—Call for tender No. 341 of 15 March 2022 of Italian Ministry of University and Research funded by the European Union—NextGenerationEU; Project code PE00000003, Concession Decree No. 1550 of 11 October 2022 adopted by the Italian Ministry of University and Research, CUP D93C22000890001, Project title "ON Foods—Research and innovation network on food and nutrition Sustainability, Safety and Security—Working ON Foods".

Data Availability Statement: No new data were created or analyzed in this study. Data sharing is not applicable to this article.

Acknowledgments: This review has been generated from the presentation delivered at the 22nd IUFoST World Congress 2024 in the special session of the International Society of Food Engineering (ISFE) Symposium on Advances on Alternative Food Processing Technologies. The authors are very grateful to Gustavo Victor Barbosa-Canovas for the invitation to be speakers in that context.

Conflicts of Interest: The authors declare no conflicts of interest.

References

1. FAO. *World Livestock 2011–Livestock in Food Security*; FAO: Rome, Italy, 2011.
2. United Nation. *World Population Prospects: The 2017 Revision*; United Nation: New York, NY, USA, 2017.
3. Corrado, S.; Sala, S. Food waste accounting along global and European food supply chains: State of the art and outlook. *Waste Manag.* **2018**, *79*, 120–131. [CrossRef] [PubMed]
4. Facchini, F.; Silvestri, B.; Digiesi, S.; Lucchese, A. Agri-food loss and waste management: Win-win strategies for edible discarded fruits and vegetables sustainable reuse. *Innov. Food Sci. Emerg. Technol.* **2023**, *83*, 103235. [CrossRef]
5. EFSA. Dietary Reference Values (DRVs) 2019. Available online: https://efsa.onlinelibrary.wiley.com/doi/toc/10.1002/(ISSN)1831-4732.021217 (accessed on 26 June 2025).
6. Harvard School of Nutrition. 5 Tips for Sustainable Eating. 2021. Available online: https://www.hsph.harvard.edu/nutritionsource/2015/06/17/5-tips-for-sustainable-eating/ (accessed on 26 June 2025).
7. Springmann, M.; Clark, M.; Mason-D'croz, D.; Wiebe, K.; Bodirsky, B.L.; Lassaletta, L.; de Vries, W.; Vermeulen, S.J.; Herrero, M.; Carlson, K.M.; et al. Options for keeping the food system within environmental limits. *Nature* **2018**, *562*, 519–525. [CrossRef] [PubMed]
8. Campbell, B.M.; Beare, D.J.; Bennett, E.M.; Hall-Spencer, J.M.; Ingram, J.S.I.; Jaramillo, F.; Ortiz, R.; Ramankutty, N.; Sayer, J.A.; Shindell, D. Agriculture production as a major driver of the Earth system exceeding planetary boundaries. *Ecol. Soc.* **2017**, *22*, 8. [CrossRef]
9. Monteiro, C.A. Nutrition and health. The issue is not food, nor nutrients, so much as processing. *Public Health Nutr.* **2009**, *12*, 729–731. [CrossRef]
10. Monteiro, C.A.; Cannon, G.; Lawrence, M.; Louzada, M.L.C.; Machado, P.P. *Ultra-Processed Foods, Diet Quality, and Health Using the NOVA Classification System*; FAO: Rome, Italy, 2019. Available online: http://www.fao.org/3/ca5644en/ca5644en.pdf (accessed on 26 June 2025).
11. Petrus, R.R.; Sobral, P.J.A.; Tadini, C.C.; Gonçalves, C.B. The NOVA classification system: A critical perspective in food science. *Trends Food Sci. Technol.* **2021**, *116*, 603–608. [CrossRef]
12. Capozzi, F.; Magkos, F.; Fava, F.; Milani, G.P.; Agostoni, C.; Astrup, A.; Saguy, I.S. A Multidisciplinary Perspective of Ultra-Processed Foods and Associated Food Processing Technologies: A View of the Sustainable Road Ahead. *Nutrients* **2021**, *13*, 3948. [CrossRef] [PubMed]
13. Sarmiento-Santos, J.; Souza, M.B.N.; Araujo, L.S.; Pion, J.M.V.; Carvalho, R.A.; Vanin, F.M. Consumers' understanding of ultra-processed foods. *Foods* **2022**, *11*, 1359. [CrossRef]
14. Ahrné, L.; Chen, H.; Henry, C.J.; Kim, H.-S.; Schneeman, B.; Windhab, E.J. Defining the role of processing in food classification systems—The IUFoST formulation & processing approach. *Npj Sci. Food* **2025**, *9*, 56. [CrossRef]
15. Sadler, C.R.; Grassby, T.; Hart, K.; Raats, M.; Sokolović, M.; Timotijevic, L. Processed food classification: Conceptualisation and challenges. *Trends Food Sci. Technol.* **2021**, *112*, 149–162. [CrossRef]
16. Tappi, S.; Tylewicz, U.; Dalla Rosa, M. Effect of nonthermal technologies on functional food compounds. In *Sustainability of the Food System: Sovereignty, Waste, and Nutrients Bioavailability*; Academic Press: Cambridge, MA, USA, 2020. [CrossRef]
17. Tylewicz, U.; Castagnini, J.M.; Tappi, S.; Romani, S.; Rocculi, P.; Rosa, M.D. Current Validation of NTP Technologies and Overview of Their Current and Potential Implementation in the Production Chain Including Agri-food Wastes. In *Nonthermal Processing in Agri-Food-Bio Sciences*; Režek Jambrak, A., Ed.; Food Engineering Series; Springer: Cham, Switzerland, 2022. [CrossRef]
18. Hati, S.; Patel, M.; Yadav, D. Food bioprocessing by non-thermal plasma technology. *Curr. Opin. Food Sci.* **2018**, *19*, 85–91. [CrossRef]
19. Pankaj, S.K.; Keener, K.M. Cold plasma: Background, applications and current trends. *Curr. Opin. Food Sci.* **2017**, *16*, 49–52. [CrossRef]
20. Thirumdas, R.; Sarangapani, C.; Annapure, U.S. Cold Plasma: A novel Non-Thermal Technology for Food Processing. *Food Biophys.* **2015**, *10*, 1–11. [CrossRef]
21. Jin, Y.; Adhikari, A. Emerging and Innovative Technologies for the Sanitization of Fresh Produce: Advances, Mechanisms, and Applications for Enhancing Food Safety and Quality. *Foods* **2025**, *14*, 1924. [CrossRef] [PubMed]

22. Maccaferri, C.; Sainz-García, A.; Capelli, F.; Gherardi, M.; Alba-Elías, F.; Laurita, R. Evaluation of the Antimicrobial Efficacy of a Large-Area Surface Dielectric Barrier Discharge on Food Contact Surfaces. *Plasma Chem. Plasma Process.* **2023**, *43*, 1773–1790. [CrossRef]
23. Li, Y.; Huang, X.; Yang, Y.; Mulati, A.; Hong, J.; Wang, J. The Effects of Cold-Plasma Technology on the Quality Properties of Fresh-Cut Produce: A Review. *Foods* **2025**, *14*, 149. [CrossRef]
24. Cozzi, L.; Vicenza, T.; Di Pasquale, S.; Maccaroni, S.; Tappi, S.; Capelli, F.; Suffredini, E. Effect of Atmospheric Cold Plasma treatment on Murine Norovirus and Hepatitis A Virus infectivity. In *Abstract Book, Proceedings of the 7th National Congress of the Italian Society for Virology, Brescia, Italy, 25–27 June 2023*; Italian Society for Virology, Ed.; Italian Society for Virology: Brescia, Italy, 2023; p. 56.
25. Velebit, B.; Milojević, L.; Baltić, T.; Grkovic, N.; Gummalla, S.; Velebit, M.; Skoka, I.; Mojsova, S.; Putnik, P. Efficacy of cold atmospheric plasma for inactivation of viruses on raspberries. *Innov. Food Sci. Emerg. Technol.* **2022**, *81*, 103121. [CrossRef]
26. Laika, J.; Sabatucii, A.; Sacchetti, G.; Michele, D.A.; Hernandez, B.M.J.; Ricci, A.; Rosa, D.M.; Lopez, C.C.; Neri, L. Cold Atmospheric Plasma Inactivation of Polyphenol Oxidase: Focus on the Protective and Boosting Effect of Mono- and Disaccharides. *J. Food Sci.* **2024**, *89*, 9283–9298. [CrossRef]
27. Chutia, H.; Kalita, D.; Mahanta, C.L.; Ojah, N.; Choudhury, A.J. Kinetics of inactivation of peroxidase and polyphenol oxidase in tender coconut water by dielectric barrier discharge plasma. *LWT* **2019**, *101*, 625–629. [CrossRef]
28. Dong, S.; Fan, L.; Ma, Y.; Du, J.; Xiang, Q. Inactivation of polyphenol oxidase by dielectric barrier discharge (DBD) plasma: Kinetics and mechanisms. *LWT* **2021**, *145*, 111322. [CrossRef]
29. Giannoglou, M.; Dimitrakellis, P.; Efthimiadou, A.; Gogolides, E.; Katsaros, G. Comparative Study on the Effect of Cold Atmospheric Plasma, Ozonation, Pulsed Electromagnetic Fields and High-Pressure Technologies on Sea Bream Fillet Quality Indices and Shelf Life. *Food Eng. Rev.* **2021**, *13*, 175–184. [CrossRef]
30. Albertos, I.; Martín-Diana, A.; Cullen, P.; Tiwari, B.; Ojha, S.; Bourke, P.; Álvarez, C.; Rico, D. Effects of dielectric barrier discharge (DBD) generated plasma on microbial reduction and quality parameters of fresh mackerel (*Scomber scombrus*) fillets. *Innov. Food Sci. Emerg. Technol.* **2017**, *44*, 117–122. [CrossRef]
31. Albertos, I.; Martin-Diana, A.B.; Cullen, P.J.; Tiwari, B.K.; Ojha, K.S.; Bourke, P.; Rico, D. Shelf-life extension of herring (*Clupea harengus*) using in-package atmospheric plasma technology. *Innov. Food Sci. Emerg. Technol.* **2019**, *53*, 85–91. [CrossRef]
32. Olatunde, O.O.; Benjakul, S.; Vongkamjan, K. High voltage cold atmospheric plasma: Antibacterial properties and its effect on quality of Asian sea bass slices. *Innov. Food Sci. Emerg. Technol.* **2019**, *52*, 305–312. [CrossRef]
33. Singh, A.; Benjakul, S. The combined effect of squid pen chitooligosaccharides and high voltage cold atmospheric plasma on the shelf-life extension of Asian sea bass slices stored at 4 °C. *Innov. Food Sci. Emerg. Technol.* **2020**, *64*, 102339. [CrossRef]
34. Tappi, S.; Nissen, L.; Casciano, F.; Antonelli, G.; Chiarello, E.; Picone, G.; Laurita, R.; Capelli, F.; Gherardi, M.; Maccaferri, C.; et al. Effect of cold plasma generated with different gas mixtures on safety, quality and nutritional aspects of fresh sea bream fillets. *Innov. Food Sci. Emerg. Technol.* **2023**, *89*, 10347. [CrossRef]
35. Molina-Hernandez, J.B.; Grande-Tovar, C.D.; Neri, L.; Delgado-Ospina, J.; Rinaldi, M.; Cordero-Bueso, G.A.; Chaves-López, C. Enhancing postharvest food safety: The essential role of non-thermal technologies in combating fungal contamination and mycotoxins. *Front. Microbiol.* **2025**, *16*, 1543716. [CrossRef]
36. Zou, F.; Yang, M.; Wu, J.; Wang, L.; Wang, H. The potential of plasma-activated water in safe and sustainable food production: A comprehensive review of recent advances and future trends. *Crit. Rev. Food Sci. Nutr.* **2025**, 1–25. [CrossRef]
37. Gebremical, G.G.; Tappi, S.; Laurita, R.; Capelli, F.; Drudi, F.; Romani, S.; Rocculi, P. Effects of incubation time of plasma activated water (PAW) combined with annealing for the modification of functional properties of potato starch. *Food Biosci.* **2024**, *59*, 104247. [CrossRef]
38. Zimmermann, U.; Pilwat, G.; Beckers, F.; Riemann, F. Effects of external electrical fields on cell membranes. *Bioelectrochem. Bioenerg.* **1976**, *3*, 58–83. [CrossRef]
39. Donsì, F.; Ferrari, G.; Pataro, G. Applications of pulsed electric field treatments for the enhancement of mass transfer from vegetable tissue. *Food Eng. Rev.* **2010**, *2*, 109–130. [CrossRef]
40. Balasa, A. Stress response of plants, metabolite production due to pulsed electric fields. In *Handbook of Electroporation*; Springer: Cham, Switzerland, 2017; Volume 4, pp. 2559–2571.
41. Kranjc, M.; Miklavcic, D. Electric field distribution and electroporation threshold. In *Handbook of Electroporation*; Springer: Cham, Switzerland, 2017; Volume 2, pp. 1043–1058.
42. Teissie, J.; Golzio, M.; Rols, M.P. Mechanisms of cell membrane electropermeabilization: A minireview of our present (lack of?) knowledge. *Biochim. Biophys. Acta* **2005**, *1724*, 270–280. [CrossRef]
43. Lytras, F.; Psakis, G.; Gatt, R.; Cebrián, G.; Raso, J.; Valdramidis, V. Exploring the efficacy of pulsed electric fields (PEF) in microbial inactivation during food processing: A deep dive into the microbial cellular and molecular mechanisms. *Innov. Food Sci. Emerg. Technol.* **2024**, *95*, 103732. [CrossRef]
44. D'Elia, F. Effect of Pulsed Electric Fields on Drying of Plant Products. Master's Thesis, University of Bologna, Cesena, Italy, 2019.

45. Wiktor, A.; Nowacka, M.; Dadan, M.; Rybak, K.; Lojkowski, W.; Chudoba, T.; Witrowa-Rajchert, D. The effect of pulsed electric field on drying kinetics, color, and microstructure of carrot. *Dry. Technol.* **2016**, *34*, 1286–1296. [CrossRef]
46. de Paula, M.; Latorres, J.M.; Martins, V.G. Potential valorization opportunities for Brewer's spent grain. *Eur. Food Res. Technol.* **2023**, *249*, 2471–2483. [CrossRef]
47. Guido, L.F.; Moreira, M.M. Techniques for Extraction of Brewer's Spent Grain Polyphenols: A Review. *Food Bioprocess Technol.* **2017**, *10*, 1192–1209. [CrossRef]
48. Al Khawli, F.; Pateiro, M.; Domínguez, R.; Lorenzo, J.M.; Gullón, P.; Kousoulaki, K.; Ferrer, E.; Berrada, H.; Barba, F.J. Innovative Green Technologies of Intensification for Valorization of Seafood and Their By-Products. *Mar. Drugs* **2019**, *17*, 689. [CrossRef]
49. Hamdi, M.; Sun, H.; Pan, L.; Wang, D.; Sun, M.; Zeng, Z.; Li, S.; Dong, Q.; Su, F. Chitosan and its derivatives as potential biomaterials for biomedical and pharmaceutical applications: A comprehensive review on green extraction approaches, recent progresses, and perspectives. *Eur. Polym. J.* **2025**, *229*, 113882. [CrossRef]
50. De Aguiar Saldanha Pinheiro, A.C.; Martí-Quijal, F.J.; Barba, F.J.; Benítez-González, A.M.; Meléndez-Martínez, A.J.; Castagnini, J.M.; Tappi, S.; Rocculi, P. Pulsed Electric Fields (PEF) and Accelerated Solvent Extraction (ASE) for Valorization of Red (*Aristeus antennatus*) and Camarote (*Melicertus kerathurus*) Shrimp Side Streams: Antioxidant and HPLC Evaluation of the Carotenoid Astaxanthin Recovery. *Antioxidants* **2023**, *12*, 406. [CrossRef] [PubMed]
51. Ngo, T.H.D.; Ngo, D.N. Effects of low–frequency ultrasound on heterogenous deacetylation of chitin. *Int. J. Biol. Macromol.* **2017**, *104*, 1604–1610. [CrossRef]
52. He, G.; Yin, Y.; Yan, X.; Wang, Y. Application of Pulsed Electric Field for Treatment of Fish and Seafood. In *Handbook of Electroporation*, 1st ed.; Miklavčič, D., Ed.; Springer: Chem, Switzerland, 2017; Volume 4, pp. 2637–2655.
53. García-Parra, J.; González-Cebrino, F.; Delgado-Adámez, J.; Cava, R.; Martín-Belloso, O.; Élez-Martínez, P.; Ramírez, R. Effect of high-hydrostatic pressure and moderate-intensity pulsed electric field on plum. *Food Sci. Technol. Int.* **2018**, *24*, 145–160. [CrossRef]
54. González-Casado, S.; Martín-Belloso, O.; Elez-Martínez, P.; Soliva-Fortuny, R. Enhancing the carotenoid content of tomato fruit with pulsed electric field treatments: Effects on respiratory activity and quality attributes. *Postharvest Biol. Technol.* **2018**, *137*, 113–118. [CrossRef]
55. Ribas-Agustí, A.; Martín-Belloso, O.; Soliva-Fortuny, R.; Elez-Martínez, P. Enhancing hydroxycinnamic acids and flavan-3-ol contents by pulsed electric fields without affecting quality attributes of apple. *Food Res. Int.* **2019**, *121*, 433–440. [CrossRef] [PubMed]
56. Guderjan, M.; Töpfl, S.; Angersbach, A.; Knorr, D. Impact of pulsed electric field treatment on the recovery and quality of plant oils. *J. Food Eng.* **2005**, *67*, 281–287. [CrossRef]
57. Rifna, E.J.; Ratish Ramanan, K.; Mahendran, R. Emerging technology applications for improving seed germination. *Trends Food Sci. Technol.* **2019**, *86*, 95–108. [CrossRef]
58. Liu, C.; Zhang, R.; Vorobiev, E.; Grimi, N. Mitigation of Acrylamide in Potato Chips by Pre-drying and Pulsed Electric Field Treatment. *Front. Nutr.* **2022**, *9*, 919634. [CrossRef]
59. Schouten, M.A.; Genovese, J.; Tappi, S.; Di Francesco, A.; Baraldi, E.; Cortese, M.; Capriolid, G.; Angelonid, S.; Vittori, S.; Rocculi, P.; et al. Effect of innovative pre-treatments on the mitigation of acrylamide formation in potato chips. *Innov. Food Sci. Emerg. Technol.* **2020**, *64*, 102397. [CrossRef]
60. Di Francesco, A.; Mari, M.; Ugolini, L.; Parisi, B.; Genovese, J.; Lazzeri, L.; Baraldi, E. Reduction of acrylamide formation in fried potato chips by *Aureobasidum pullulans* L1 strain. *Int. J. Food Microbiol.* **2019**, *289*, 168–173. [CrossRef]
61. Witrowa-Rajchert, D.; Wiktor, A.; Sledz, M.; Nowacka, M. Selected Emerging Technologies to Enhance the Drying Process: A Review. *Dry. Technol.* **2014**, *32*, 1386–1396. [CrossRef]
62. Soltani Firouz, M.; Farahmandi, A.; Hosseinpour, S. Recent advances in ultrasound application as a novel technique in analysis, processing and quality control of fruits, juices and dairy products industries: A review. *Ultrason. Sonochem.* **2019**, *57*, 73–88. [CrossRef]
63. Knorr, D.; Zenker, M.; Heinz, V.; Lee, D.U. Applications and potential of ultrasonics in food processing. *Trends Food Sci. Technol.* **2004**, *15*, 261–266. [CrossRef]
64. Virot, M.; Tomao, V.; Le Bourvellec, C.; Renard, C.M.; Chemat, F. Towards the industrial production of antioxidants from food processing by-products with ultrasound-assisted extraction. *Ultrason. Sonochem.* **2010**, *17*, 1066–1074. [CrossRef] [PubMed]
65. Hu, Q.; Ma, X.; Cai, T.; Li, Y. Flavonoid intake, inflammation, and atherosclerotic cardiovascular disease risk in U.S. adults: A cross-sectional study. *Nutr. Metab.* **2025**, *22*, 24. [CrossRef]
66. Jomova, K.; Alomar, S.Y.; Valko, R.; Liska, J.; Nepovimova, E.; Kuca, K.; Valko, M. Flavonoids and their role in oxidative stress, inflammation, and human diseases. *Chem.-Biol. Interact.* **2025**, *413*, 111489. [CrossRef]
67. Li, L.; Wang, Z.; Yu, Z.; Niu, T. Dietary Flavonoid Intake and Anemia Risk in Children and Adolescents: Insights from National Health and Nutrition Examination Survey. *Antioxidants* **2025**, *14*, 395. [CrossRef]

68. Santos, V.O.; Rodrigues, S.; Fernandes, F.A.N. Improvements on the Stability and Vitamin Content of Acerola Juice Obtained by Ultrasonic Processing. *Foods* **2018**, *7*, 68. [CrossRef]
69. Frias, J.; Peñas, E.; Ullate, M.; Vidal-Valverde, C. Influence of Drying by Convective Air Dryer or Power Ultrasound on the Vitamin C and β-Carotene Content of Carrots. *J. Agric. Food Chem.* **2010**, *58*, 10539–10544. [CrossRef]
70. Nowacka, M.; Wedzik, M. Effect of ultrasound treatment on microstructure, colour and carotenoid content in fresh and dried carrot tissue. *Appl. Acoust.* **2016**, *103 Pt B*, 163–171. [CrossRef]
71. Fonteles, T.V.; Leite, A.K.F.; da Silva, A.R.A.; Fernandes, F.A.N.; Rodrigues, S. Sonication Effect on Bioactive Compounds of Cashew Apple Bagasse. *Food Bioprocess Technol.* **2017**, *10*, 1854–1864. [CrossRef]
72. Nowacka, M.; Fijalkowska, A.; Dadan, M.; Rybak, K.; Wiktor, A.; Witrowa-Rajchert, D. Effect of ultrasound treatment during osmotic dehydration on bioactive compounds of cranberries. *Ultrasonics* **2018**, *83*, 18–25. [CrossRef] [PubMed]
73. Bindes, M.M.M.; Reis, H.M.M.; Cardoso, V.L.; Boffito, D.C. Ultrasound-assisted extraction of bioactive compounds from green tea leaves and clarification with natural coagulants (chitosan and *Moringa oleífera* seeds). *Ultrason. Sonochem.* **2019**, *51*, 111–119. [CrossRef]
74. Díaz-de-Cerio, E.; Tylewicz, U.; Verardo, V.; Fernández-Gutiérrez, A.; Segura-Carretero, A.; Romani, S. Design of Sonotrode Ultrasound-Assisted Extraction of Phenolic Compounds from *Psidium guajava* L. Leaves. *Food Anal. Methods* **2017**, *10*, 2781–2791. [CrossRef]
75. Papoutsis, K.; Pristijono, P.; Golding, J.B.; Stathopoulos, C.E.; Bowyer, M.C.; Scarlett, C.J.; Vuong, Q.V. Optimizing a sustainable ultrasound-assisted extraction method for the recovery of polyphenols from lemon by-products: Comparison with hot water and organic solvent extractions. *Eur. Food Res. Technol.* **2018**, *244*, 1353–1365. [CrossRef]
76. Altemimi, A.; Watson, D.G.; Choudhary, R.; Dasari, M.R.; Lightfoot, D.A. Ultrasound assisted extraction of phenolic compounds from peaches and pumpkins. *PLoS ONE* **2016**, *11*, e0148758. [CrossRef]
77. Siucińska, K.; Mieszczakowska-Frąc, M.; Połubok, A.; Konopacka, D. Effects of Ultrasound Assistance on Dehydration Processes and Bioactive Component Retention of Osmo-Dried Sour Cherries. *J. Food Sci.* **2016**, *81*, C1654–C1661. [CrossRef] [PubMed]
78. Altemimi, A.; Choudhary, R.; Watson, D.G.; Lightfoot, D.A. Effects of ultrasonic treatments on the polyphenol and antioxidant content of spinach extracts. *Ultrason. Sonochem.* **2015**, *24*, 247–255. [CrossRef]
79. Gamboa-Santos, J.; Montilla, A.; Soria, A.C.; Cárcel, J.A.; García-Pérez, J.V.; Villamiel, M. Impact of power ultrasound on chemical and physicochemical quality indicators of strawberries dried by convection. *Food Chem.* **2014**, *161*, 40–46. [CrossRef]
80. Oladejo, A.O.; Ma, H.; Qu, W.; Zhou, C.; Wu, B. Effects of Ultrasound on Mass Transfer Kinetics, Structure, Carotenoid and Vitamin C Content of Osmodehydrated Sweet Potato (*Ipomea batatas*). *Food Bioprocess Technol.* **2017**, *10*, 1162–1172. [CrossRef]
81. Nowacka, M.; Dadan, M.; Tylewicz, U. Current Applications of Ultrasound in Fruit and Vegetables Osmotic Dehydration Processes. *Appl. Sci.* **2021**, *11*, 1269. [CrossRef]
82. Shen, L.; Pang, S.; Zhong, M.; Sun, Y.; Qayum, A.; Liu, Y.; Rashid, A.; Xu, B.; Liang, Q.; Ma, H.; et al. A comprehensive review of ultrasonic assisted extraction (UAE) for bioactive components: Principles, advantages, equipment, and combined technologies. *Ultrason. Sonochem.* **2023**, *101*, 106646. [CrossRef] [PubMed]
83. Rostamabadi, H.; Karaca, A.C.; Nowacka, M.; Mulla, M.Z.; Al-Attar, H.; Rathnakumar, K.; Subasi, B.G.; Sehrawat, R.; Kheto, A.; Falsafi, S.R. How high hydrostatic pressure treatment modifies the physicochemical and nutritional attributes of polysaccharides? *Food Hydrocoll.* **2023**, *137*, 108375. [CrossRef]
84. Buzrul, S.; Alpas, H. Treatment of foods using high hydrostatic pressure. In *Progress in Food Preservation*, 1st ed.; Bhat, R., Alias, A.K., Paliyath, G., Eds.; Wiley-Blackwell: Hoboken, NJ, USA, 2012. [CrossRef]
85. Patrignani, F.; Lanciotti, R. Applications of High and Ultra High Pressure Homogenization for Food Safety. *Front. Microbiol.* **2016**, *7*, 1132. [CrossRef] [PubMed]
86. Lacroix, N.; Fliss, I.; Makhlouf, J. Inactivation of pectin methylesterase and stabilization of opalescence in orange juice by dynamic high pressure. *Food Res. Int.* **2005**, *38*, 569–576. [CrossRef]
87. Tribst, A.A.L.; Franchi, M.A.; Cristianini, M. Ultra-high pressure homogenization treatment combined with lysozyme for controlling Lactobacillus brevis contamination in model system. *Innov. Food Sci. Emerg. Technol.* **2008**, *9*, 265–271. [CrossRef]
88. Bonfim, R.C.; de Oliveira, F.A.; Godoy, R.L.d.O.; Rosenthal, A. A review on high hydrostatic pressure for bivalve mollusk processing: Relevant aspects concerning safety and quality. *Food Sci. Technol. Braz.* **2019**, *39*, 515–523. [CrossRef]
89. Wang, Y.; Ma C-m Yang, Y.; Wang, B.; Liu X-f Wang, Y.; Bian, X.; Zhang, G.; Zhang, N. Effect of high hydrostatic pressure treatment on food composition and applications in food industry: A review. *Food Res. Int.* **2024**, *195*, 114991. [CrossRef]
90. Roobab, U.; Fidalgo, L.G.; Arshad, R.N.; Khan, A.W.; Zeng, X.A.; Bhat, Z.F.; Bekhit, A.E.A.; Batool, Z.; Aadil, R.M. High-pressure processing of fish and shellfish products: Safety, quality, and research prospects. *Compr. Rev. Food Sci. Food Saf.* **2022**, *21*, 3297–3325. [CrossRef]
91. Lan, Q.; Tappi, S.; Braschi, G.; Picone, G.; Rocculi, P.; Laghi, L. Comparative metabolomic analysis of minced grey mullet (*Mugil cephalus*) pasteurized by high hydrostatic pressure (HHP) during chilled storage. *Food Biosci.* **2024**, *61*, 104539. [CrossRef]

92. Liu, Y.; Deng, J.; Zhao, T.; Yang, X.; Zhang, J.; Yang, H. Bioavailability and mechanisms of dietary polyphenols affected by non-thermal processing technology in fruits and vegetables. *Curr. Res. Food Sci.* **2024**, *8*, 100715. [CrossRef] [PubMed]
93. Betoret, E.; Betoret, N.; Calabuig-JImenez, L.; Patrignani, F.; Barrera, C.; Lanciotti, R.; Dalla Rosa, M. Probiotic survival and in vitro digestion of *L. salivarius* spp. *salivarius* encapsulated by high homogenization pressures and incorporated into a fruit matrix. *LWT* **2019**, *111*, 883–888. [CrossRef]
94. Lima, M.A.; Rosenthal, A. High pressure homogenization applied to fruit juices: Effects on microbial inactivation and on maintenance of bioactive components. *Food Sci. Technol. Int.* **2022**, *29*, 857–870. [CrossRef] [PubMed]
95. Yudhistira, B.; Punthi, F.; Gavahian, M.; Chang, C.-K.; Hazeena, S.H.; Hou, C.-Y.; Hsieh, C.-W. Nonthermal technologies to maintain food quality and carbon footprint minimization in food processing: A review. *Trends Food Sci. Technol.* **2023**, *141*, 104205. [CrossRef]

MDPI AG
Grosspeteranlage 5
4052 Basel
Switzerland
Tel.: +41 61 683 77 34

Foods Editorial Office
E-mail: foods@mdpi.com
www.mdpi.com/journal/foods

www.ingramcontent.com/pod-product-compliance
Lightning Source LLC
Chambersburg PA
CBHW061555210926
51850CB00009B/2107